Building Surveys and Reports

Edward A. Noy FASI, ARSH

Third Edition
revised by

James Douglas BSc, MRICS

Blackwell
Publishing

Blackwell Publishing Ltd
Editorial offices:
Blackwell Publishing Ltd, 9600 Garsington Road, Oxford OX4 2DQ, UK
 Tel: +44 (0)1865 776868
Blackwell Publishing Inc., 350 Main Street, Malden, MA 02148-5020, USA
 Tel: +1 781 388 8250
Blackwell Publishing Asia Pty Ltd, 550 Swanston Street, Carlton, Victoria 3053, Australia
 Tel: +61 (0)3 8359 1011

First edition published 1990 by BSP Professional Books
Second edition published 1995 by Blackwell Science Ltd
Third edition published 2005 by Blackwell Publishing Ltd

Library of Congress Cataloging-in-Publication Data
Noy, Edward A.
  Building surveys and reports / Edward A. Noy. – 3rd ed. / revised by James Douglas.
    p. cm.
  Includes bibliographical references and index.
  ISBN-13: 978-1-4051-2147-7 (alk. paper)
  ISBN-10: 1-4051-2147-5 (alk. paper)
  1. Building inspection. 2. Buildings–Defects. I. Douglas, James. II. Title.

TH439.N. 68   2005
690'.21–dc22
2005041028

ISBN-10: 1-4051-2147-5
ISBN-13: 978-14051-2147-7

A catalogue record for this title is available from the British Library.

Set in 10/12pt Palatino
by SNP Best-set Typesetter Ltd., Hong Kong
Printed and bound in India
by Replika Press Pvt. Ltd.

The publisher's policy is to use permanent paper from mills that operate a sustainable
forestry policy, and which has been manufactured from pulp processed using acid-free
and elementary chlorine-free practices. Furthermore, the publisher ensures that the text
paper and cover board used have met acceptable environmental accreditation standards.

For further information on Blackwell Publishing, visit our website:
www.thatconstructionsite.com

Contents

Internal and external finishes, 322 – 16.13 Services, 323 –
16.14 Recording the defects, 323
FLOOD DAMAGE
16.15 Causes, 323 – 16.16 Preliminary examination, 323 –
16.17 General effects of flooding, 324 – 16.18 Foundations, 324
– 16.19 Ground floors, 325 – 16.20 Suspended ground floors,
325 – 16.21 Floor finishes, 325 – 16.22 Wall finishes, 325 – 16.23
Metal finishes and fastenings, 326 – 16.24 Drainage systems,
326 – 16.25 Pavings, 326 – 16.26 Recording defects, 326

Preface to Second Edition

This book provides a comprehensive guide for surveyors and architects on the steps to take when approached by a client asking for a structural survey. It deals with all types of buildings: domestic, commercial and industrial. Advice is given on how to diagnose faults, with many detailed sketches and photographs to illustrate the text. Examples of various types of reports are given in the appendices.

We are living in an era of change. Adaptation of buildings for different uses and extensions to existing buildings are commonplace. In each of these cases measured and structural surveys are necessary. Some of the difficulties which are met with are described in chapters 3 and 4, and advice is given as to how to avoid mistakes.

The book covers both old and new methods of construction. The subject has been treated basically under the elements of construction, most of which are interrelated. Much of the information is the result of considerable practical experience over the past fifty years and is written mainly for the newly qualified surveyor or architect. However, it may be that the book will also prove of interest to the experienced professional surveyor. I have, therefore, assumed that the reader has some knowledge of building techniques.

Flood and fire damage has been given a separate chapter since it involves different structural problems in diagnosing the cause, as well as negotiations with insurance assessors before steps for reinstatement can be put in hand.

During the past forty years there have been many new materials and construction techniques using new and traditional materials. The surveyor can no longer be dependent on a limited range of materials, but must exercise his judgment in a widening realm of alternatives. The fabric of a building has to satisfy different user needs and occupational factors. The surveyor's duty is to identify what performance is required from the fabric in terms of durability and weathertightness. It is therefore essential that he must have a sound knowledge of not only building construction, but also the performance of materials in use.

I have been encouraged by the fact that the first edition of my book was well received. Preparing a second edition has given me the opportunity to revise and update the text, thereby covering the latest procedures and research. New material and additional illustrations have been included. The chapter dealing with legal aspects has been expanded and refers to recent case law now available.

Reference has been made to the various publications by the Building Research Establishment which I duly acknowledge. Grateful thanks are due to Mr C.I. Berry, BArch, ARIBA for reading my manuscript and making valuable suggestions, and my wife Audrey for her invaluable support. Finally, my thanks are due to the publishers for their help and advice throughout the production of the book.

Edward A. Noy

Preface to Third Edition

This book focuses primarily on traditional construction of residential and non-residential buildings. It aims to provide the reader with guidance on the methodology and risks of inspecting and surveying buildings generally. What makes this revised edition different from most of its competitors is that it includes a comparison of the various surveys available. In this regard reference has been made to the Construction Industry Council's 1997 guidance note on the definitions of building inspections and surveys (see Appendix I).

Chapter 7 includes a new section on no-fines concrete construction. This is an important form of non-traditional domestic construction of which surveyors should be aware.

Also contained in this edition is a second new appendix, which describes typical schedules used in condition and dilapidations surveys. This supplements the other sample survey checklists described in Appendices III and IV.

The importance given to construction safety has increased since the early 1990s. The third new appendix contains guidance from the Royal Institution of Chartered Surveyors (RICS) on surveying properties safely.

The fourth new appendix contains a glossary of terms relevant to building inspections and surveys.

James Douglas

Acknowledgements

A major advantage of revising someone else's work is that there is already an extensive body of material to work from. Indeed, the bulk of this book did not require any drastic overhauling, which clearly reflects well on Edward Noy's previous editions. Naturally, therefore, I would like to thank Edward for allowing me to revise his book to produce this third edition. I hope that he is pleased with the result.

I would, of course, like to express my appreciation to Blackwell Publishing for their confidence in allowing me to revise this work. It is gratifying to know that publishers are keen to continue promoting books on building surveying and related construction topics. The construction industry needs to keep expanding its body of literature so that the knowledge of and guidance on best practice of inspecting and surveying buildings can be disseminated more widely.

In particular, my thanks go to Julia Burden, Blackwell's deputy divisional director, for giving me the opportunity to revise this edition. I also wish to express my appreciation to her colleague Emma Moss for the reminders and friendly cajoling throughout the revision of this work.

My gratitude, too, goes to my colleague Edith Bowman FRICS for her guidance on the legal implications of undertaking surveys.

I would also like to thank Tony Condron of Delvemade Ltd for his feedback and technical guidance on the problem of cut edge corrosion and its solution.

Wherever possible every attempt has been made to acknowledge the numerous sources used in this book. The reviser would like to apologise in advance if there is any case where this has not been achieved. Due acknowledgement will be made in any subsequent edition.

James Douglas

1 General Principles and Responsibilities

1.1 What is a building survey?

Definitions

In 1997 the Construction Industry Council (CIC) published a leaflet entitled *Definitions of Inspections and Surveys of Buildings* (see Appendix I). Although the definitions specifically apply to England and Wales, they are also relevant to the rest of the UK. The CIC is the organisation representing the main professional bodies in construction and property, such as the Royal Institution of Chartered Surveyors (RICS), the Royal Institute of British Architects, the Chartered Institute of Building, the Association of Building Engineers, the Architecture and Surveying Institute, the Chartered Institution of Building Services Engineers, the Institution of Civil Engineers and the Institution of Structural Engineers.

One of the most significant consequences of the CIC list of definitions was the scrapping of the term 'structural survey'. Up until 1997 'structural survey' was the commonly accepted term for a Scheme 3 survey – the full building survey (Staveley 1998). Although surveyors and lawyers in the UK had been using the term 'structural survey' for decades, many professionals, particularly consulting engineers, felt that it was misleading. It implied that the survey focused on structural issues relating to the property being surveyed – in other words, that it would only deal with the loadbearing characteristics of the building. This of course was not the case as 'structural surveys' assessed the property's fabric and services as well as addressed its stability. Any major 'structural' findings were then referred to an engineer for more detailed analysis.

Nowadays, therefore, either 'structural inspection' or 'structural assessment' is the more accurate term to describe a building-related investigation undertaken by consulting engineers (IStructE 1991). It is essentially a specialist investigation that often follows a condition/building survey, to assess in more detail a problem or requirement relating to the property's loadbearing elements – such as foundations, walls, floors, beams and columns and roofs.

See Appendix VIII for the definition of 'building survey' and other related terms.

Categories of property survey

As indicated in the CIC list, there is a wide range of property surveys. Table 1.1 categorises property surveys into five main groups and shows their relationship to one another.

Table 1.1   Types of property surveys.

Type of Surveys	Examples (a)	Scheme
Acquisition surveys (b)	Mortgage valuation (1)	1
	Homebuyers survey and valuation (2)	2
	Home inspection report (2) (c)	2
	Building survey (3)	3
Lease-related surveys (d)	Schedule of dilapidations survey (9)	2 (e)
	Schedule of condition survey (8)	2 (e, f)
Record surveys (g)	Schedule of condition survey (8)	2 (e)
	Measurement/dimensional survey (10)	2 (e)
	Inspection of building prior to alteration (5)	2 (e)
	Conservation plan inspection (5)	3 (e)
	Inspection of buildings under construction (11)	2 (e)
	Stock condition survey (7)	2 (e)
	Maintenance survey (7)	2 (e)
Reinstatement surveys (h)	Fire damage survey (6)	2 (e)
	Flood damage survey (6)	2 (e)
	Other damage/insurance-related survey (e.g. following a burglary) (6)	2 (e)
Specialist surveys (i)	Access audit (4)	2 or 3 (e)
	Defect assessment or diagnostic survey (4)	2 or 3 (e)
	Elemental investigation (4)	2 or 3 (e)
	Sanitary survey (4) (j)	2 or 3 (e)
	Structural inspection/assessment (4)	2 or 3 (e)
	Post-occupancy evaluation (4) (k)	2 or 3 (e)

Notes:
(a) The equivalent CIC type of survey number is shown in brackets.
(b) These are surveys required as a result of the intended purchase of a property and account for the majority of such commissions.
(c) This forms part of the proposed home information pack.
(d) These surveys are required under the terms of a property lease.
(e) The level of detail of these surveys is approximate to the Scheme indicated, even though they are not officially classed as such.
(f) A Schedule of Condition can be undertaken outwith the context or requirements of a lease (e.g. before adaptation works or demolition of an adjacent/nearby building).
(g) This type of survey is usually related to adaptation, conservation or maintenance work.
(h) Insurance claims usually prompt this kind of survey.
(i) This type of survey sometimes follows as a result of a Scheme 1 or 2 survey.
(j) The test and examination of the drainage system is the most important single item in this type of survey (Moseley and Curtis 1973).
(k) This type of survey aims to assess a building's performance and is usually undertaken at least six months after it has been built or adapted (Preiser 1989).

Synchronic and diachronic surveys

Another way of categorising property surveys is to consider them either synchronically or diachronically (Brand 1994). A synchronic survey is a snapshot assessment of a building and the way it all fits together at a particular moment in time. This usually means the present, but buildings can be studied as regards how they worked at one time in the past. In other words, it is about studying buildings in terms of immediacy and is the preference of building surveyors as well as 'city planners and architects looking for design ideas' (Brand 1993). Building surveys, condition surveys and dilapidation surveys are typical examples of this kind of appraisal.

A diachronic survey, on the other hand, is a way of studying buildings in terms of how they change or evolve over time. This is the way architectural historians (and building maintenance surveyors) appraise buildings (Brand 1993). Maintenance surveys as well as conservation plan inspections and other record surveys of older properties are typical methods of studying buildings diachronically.

Stock condition surveys

These are surveys that are undertaken on a large number of properties one after the other, or simultaneously if more than one surveyor is being used. They are most common for determining the state of repair of housing. However, the same approach can be used when assessing the condition of other large property stock such as warehouses and other industrial or commercial buildings.

The reader is referred to the relevant RICS guidance note on this type of survey (RICS 1995). It is usually carried out on a regular (e.g. quinquennial) basis on ecclesiastical buildings as well as housing stocks. Data on the most recent English and Scottish house condition surveys undertaken in 2001 and 2002 respectively can be obtained from the research departments of the Research, Analysis & Evaluation Division (2003) and Communities Scotland (2004) respectively.

1.2   Housing quality initiatives

Home condition reports

The new housing legislation proposed by the Government will require sellers of dwellings to supply a standard set of information referred to as a Home Information Pack (HIP) (ODPM 2003a). This will need to be provided before marketing a property for sale and made available to prospective purchasers (Melville and Gordon 2004). It will require a home 'condition report based on a professional survey of the property, including an energy efficiency assessment' (ODPM 2003a). It is analogous to a 'homesellers survey and valuation', which would be similar to the well-established homebuyers survey and valuation (see Parnam and Rispin 2001).

The HIP is deemed necessary for three main reasons (ODPM 2003a):

- To improve the home purchase process.
- To enable compliance with EU Directive 2002/91/EC, Article 6 of which requires that member states ensure an energy performance certificate is made available to the prospective buyer/tenant when buildings are constructed, sold or let.
- To improve household energy efficiency.

Fitness standard

The current fitness standard for England and Wales was introduced through the Local Government and Housing Act 1989 which inserted a new Section 604 in the Housing Act 1985. According to the DETR (1998) 'a dwelling is unfit if, in the opinion of the authority, it fails to meet one of the requirements set out in paragraphs (a) to (i) of s.604 (1) and, by reason of that failure, is not reasonably suitable for occupation'. The requirements constitute the minimum deemed necessary for a dwelling house (including a house in multiple occupation) to be fit for human habitation (Douglas 2002).

These fitness standards require that a dwelling house should:

- Be free from serious disrepair.
- Be structurally stable.
- Be free from dampness prejudicial to the health of the occupants.
- Have adequate provision for lighting, heating and ventilation.
- Have an adequate piped supply of wholesome water.
- Have an effective system for the drainage of foul, waste and surface water.
- Have a suitably located WC for exclusive use of the occupants.
- Have a bath or shower and wash-hand basin, with hot and cold water.
- Have satisfactory facilities for the preparation and cooking of food including a sink with hot and cold water.

Tolerable standard

According to the Scottish Executive (2003) 'The Tolerable Standard (which is equivalent to the Fitness Standard in England) as amended by the Housing (Scotland) Act 2001 was introduced in the 1969 Housing (Scotland) Act following recommendations made in the 1967 Cullingworth report'. Other than the incorporation of the 'basic/standard amenities' (e.g. hot and cold running water) by the Housing (Scotland) Act 2001, it has remained largely unchanged.

The Scottish Executive (2003) emphasises that the standard is *not* intended to be a measure of acceptable housing conditions. It is distinct from the Building Regulations for example, which provide minimum standards for new construction and reflect modern expectations of the facilities and amenities to be provided in modern homes. The standard sets the base line below which houses should not be allowed to continue in occupation.

A house meets the tolerable standard for the purposes of this Act according to the Scottish Executive (2003) if it:

- Is structurally stable.
- Is substantially free from rising or penetrating damp.
- Has satisfactory provision for natural and artificial lighting, for ventilation and for heating.
- Has an adequate piped supply of wholesome water available within the house.
- Has a sink provided with a satisfactory supply of both hot and cold water within the house.
- Has a water closet available for the exclusive use of the occupants of the house and suitably located within the house.
- Has a fixed bath or shower and a wash-hand basin, each provided with a satisfactory supply of both hot and cold water and suitably located within the house.
- Has an effective system for the drainage and disposal of foul and surface water.
- Has satisfactory facilities for the cooking of food within the house.
- Has satisfactory access to all external doors and outbuildings.
- Any reference to a house not meeting the tolerable standard or being brought up to the tolerable standard shall be construed accordingly.

Decent homes and quality housing initiatives

It is the Government's aim to 'By 2010, bring all social housing (in England) into decent condition with most of the improvement taking place in deprived areas, and increase the proportion of private housing in decent condition occupied by vulnerable groups'.

A home is classified as decent if it:

- Meets the current statutory minimum standard.
- Is in reasonable repair.
- Has reasonably modern facilities and services.
- Provides a reasonable degree of thermal comfort.

Similar schemes are being proposed for Wales, Northern Ireland and Scotland. For example, the Scottish Executive (2003) set out proposals for a national standard based on a minimum set of quality measures for all houses in the social rented sector in Scotland. In February 2004 the Minister for Communities launched The Scottish Housing Quality Standard (here referred to as 'the Standard'). The announcement set out a range of measures that local authority and Registered Social Landlord (RSL) stock had to reach by March 2015 and required all social landlords to draw up Standard Delivery Plans (SDPs) to show how they were going to reach that target. This is similar to the Decent Home Strategy for social housing in England.

In Scotland the Government's intention has been 'to define a standard that is relevant to the twenty-first century and is consistent with views on what constitutes acceptable, good quality housing. It differs from the statutory Tolerable Standard (which is considered a very basic standard of acceptability) and the Building Standards as they apply to new housing' (Scottish Executive 2003).

As initially proposed by the Scottish Executive (2003) the standard is based on a number of broad quality criteria. To meet the standard the house must be:

- Compliant with the tolerable standard (as described above).
- Free from serious disrepair (such as dilapidation or structural instability).
- Energy efficient (i.e. has a National Home Energy Rating (NHER) of at least 5).
- Provided with modern facilities and services (e.g. indoor WC, disabled access, etc).
- Healthy, safe and secure (e.g. free from mould and contains no faulty electrical or gas installations).

Domestic survey implications

Given these statutory influences any surveyor inspecting a residential property should be cognisant of them. The proforma survey checklist in Appendix III includes a brief reference to fitness and amenities. The degree of compliance with these basic housing requirements should be noted when inspecting domestic properties.

1.3   The purpose of the survey

There are several conditions under which a surveyor may be required to survey or examine a building and the first point to ascertain is the reason for which the advice is being sought. The following is a list of the most usual reasons:

- To prepare a measured drawing of the building to enable a scheme for alterations, improvements or extensions to be prepared.

- To prepare a report on the condition of a property to be purchased.
- To prepare a schedule of condition for a property to be taken on long lease.
- To advise on the repair and preservation of a building (including 'listed' buildings).
- Work to be carried out to satisfy the requirements of the local or other authority, i.e. dangerous structure notices, public health notices or a factory inspector's notice.
- To prepare plans in connection with party wall agreements. This is usually required where alterations to a party wall are contemplated (see Anstey 1998).
- To advise on the repair of a building damaged by fire or flood.
- To make a structural appraisal of existing buildings for 'change of use'.

No doubt it would be readily understood that several of the surveys mentioned above would be carried out simultaneously. For instance, a surveyor is often asked to report on the condition of a property and at the same time prepare a scheme for an extension or alterations. The surveyor's report on the items to be examined will also vary with each building; a lot will depend on whether or not he is being asked to report on the general condition of a building or examine specific defects. In certain cases the surveyor may consider it advisable to ask if there is any particular point which the client has noticed and which might be giving them reason for concern. Initially, the client will almost certainly be worried by the question of structural stability and will wish to have the surveyor's advice on this matter as soon as possible. On accepting instructions, the surveyor must therefore arrange an early date to examine the premises with this object in mind. Owners seldom realise that a building ten or more years of age is unlikely to be in perfect condition; even in quite small properties expenditure may have to be incurred in order to put the property in sound condition.

Reference to the early history of the building is often important. Very few owners can provide clear details about old buildings. Local authorities or local builders can often produce the original plans, but it is well to remember that alterations were often carried out in the past without submitting plans to the authorities concerned, so any drawings produced should be checked carefully. It was also quite common for details to be altered at the time of building but not amended on the plans.

As soon as a commission has been received to survey a building for alterations or extensions, it is important to consider the nature of the proposed scheme and to ensure that adequate information is obtained on site (Douglas 2002). It is therefore advisable to discuss the proposals with the client before commencing the survey and reach an agreement as to what precisely they require by way of advice and the specific parts of the building which are to be examined. This procedure enables the surveyor to make notes and sketches during his examination of the area concerned which will assist him when preparing the scheme. For example, an extension may necessitate repositioning some essential services or breaking into a party wall where the interests of the adjoining owner would be affected. If the work envisaged involves an extension at the front or rear of a property, close to an adjoining building, then information should be obtained as to the interests of the adjoining owner, i.e. rights of light, air, drainage or other easements. The omission of these particulars may seriously affect the work and cause unnecessary delays. If this is the case it is advisable to contact the adjoining owner at the earliest opportunity and let him/her know what your client proposes to do.

Surveyors are often asked to advise a client as to the desirability of taking a property on lease. One of the clauses in a lease agreement usually states that the property is to be given up at the end of the term in a condition similar to that when the new agreement was signed. In such cases the surveyor should carefully examine the property and prepare a detailed schedule of condition in order that at the termination of the lease there can be no dispute as to its condition (see Appendix II). Special care should be taken to identify all the rooms referred to in the report, and when dealing with a large property it is advisable to attach a plan and number all the rooms.

When asked to investigate a specific defect, such as a dangerously bulged wall or settlement, surveyors would be unwise to commit themselves to a definite opinion derived from one examination especially if they are not entirely familiar with the district and that type of property. Long term observation is usually required in order to establish with any degree of certainty the exact cause of the failure. The investigation would probably necessitate the use of tell-tales or plumbing of the walls. The client should be advised that it may take several months to reach a decision together with some brief details of the measures that have been taken. Appendix VI is an example of a report on a defective roof to a village hall.

In the case of a 'material change of use', that is e.g. a change of use from a large private house to a hotel, the building must comply with the current building regulations and it is advisable to involve the local authority throughout the process. Apart from the details required for the alterations or extensions, the local authority may require details of the existing structure, services and fittings so that they can deal with the whole building and not merely with the new work. There are several points that the surveyor may have to consider when dealing with a change of use such as Part E of the building regulations: resistance to the passage of sound.

This part of the building regulations was extended in June 1992 to include any material change of use of a building into a dwelling including conversion into flats. The requirement incorporates such works as sound insulation to floors or ceilings to prevent sound passing via the existing structure to adjacent rooms. When undertaking work of this nature it is advisable to remember that the local authorities have the power to relax the requirements where implementation would be unreasonable. In such cases the surveyor should ensure that the instructions he receives from his client are clearly defined. It is also important to explain to the client at the outset what is to be done and the information the report will contain.

The structural appraisal may require the surveyor to check the ability of the building to sustain increased floor loads, and the upgrading or structural fire protection (see Douglas 2002). Other important evidence which may have to be considered is that the building may have undergone several alterations over the years. The owner may have documentary material which should be considered in the appraisal.

1.4 Building condition assessment

As part of the preliminary assessment of a building or stock of properties it may be useful, if not necessary, to categorise them into broad condition ratings. Table 1.2 shows a convenient way of doing this.

1.5 Surveyor's responsibilities

Introductory

Surveyors are expected to have a working knowledge of the law relating to their profession to enable them to perform their duties adequately. It is not for the surveyor to assume the role of a solicitor. If a surveyor is confronted with a problem that exceeds the knowledge that he can reasonably be expected to have, then he would be wise to discuss the matter with his solicitor who will have a much wider knowledge of legal matters. The surveyor's legal liabilities, particularly the subject of negligence, are dealt with in Chapter 18. In the following paragraphs emphasis will be placed on matters of common occurrence concerning contracts and fees, together with examples of letters of contract.

Establishing the client's requirements

The following points as highlighted by Watts and Partners (2004) should be used as a checklist for establishing the client's brief and the extent and method of survey, particularly for commercial/industrial buildings:

- Identify the reasons for and scope of survey (as per Table 1.2).
- Ascertain the scope and degree of detail required (Residential Property Faculty 2004).
- Identify energy conservation requirements.

- Ascertain access for inspection.
- Ascertain tenure and request relevant documents.
- Check if the property is on contaminated land.
- Ascertain information available (at the desktop stage).
- Consider impact of statutory requirements (e.g. Disability Discrimination Act 1995).
- Establish if costings are required.
- Establish and confirm instructions (RICS 1998; Residential Property Faculty 2004).

Table 1.2   The spectrum of building condition (based on Douglas 2002).

Condition	Rating			
	Bad	**Poor**	**Fair**	**Good**
Category	Ruinous	Dilapidated	Wind and watertight	Optimal
Description	Only some walls present. Little or no roof structure remaining. No windows and doors left. Vegetation growing throughout building.	Extensive defects to the structure and fabric. Crumbling fabric nearing dereliction and redundancy.	Showing signs of wear and tear. Potentially neglected and approaching being run down.	Reasonable state of repair. No major defects. Satisfies most user requirements.
Typical risks	Instability. Partial/full collapse. Infestation by vermin and vegetation. Squatters. Redundancy. Obsolescence. Unsafe areas. Flooding. Saturation of walls and floors.	Ruination. Squatters. Redundancy. Obsolescence. Vandalism. Unsafe parts. Deleterious materials. Flooding/fire. Significant levels of dampness, timber decay and movement.	Accessibility problems. Redundancy. Obsolescence. Deleterious materials. Vandalism. Flooding. Dampness, timber decay and movement.	Minor dampness, timber decay and movement. Redundancy. Obsolescence. Vandalism. Flooding.
Response	Demolition? Restoration? Conservation? Conversion to other use?	Demolition? Renovation? Conversion to same or modified use?	Maintenance. Modernisation? Refurbishment? Rehabilitation? Alteration? Extension?	Maintenance and minor improvements. Minor alterations? Extension?

Due diligence

Surveyors must demonstrate due diligence before, during and after undertaking a building survey. Watts and Partners (2004) suggest the following issues are relevant to the due diligence process:

- Occupation considerations (e.g. fit out, sub-division, etc.).
- Repairs and defects (e.g. patent or latent defects, etc.).
- Environmental considerations (e.g. risk of contamination).
- Legal issues (e.g. restrictive covenants, party wall matters, road to property adopted?).
- Due diligence team members (e.g. solicitor, town planner, etc.).

Keeping records

All surveyors should retain their entire records concerning property surveys and reports (Glover 2003). Such documentation should be neatly filed and stored in chronological order for easy retrieval in the event of a query, dispute, or worse still, court case. A poorly kept records file on a survey commission will undermine the surveyor's credibility if the case goes to court (Vegoda 1993).

Health and safety

Inspecting and surveying buildings is a risky business physically as well as financially. Buildings can pose a wide range of physical dangers for the unsuspecting and inadequately prepared surveyor (RICS 2004). To minimise if not avoid such problems surveyors should always undertake a risk assessment before embarking on a property survey.

The risk assessment procedure follows three basic steps: risk identification, risk measurement and risk control.

Risk identification

This involves pinpointing the likely risks associated with the proposed survey. The RICS (2004) guidance in Appendix V lists the main physical risks associated with surveying buildings. (Other risks are dealt with in the next chapter.) Typical health risks include falls from heights, falling through defective floors and roofs, stepping on or hitting sharp objects, exposure to deleterious substances such as asbestos and silica dusts, contaminated air, etc. For example, inspecting roof voids and other enclosed spaces poses a number of dangers for any surveyor. Falling through the space between the ceiling ties or receiving a head injury from the protruding slate nails are common risks when inspecting roof spaces – particularly if the surveyor has an insufficient level of temporary lighting.

Risk measurement

This includes the attempt to evaluate the degree of risk. The two main methods are qualitative (e.g. low/medium/high) and quantitative (e.g. 1, 2, 3, 4 and 5, with 5 being the highest level of risk). For most straight-forward building inspections a qualitative approach to measuring risk should suffice.

Risk control

Once the risks have been assessed qualitatively the extent of precautions required can be determined. Normally only risks identified as medium and high would involve or necessitate some controls. For example, inspecting a roof space could be classed as a medium risk. This would necessitate controls such as providing PPE (personal protective equipment) such as hard hat (with front mounted torch), overalls (with high visibility stripes) and facemask. Crawlboards may be required to provide safe access across the ceiling ties if there is no decking in the roof space. Other health and safety precautions that the surveyor should consider taking, especially in case of an emergency, are:

- First aid kit.
- Mobile phone (with charger in car).
- Personal attack alarm.
- Safety whistle.
- Safety spectacles/goggles.
- High visibility jacket.
- Safety shoes/boots with steel toe-caps and steel soles – to protect the surveyor in dilapidated property or on a building site where there may be sharp objects on the floor, such as nails or spikes, which can puncture the foot.

Expert witness

A surveyor may be called upon to act as an expert witness in relation to a court case involving the survey of a property. Before volunteering for such a role, however, surveyors should follow Part 35 of the Civil Procedures Rules (CPR) which deals with expert witnesses and came into force on 26 April 1999 (Watt and Partners 2004).

Most importantly surveyors acting as expert witnesses must bear in mind where their ultimate loyalty lies – with the court. According to the CPR (Watts and Partners 2004) 'it is the duty of an expert to help the court on matters within his/her expertise. This duty overrides any obligation to the person from whom he/she has received instructions or by whom he/she is paid'.

As highlighted by Watts and Partners (2004), the expert witness's report should follow CPR requirements and contain a number of key points:

- The report should contain a statement setting out the brief and instructions given.
- It should be written in the first person (unlike most building survey reports, which are usually written in the third person).
- Any assertions or findings must be substantiated.
- The report should be addressed to the court.
- It should contain the expert's curriculum vitae.
- The report should follow a coherent structure (such as the model form of report produced by the Academy of Experts).
- It should follow the criteria for report writing discussed in Chapter 17.

Professional practice

In the realm of professional practice, particularly for undertaking building surveys, it is important to differentiate between personal conduct, professional conduct and professional competence.

Personal conduct

Personal conduct covers deficient or unacceptable performance or behaviour due to factors other than those associated with the exercise of professional skills and may include:

- Bullying.
- Sexual or racial harassment.
- Lack of probity (i.e. unscrupulous behaviour).
- Lack of reliability and poor timekeeping.
- Acting on duty under the influence of drugs or alcohol.
- Inappropriate or criminal behaviour.
- Inappropriate use of employer's facilities.
- Failure to follow organisational policies and procedures.

Professional conduct

This covers deficient/poor/unacceptable performance or behaviour arising from the exercise of professional skills and may include:

- Neglect or disregard of professional responsibilities to clients.
- Failure to comply with the relevant professional guidance (e.g. on property surveys – Residential Property Faculty 2004; RICS 1998).

- Any abuse by the surveyor of his or her position of trust, including a breach of professional confidence, or any form of indecency towards a client.
- Inappropriate or unacceptable attitudes and behaviour towards clients and their employees/colleagues, and the public.
- Unprofessional and inappropriate attitudes and approach to work – such as failure to comply with the RICS's nine core values: (1) Act with integrity. (2) Always be honest. (3) Be open and transparent in your dealings. (4) Be accountable for all your actions. (5) Know and act within your limitations. (6) Be objective at all times. (7) Never discriminate against others. (8) Set a good example. (9) Have the courage to make a stand.

Professional competence

Performance that is deficient over a period of time is usually considered 'poor performance'. On the other hand poor performance might be the committing of a specific, though isolated, serious incident. Poor performance may encompass:

- Failure to keep professional skills and knowledge up-to-date (e.g. not fulfilling the profession's Continuing Professional Development (CPD) requirements; or in the case of trainee surveyors, failure to develop professional skills and knowledge appropriately).
- Failure to work effectively with colleagues.
- Failure to recognise the limits of professional competence (e.g. undertaking work that is beyond one's area of skill – such as undertaking detailed inspection and testing of electrical or lift installations).
- Failure to consult senior colleagues as appropriate.
- Attempting to undertake techniques in which the practitioner has not been appropriately trained.
- Failure to communicate effectively with clients.

1.6   Contracts and fees

Agreement and letters of confirmation

The surveyor's duty is a contractual one and, therefore, depends on the terms of the agreement made between the surveyor and client. The client employs the surveyor to look after his interests and thus the surveyor becomes his agent for all purposes relating to the examination of the building. To avoid misunderstanding a contract for professional services should always be put in writing (RICS 2004).

A simple contract requires:

- An offer.
- An acceptance.
- An intention to create legal relations.
- Consideration – which is the bargain element.

Thus, if a surveyor agrees either verbally or in writing to provide a report a simple contract will be formed. In a professional situation the courts will assume that there is an intention to create legal relations. The consideration, that is the bargain element, will be the agreed fee. The letter should express clearly what the two parties have agreed and the duties should be clearly enumerated, together with any exclusions such as the examination of inaccessible areas. This is important for the protection of the interests of both parties. In the event of a dispute it is important that a formal agreement showing the intention of both parties is produced. If no written agreement exists the surveyor will have difficulty in suing his client for fees. If the surveyor is engaged by a limited liability company or public body it is advisable to obtain a formal letter of appointment.

Experience shows that many surveyors have difficulties over fees with their clients. This is usually due to the lack of a clear arrangement at the outset. In order to avoid possible problems it is strongly recommended that the surveyor's fees are confirmed in writing including those of any consultants he considers necessary to examine the various services. This matter can then be included in the letter confirming the client's instructions.

There is no fixed scale of professional charges for building surveys. The various professional institutions simply state 'a fee by arrangement according to circumstances'. Various methods have been tried for assessing fees, but it is now generally agreed that the fairest means of calculating should be on an hourly basis at an agreed rate per hour, plus travelling expenses, and if necessary hotel expenses unless other arrangements have been made. The hourly rate should naturally include all office overheads and expenses (see Table 1.3).

It often happens that a client desires to know the amount he will have to pay his surveyor before giving instructions to proceed. It is not easy to assess charges from a client's letter or a description given over the telephone. A preliminary visit to the property, preferably with the client will soon provide the surveyor with a good guide as to the approximate number of hours which the survey will take. This preliminary 'walkabout' is especially important if the property is extensive, i.e. a large commercial or industrial complex. The surveyor will then be in a position to ascertain whether or not a professional consultant is required to examine and report on a particular service or a builder is required to test drains and provide general attendance. Here again a clear understanding should be reached with the client that a consultant or builder is necessary and that the client will be responsible for the fees. However, it sometimes happens that the survey is being undertaken for a client for whom the surveyor regularly

Table 1.3   Typical costs (at 2005 prices) and timings for property surveys of low-rise dwellings.

Level of survey	Indicative cost range (a)	Average time to complete survey (a)
Scheme 1	£100–£300 (b)	<1 hour
Scheme 2	£300–£700 (c)	Between 2 and 4 hours
Scheme 3	£500+ (c)	Between 4 and 8 hours

Notes
(a) For a typical size dwelling up to two storeys in height.
(b) Alternatively, the required fee for a typical mortgage valuation survey may be based on around 0.1–0.15% of the property's capital value.
(c) An agreed hourly rate is usually the basis for determining the level of fee required for these surveys.

acts. In such cases the client will usually instruct the surveyor to proceed with the survey on the hourly rate previously agreed. Surveyors are often instructed to carry out a building survey and at the same time prepare details of alterations required by the client. As far as fees are concerned it is recommended that the two matters are dealt with separately. This point should be clearly established with the client at the outset.

Six examples of letters of contract are given below. They are given as a guide and will naturally need to be adapted to suit each individual case.

Letter A

Surveyor confirming appointment with details of fees.

_____Date

Dear Sir,

Shop and storage property: <u>No. 14 Blank Street, Sevenoaks, Kent</u>.

Further to our telephone conversation today, I am pleased to hear that you intend to appoint me as your surveyor and that you wish me to carry out a building survey at the above-mentioned property.

I also confirm that my fees are calculated on an hourly basis. As agreed during our telephone conversation we will meet on the site at (insert time) on the (insert date) for a preliminary walk round and discussion. At this meeting I shall have the opportunity to assess the time required to carry out the survey and prepare the report. I shall then be in a position to submit an approximate fee for your consideration.

I understand that the property is vacant and that the keys are in your possession. In the meantime if you have any queries please do not hesitate to telephone me.

Yours faithfully

Letter B

Follow up to letter A confirming fees and stating the areas which are inaccessible.

                              _____Date

Dear Sir,

Shop and storage property: <u>No. 14 Blank Street, Sevenoaks, Kent</u>.

I refer to our meeting at the above and our subsequent telephone conversation, and confirm that you wish me to proceed with the survey and that you agree to my fee amounting to approximately (insert fee).

    During my preliminary examination I found that there is no access to the roof space in the single storey building at the rear of the property. I am therefore unable to examine the internal roof structure and report on it. However, I will examine the slate covering and report any defects found. I will let you have my report by (insert date).

Yours faithfully

Letter C

Surveyor confirming appointment with client previously known to him.

                              _____Date

Dear Mr Jones,

Warehouse property: <u>Blank Street, Tunbridge Wells, Kent</u>.

I thank you for your letter dated (insert date) instructing me to carry out a building survey and valuation of the above property. I will contact the agents (insert name) immediately and arrange to collect the keys. I understand that the property is approximately 80 years old and has been vacant for some considerable time. If I consider that the electrical and heating services require an examination by qualified engineers I will inform you immediately with details of their fees. My fees will be on an hourly basis as previously agreed.

    I will let you have my report and valuation in about six days' time.

Yours sincerely

Letter D

Follow up to letter C confirming appointment of consultants.

_____ Date

Dear Mr Jones,

Warehouse property: <u>Blank Street, Tunbridge Wells, Kent</u>.

Further to my letter dated (insert date of first letter) and our subsequent telephone conversation on the (insert date), I confirm that the electrical and heating installations appear to be in poor condition. I note that you agree that I appoint consultants to examine and report on the two installations. I have worked with the following firms on other surveys and am confident that they will submit a reliable report.

(insert names and addresses of the two consultants)

I have approached the two firms and explained the extent of the service required and ascertained their fees. They will collect the keys from the estate agent's office. Their combined fees for this service amount to approximately (insert amount) and their reports will be submitted to me in about six days' time. By that time my report and valuation should be completed.

Yours sincerely

Letter E

Surveyor confirming appointment with a client who requires a Schedule of Condition on a Leasehold property (see sample in Appendix II).

_____ Date

Dear Mr Brown,

Detached house: <u>10, Blank Street, Eastbourne, Sussex</u>.

I was interested to hear that you propose to take the above property on a (insert number of years) year lease and that you require a report as to the condition of the building. I thank you for letting me have a copy of the lease and I have made a note of the repair clause which sets out your obligations.

I note that you are anxious to sign the lease and occupy the house. I will therefore proceed with the survey as quickly as possible. I understand that the keys are with the freeholder (insert name) and that they have agreed to meet me at the property on (insert date) at (insert time).

My fees are calculated on an hourly basis and amount to approximately (insert amount) per hour plus travelling expenses.

Yours sincerely

Letter F

Follow up to letter E enclosing Schedule of Condition and fee account.

_____Date

Dear Mr Brown,

Detached house: <u>10, Blank Street, Eastbourne, Sussex</u>.

Further to my letter dated (insert date) I have to inform you that I have completed my survey and enclose two copies of the Schedule of Condition. The Schedule should of course be signed by yourself and the landlord and a copy kept with the lease. If you have any queries I shall be pleased to give you any further information you may require.
 I also enclose my account for fees which I trust you will find satisfactory.

Yours sincerely

Cost and timing of surveys

The cost of a property survey will depend on three key factors: (1) the type of survey required, (2) the size and location of property being surveyed, and (3) the risks involved. All of these factors will impact on the time taken to carry out the survey effectively. The surveyor spending an inadequate amount of time to undertake the inspection is one of the main causes of poor quality property survey reports.

Given the wide range of property types it would be difficult as well as unwise to prescribe a universal scale of fees and times for all building surveys. However, Table 1.3 gives an approximate indication of the likely costs and timescales for the three main levels of property surveys of a three-apartment detached/semi-detached dwelling not exceeding two storeys. A building survey of a larger dwelling (e.g. three-storey mansion) or multi-storey commercial property would clearly be more time consuming and thus more expensive.

2 Procedure and Equipment

2.1 Basic survey methodology

Criteria

Any building inspection or survey should be undertaken in a systematic manner. This will help minimise the risk of errors or omissions during the survey process. Complying with the following criteria should help in this regard.

Expertise

The qualified surveyor should have the adequate level of skill and knowledge to undertake the survey. A sound knowledge of building construction along with experience in carrying out the type of building inspection or survey required is essential. An awareness of obsolete construction methods (e.g. wattle and daub, Mundic, crux frame, etc.) is particularly important for investigating very old buildings.

Care

Use appropriate PPE and take sufficient time to carry out the survey, which should be done in a cautious, safety-conscious manner. For safety and security reasons two surveyors may be needed to undertake the survey – particularly if it is of a large empty building.

Being methodical

A systematic, step-by-step approach starting from the outside of the property, going inside and then outside, should be adopted to minimise oversights and errors. This approach is illustrated in Fig. 2.1.

Main stages

In summary the main stages of the process of surveying a building are shown in Fig. 2.1:

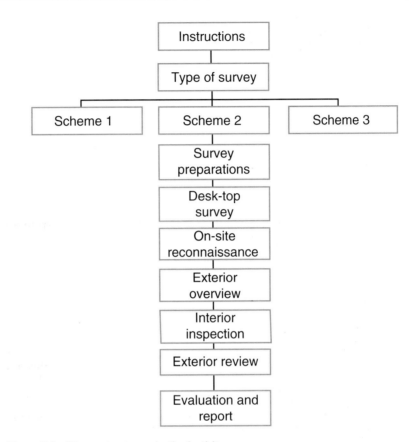

Figure 2.1 The main stages in the building survey process.

Instructions

A rigorous evaluation of the client's requirements needs to be made at this stage to determine the type of survey that would best suit their needs. It is important to follow the guidance of a relevant professional body such as the Royal Institution of Chartered Surveyors (RICS 2003, 2004). Any instructions must be confirmed in writing.

Preparations

Arranging access to the property being surveyed is important especially if it is vacant. This may mean having someone on site at the property with the keys to facilitate entry. It would also mean that the surveyor would not be alone when the survey commences.

The type of survey will determine the extent of equipment needed. Surveyors should in any event have a standard set of PPE (referred to in Chapter 1) as well as a basic range of building surveying equipment (see later in this chapter).

Desk-top survey

An examination of all documentation and other information related to the building being surveyed is advisable. This is best done before the building survey proper is undertaken as it prepares the surveyor more fully for the task at hand (Douglas 2002).

Typical sources of information are:

- As-built plans and drawings to show the building's original construction. In the absence of these a measured or dimensional survey may be required (see next chapter).
- Title deeds or other property documents in the local Record Office (e.g. to confirm the building's age – see next main section on dating of buildings – and any restrictive covenants).
- Ordnance survey maps of the area (e.g. to indicate the property's precise location and physical context).
- Geological maps of the area (e.g. Coal Board plans or British Geological Survey plans to give information on the presence of previous or existing mineworkings and typical ground conditions).
- The various Royal Commissions on Ancient and Historic Buildings – for photographic and other records of the property.
- Previous building control approvals (called building warrants in Scotland) or planning permissions – to indicate any past major changes or interventions to the property. The absence of these would indicate that any structural alterations etc. to the property had been undertaken unlawfully (see the section 'Illicit alterations' later in this chapter).
- Other local authority records – statutory notices or previously objected schemes by neighbours.
- Ecclesiastic records – particularly parish church documents relating to local inhabitants and their residencies.

Reconnaissance survey

Establishing the physical and spatial contexts of the building on site is critical. A number of issues may need to be considered at this stage:

- What is the actual physical context of the property? Is it different from the map/plan? Is the property detached/semi-detached/terraced etc?
- Are there any major obstacles or natural formations that might have an impact on the property? For example, are there any trees closer to the property than their mature height? If so there is a higher risk of tree root damage.
- Is the property in a mining area? Is the mine redundant or still active? If active, this would increase the risk of mining subsidence.
- Do any of the surrounding properties display common faults that would suggest subsidence or other problems with the topography?

- Are there any access problems that may adversely affect the use or worth of the property?
- Are any of the adjoining/approach roads in a poor condition (which might suggest that they have not yet been adopted)?
- Is there space for future lateral extensions to the building being surveyed?
- Is the property in a low-lying area that is susceptible to flooding?

On-site survey

An on-site survey comprises three stages: exterior overview, interior inspection and review of exterior and surrounding site (Douglas 2002). The external appraisal focuses on the main elements of the building and records any significant defects. Usually this is done using a top-down approach – inspecting the roof, then the walls of the upper floors down to the ground or basement level. The internal inspection is done on a room-by-room basis, normally working clockwise. It should record any defects in the floors, walls, ceilings and fittings. Any problems relating to the use and condition of the building as well as to what extent any defects found externally are affecting it internally should be recorded. In the absence of the original drawings, the on-site stage may require a measured or dimensional survey of the building. A building survey might prompt the need for an appraisal of the property's structural capacity – thus requiring a specialist structural inspection.

Evaluation and report

Once the survey has been completed and the evidence collected, the surveyor should then evaluate his findings. These will be articulated in the building survey report (see Chapter 17). Any verbal feedback given by the surveyor to the client before submitting the report should be kept brief. It should also be consistent with any subsequent written feedback. The latter should not be so concise as to overlook any important issues, which might expose the surveyor to a negligence claim (see Chapter 18).

Dating buildings

One important piece of information that should be established either before or during a survey of a building is its date of construction. This is required for a variety of reasons. First, in most Scheme 1 to 3 surveys it is a normal requirement to establish the property's age as a determining factor in a loan application. All home condition reports require what is now called an 'age assessment' (Melville and Gordon 2004). The expression 'the property was built around the late 1890s', however, is normally sufficient for the purposes of a Scheme 1 survey report.

Second, the owner will usually want to know how old the property is for personal as well as general interest reasons (Cunnington 1988). This information will also alert them to potential problems or replacements ahead.

Third, the age of the property may need to be established for conservation reasons. Its date of construction would indicate the building's likelihood or potential of being listed – particularly if it predates the nineteenth century. This could also have maintenance and adaptation implications for the property (Douglas 2002). Moreover, dating is vital in formulating proposals for what should happen to a building, and will therefore form a vital part in its investigation. Guidance on conservation plans from various agencies such as the Heritage Lottery Fund, Historic Scotland (2000a), etc. has emphasised the importance of demonstrating an understanding of the building when applying for listed building consent.

Fourth, the age of a building could indicate its form of construction and suggest the materials that were likely to have been used in its structure and fabric. In addition, defects are usually related to forms of construction commonly adopted at particular times (Melville and Gordon 2004). For example, if a property was built before 1875 it is unlikely to have a damp-proof course. Thus this could indicate the risk of bridging or rising dampness in the building.

Finally, an age assessment is a necessary prerequisite for calculating reinstatement costs for fire insurance purposes (Melville and Gordon 2004). The age of the building will indicate its fire risk for the insurers and help them establish the extent of any uncommon remedial works.

Dating is an essential first step in assembling evidence on a building and interpreting it (Kelsall 2001). External evidence is derived mainly from written and graphical sources of information about the property (see below). Internal evidence is obtained from the building itself – its style, size and shape, morphology, form of construction and materials used.

However, many buildings can present a variety of difficulties when trying to establish their date of construction. No guidance therefore can conclusively demonstrate how to achieve this objective in all cases. Even dating based solely on external inspections of the believed age of a building can deceive the experienced surveyor (Watts and Partners 2004). For example, some owners may undertake alterations to make their building look older or younger than it is by installing Georgian-style timber windows in a Victorian property or vice versa. Others may eliminate much of the physical evidence of their property – such as removing a timber-framed building's physical evidence by stripping it back to the original and covering the main members (Kelsall 2001).

Establishing the date of a building's construction can be helped using the main methods summarised below (Douglas 2002).

Building inscriptions

For example, confirming foundation stone laying ceremony; date plaque – usually above the front doorway; plaque containing other data such as the designer and/or builder's name. Care has to be taken with such inscriptions, however, to ensure that they are accurate. One cannot be certain, for example, that a later owner did not add an inscription or that it dates more than a specific phase of development (Kelsall 2001).

Documentary evidence

For example, legal documents such as title deeds relating to the property; drawings and illustrations of the building as proposed, which even if undated might still suggest a distinct time period because of their style or quality; as-built drawings showing date of building control approval; local ecclesiastical records; local building guides such as Pevsner's (e.g. Forsyth 2003). Archival material such as land registration documents or estate records deposited at the local Record Office indicating when the property was registered can provide good primary evidence of dating (Cunnington 1988). Books on the designs of notable architects may prove useful sources of information about distinctive buildings that they were associated with (Kelsall 2001).

Maps and plans

A dated Ordnance Survey plan, for example, showing the presence of the property can indicate its likely date of construction. By working backwards through OS maps or other archival material such as street directories, statutory lists, valuation rolls or rates books, one can determine when a building first appeared. If the property is absent from earlier plans this suggests that it was built after the date indicated. In addition, gaps in the series or significant changes in value can be a clue to adaptations such as extensions or rebuilding (Kelsall 2001).

Architectural style

For example, Art Deco style emerged during the 1930s; bay windows are a common feature of Victorian buildings from the 1880s onwards. Comparative analysis with other known and dated examples is usually required when attempting to date an old building (Kelsall 2001).

Form of construction

Cavity walls, for example, did not become popular until the early 1920s; no-fines concrete started being used in residential construction in the late

1940s (see Chapter 7). Measuring the thickness of the walls will help identify if they are solid or cavity construction – 225 mm for solid construction, whereas cavity wall construction is at least 250 mm thick.

Radiocarbon dating

This technique is used by above-ground/buildings archaeologists investigating old properties to determine the date of organic materials. This scientific method will only indicate the likely age of organic materials used in the building, such as timbers, to within ±50 years. The analysis of replacement timbers would give a misleading result if they are of a different period from the original timbers.

Dendrochronology

This technique involves counting the growth rings of timber members to determine their age by reference to a database of tree types. Again, this is another specialist archaeological technique (Switsur 2001), which would probably only be used in rare circumstances relating to a building survey.

Thermoluminescence (TL)

This scientific method is used for dating pottery and other fired clay and quartz-based materials. It often provides the only way to distinguish medieval or Tudor bricks and chimney pots from Victorian reproductions (Switsur 2001).

Local place names

For example, Robert Burns Street in Ayrshire is unlikely to have existed before the Scottish bard was born in 1759, unless of course it was renamed at a later date or relates to another person of the same name.

Data capture

There is a variety of ways for surveyors to record their survey findings, such as a checklist, dictating machine, digital still/video camera, etc. A proforma checklist in a notepad is a useful *aide-mémoire* for taking notes on site. Its format should be flexible enough to accommodate most building types yet allow space for the recording of novel or unconventional details relating to the property. The big drawback with checklists is that they can inhibit originality in the report format and fail to cover all contingencies.

Care should be taken when using a dictating machine to record the survey findings. The prose in a spoken word account does not directly translate well into a written form. It must be rephrased to a professionally written style of prose – usually in the third person. For example, Glover (2003) reported a case where a building surveyor was found negligent in undertaking a building survey of a farmhouse after relying on a dictating machine to record his findings. Appendices III and IV show two examples of proforma checklists that can be used by surveyors on site.

2.2 Preliminary operations

Various methods are used to carry out a building survey, depending on the experience and skill of the surveyor. Most surveyors have their own methods of examining existing buildings. The important thing is that it is essential that some system is adopted which will ensure that the building is thoroughly examined and all defects investigated. If the surveyor is unacquainted with the locality it is worth his while to make a preliminary inspection. In the case of domestic property he will be able to appreciate its general condition and form an idea of its age, type of design and construction. There may be other special considerations: for example in a mining area, the provisions of the Coal Mining Subsidence Act 1991 must be borne in mind.

It is generally convenient to inspect the outside of a building first and then examine the inside. However, a preliminary 'walk round' the building should be made first. This will give the surveyor a general idea of the condition, age and layout of the building. When dealing with commercial and industrial buildings several storeys high, there are often inaccessible areas which require examination. In such cases it is advisable to make arrangements with a local builder to provide ladders or scaffolding and at the same time carry out drain tests if the surveyor considers it necessary. These arrangements should be made as early as possible and confirmed in writing to the builder giving dates and times, and the address of the property. In addition the builder will require some guidance concerning the height of the building and details of the type of drain test required. The client should be informed that a 'builder's attendance' is necessary and that he will require payment for his services.

If it has not been possible to make a preliminary visit to the property then it is advisable for the surveyor to arrive an hour or so before the appointed time in order to familiarise himself with the property before the builder arrives. This early visit will give the surveyor time to make himself known to the occupier. He will also have the opportunity to 'walk round' the premises and locate entrances, roof access, and the best position for ladders etc.

If the building is a commercial or industrial property with extensive heating, electrical, gas and lift installations then a specialist report on the condition and efficiency of the services will be required. The specialist will require a certain amount of information from the surveyor concerning the following:

- A brief description of the installation and equipment which requires examination and testing.
- Position of intake cables and meter etc.
- Position of lifts and motor room. A simple free-hand sketch showing the location of the lift shaft and motor room would be useful.
- Type of heating installation to be examined and location of boiler room etc.

The specialists will charge a fee for their services and the surveyor should make sure that he explains this matter when confirming his instructions to the client.

If extensive repairs and maintenance works are envisaged, then measured drawings should first be prepared. These drawings need not be elaborate, the object being to attach a copy to the surveyor's report clarifying the position of fractured walls, leaning chimneys and defective floors etc. For surveys dealing with the structural condition of the buildings, it is expedient to itemise the elements to be examined. This form or 'check list' reduces the risk of omissions and forms a sound basis for the preparation of a report. The writer has used the checklists shown in Appendices III and IV for many years and has found that they are a good guide to the various items a surveyor should take into consideration during his inspection. Naturally, some of the items will require adjustment to provide for certain peculiarities of construction or perhaps be deleted to suit different types of building. The writer does not claim that the lists contain every item a surveyor may find during his examination, but it is considered that they provide a methodical approach to a building survey.

2.3   Property risks

Three key defect risk areas

Deterioration and failure of a building's structure and fabric are generally associated with one or a combination of the following three principal causes (Addleson and Rice, 1992):

- Dampness: see Chapter 8 and Burkinshaw and Parrett (2003)
- Timber decay: see Chapter 9 and Ridout (1999)
- Cracking and movement: see Chapters 5–7 and Dickinson and Thornton (2004).

All buildings to a greater or lesser degree are exposed to these major influences (see Fig. 2.2). Within each of these three basic causes, however, are numerous agencies or sources such as condensation (i.e. dampness), dry rot (i.e. timber decay) and subsidence (i.e. movement). As pointed out by

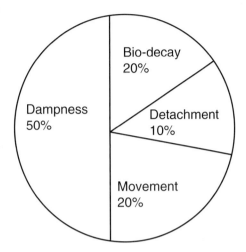

Figure 2.2   Analysis of building defects (Ransom 1987).

Addleson (1992), the causes are few but the specific agencies (or sources) are many. This often makes it tricky to achieve the correct diagnosis without a thorough investigation.

Thus 'the process of diagnosing faults in buildings is often made difficult in a specific situation because of the presence of relatively few anomalies (or symptoms) but with a wide range of possible causes and defects' (Addleson 1992). Moreover, the precise trigger mechanism (i.e. agency) of a building fault is not always apparent even though the cause (e.g. dampness) may be obvious. The same symptoms can be caused by different agencies. A variety of agencies working in combination can trigger a defect. Rainwater penetration, for example, can lead to dry rot, which is the most destructive form of timber decay in buildings (Ridout 2001). Dampness is of course the 'cause' of the problem in this case, but the source of the moisture may not be so obvious. In any therapy to eradicate fungal attack, identifying and eliminating the source of moisture is a primary control measure (Palfreyman and Urquhart 2002).

Similarly, a single cause can result in several symptoms. For example, penetrating dampness through a blocked cavity wall can cause corrosion of the wall ties (leading to horizontal expansion of the mortar joints) and moisture staining or soiling on the inside face of the inner leaf of brickwork.

It is no surprise that these three key risks are the main defect categories specifically addressed in the RICS's Homebuyers Survey and Valuation pro-forma. Generally speaking, if a building were unaffected by these three defects, prima facie its condition could be rated as being reasonably satisfactory. However, the building may contain risks such as deleterious materials or others as listed above, which could adversely affect its value, potential or safety. One must, therefore, be careful about jumping to the conclusion that such a building is defect free (Curtin 1993).

Other technical risk areas

The above focus on the three defect risk areas, however, should not distract the surveyor from considering other causes of building failures. For further guidance on general property defects, consult Hinks and Cook (1997), Son and Yuen (1996) and Carillion (2001).

Buildings are complex artefacts that are vulnerable to a wide range of risks. Any of those risks can compromise a property's functionality, stability or value. Every surveyor, therefore, should check if the following typical risks are present or potential in the property being surveyed:

- Accessibility: Lack of accessibility in the premises in terms of compliance with the Disability Discrimination Act?
- Subsidence: Shrinkable clay subsoil? Trees nearby? Mineworking area?
- Settlement: Overloading? Differential settlement at lateral extension?
- Heave: Expansion of clay subsoil likely? There is a greater risk of this if flooding or other source increases the level of the site's water table.
- Flooding: River/stream/sea nearby? Property in flood plane area?
- Soil erosion: Particularly in areas at the seaside where the coastline is eroded.
- Pollution: Noise/disturbance from adjacent/nearby properties – anti-social neighbours, industrial processes?
- Deleterious materials: Asbestos, lead, Mundic, woodwool slabs, high alumina cement concrete (HACC), etc. (see Douglas 2002).
- Fire: Flammable materials in the construction or stored in the property? Fire caused by accidental ignition of flammable materials or arson? Overheating of loose or faulty electrical connections is another common cause of fires in buildings.
- Instability: Bomb or gas blast, earthquake, storm damage, settlement, dilapidation, etc.
- Insecurity: Vulnerability to break-ins, ram-raiding, etc.
- Obsolescence: See Douglas (2002).

- Redundancy: See Douglas (2002).
- Sick building syndrome: Particularly air-conditioned buildings built or refurbished since the early 1970s.
- Deficiency: Lacking in performance – acoustic, fire, thermal, structural, etc.).
- Degradation and deterioration: See Richardson (2001).
- Non-traditional construction: See Chandler (1991) and Douglas (2002).
- Illicit alterations: See next section.

Illicit alterations

The surveyor should be on the lookout for alterations to the property being surveyed. Some property owners often undertake many such changes without obtaining the proper consents, e.g. planning permission or building control approval. Even where such approvals have been obtained, the work might have been carried out either incorrectly or not in accordance with the proper specification.

In a wide variety of cases such changes to the structure and fabric of a building are obvious. Sometimes, however, they are not. Typical illicit alterations made by householders are:

- Removal of chimneystack (above roof): without providing background ventilation to the remaining section of the now redundant flue.
- Removal of chimneybreast: without providing adequate support to the upper section of the chimneystack that is left, which would make the remaining section of the chimney unstable.
- Removal of other elements: removal of stone mullions to provide larger window opening can overstress the lintel supporting the opening leading to it sagging; or removal of partitions or loadbearing walls without checking for support of structure above.
- Forming openings in loadbearing walls: without installing an appropriate lintel to support the masonry above; or making good the walling around the opening.
- Changes in layout: such as modifying the layout of the kitchen/bathroom without complying with the Building Regulation requirements for circulation space, etc.
- Modifications to the electrical installation: adding power circuits (such as for a shower unit) but not installing the correct fuse or wire rating will increase the fire risk as a result of possible overheating in the wiring connections (see Chapter 14).
- Loft conversions: decking the roof space to provide informal habitable floor surface but retaining a 'Ramsey' ladder access instead of providing a proper staircase.

- Window replacements: replacement of existing windows with inadequate or different windows – for example, windows with only deadpanes and hopper sashes in upper floor rooms provide no means of escape; window frames without adequate trickle ventilation; or window at first floor and above with the sill below the 1.05 m minimum safety height.

2.4   Equipment for measured drawing surveys

The principal instruments necessary for a survey requiring a measured drawing are as follows:

- A 30 m steel tape, 2 m folding survey rod and a 2 m flexible steel tape.
- If mouldings are to be measured and reproduced, a length of Code 3 sheet lead about 25 mm wide and about 800 mm long should be included.
- A bricklayer's level and straight edge about 2 m long.
- An A4 tracing paper sketch pad could be used for the sketches. A metal board with a stout spring clip will hold the pad secure and provide a firm surface.
- When no builder is involved it is often necessary to gain access to high ceilings and roof spaces. A telescopic aluminium ladder capable of extending to about 3 m will meet this need, and will be found useful for most surveys. This type of ladder is usually about 1 m long when retracted and can easily fit into a car (see Fig. 2.4).
- For surveys of existing buildings never wear your best suit! Some old clothing or preferably a boiler suit or nylon slip-over trousers with an elastic grip top will be found most useful.
- A builder's safety helmet should always be worn when surveying empty old buildings, the inside of pitched roofs and confined spaces.
- Photographs can be of great help when setting out elevations, particularly elevations containing ornamental stonework and elaborate joinery details. The photographs can then be checked against the measured drawing details. Photographs are also useful where a 'listed' building is being altered or an application for demolition is made in which a dispute may arise. A digital camera is probably the most suitable for this type of work, but for large, complex elevations a whole plate print 16.5 × 21.6 cm or 20.5 × 25.4 cm will convey more detail. This type of print is better if a coloured picture is required to be exhibited or published. If the property is extensive and the surveyor's photographic knowledge is slight, it is often wise to engage the services of a professional photographer. If the client is proposing to carry out an elaborate conversion job, the extra cost would no doubt be justified.

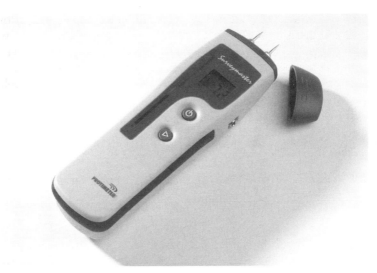

Figure 2.3 'Surveymaster SM' moisture-reading meter (courtesy of Protimeter Ltd).

- A pair of binoculars is useful when examining roofs, parapets and chimney stacks.
- A small pocket compass should also be carried to check the orientation of the property.
- An electric torch of the rubber cased type is preferable. They have a better 'grip' and stand up to a fair amount of knocking about. A set of spare bulbs and batteries are useful additions.

2.5   Equipment for examining defects

The following paragraphs describe the equipment necessary for the examination of structural defects. Two sets of equipment are outlined: a general purpose kit for all types of investigations, and a specialised kit for more complex work where a more extensive investigation is required. The general purpose one should include all the items described under 'Equipment for measured drawings' and these are, therefore, not repeated.

General purpose kits

These should include all the items described in section 2.4, plus the items listed here.

(1)   Electronic moisture-reading meter (around £350 at 2004 prices) (see Fig. 2.3). There are two main types currently in use, the conductiv-

Figure 2.4 Telesteps® telescopic ladder (courtesy of The Telescopic Ladder Company Ltd).

     ity and the capacitance type. These instruments measure conductivity and do not actually measure content.

(2)   Telescopic ladder (around £250 at 2004 prices) (see Fig. 2.4) A fairly recent innovation is the use of a telescopic ladder for building inspection purposes. The ladder is only 78 cm when closed, and 3.3 m when fully extended, locking at ten different intermittent heights in between.

(3)   A bradawl, pocket knife, cold chisel, small hammer and pliers are necessary for exploring on a limited scale. The bradawl and pocket knife have many uses, such as testing timber for rot and mortar joints for sulphate attack.

(4)   A vernier scale is useful for measuring the width of joints or cracks, frames and pipes etc.

(5)   An adhesive material such as Blu-Tack™ can be most useful as a temporary fixing for measuring tapes etc.

(6)   For recording the movement of cracks various methods have been tried out over the years. Strips of glass bedded in cement fixed across

a crack will break at the first sign of movement but with this method the amount cannot be measured. There is also the danger that the cement patch will shrink and show a hair crack or that the glass may have been broken by vandals. A proprietary sliding acrylic plastic tell-tale has now been developed (e.g. by Avonguard). The tell-tale consists of two plates which overlap for part of their length. One part is calibrated in millimetres and the transparent overlapping plate is marked with a hair line cursor. As the crack opens or closes so one plate moves relative to the other. Thus the difference between the hair line cursor and the calibration represents the amount of movement and is capable of monitoring 1 to 10 mm. This instrument can also monitor cracks at internal corners of walls and partitions. The instrument can be glued in position with a rapid setting epoxy resin. However, it does require a fairly flat surface. A very satisfactory method is to use crack markers at frequent intervals at about 30 mm either side of the crack. Small holes are cut into the brickwork and non ferrous pegs inserted set in cement mortar and left projecting about 5 mm. This method enables accurate dimensions to be taken across the pegs with vernier calipers. Alternatively, small centre-punched non-ferrous discs set in epoxy resin glued on either side of the crack are most efficient. Measurements are taken between the centre punch points with the vernier calipers. These two methods have the advantage of being accurate and can be fixed to irregular masonry surfaces.

(7) Mirrors (approx. 100 mm × 100 mm) can be very useful for inspection of otherwise inaccessible parts of a building, such as timbers in a roof space and the underside of parapet wall copings. Inspection mirrors attached to a telescopic handle are especially useful.

(8) In order to examine straight lengths of drain, reflectors are useful. They are placed in the channels of the inspection chambers and reflect the light along the drain.

(9) A magnifying glass is often helpful to identify the nature of fungi, small holes and the condition of surface finishes.

(10) Two or three sticks of wax-coated chalk will often be found useful for marking out defects.

(11) A plumb bob and line is essential when the plumb of a building is necessary.

(12) A combined lifting and cleaning key is useful for carrying out inspection chamber examinations.

(13) I have not included the various grades of pencils in the above list of equipment as this is obvious, but it is advisable to carry two or three. They are easily mislaid during the walkabouts and it is far easier to pick up a new one than go searching for the old one.

(14) Finally, although water is usually available on the property, a towel and soap or a pack of hand wipes are useful items to carry.

Figure 2.5   Surveyor using a borescope (courtesy of Olympus Industrial).

Specialised kits

Much of the equipment used for these non-destructive tests is very expensive to purchase. Most of them can, however, be hired from an appropriate supplier. Non-destructive investigations that may form part of a diagnostic or general survey can involve a variety of techniques and equipment (see GB Geotechnics Ltd 2001). Typical examples of these are as follows:

(1) Endoscopy: A fibre optic borescope (shown in Fig. 2.5) is a useful tool with which building surveyors can inspect hidden voids and cavities for problems such as bridging, wall-tie corrosion and fungal attack (Hollis 2000) (Around £1,000–£3,000 at 2004 prices).

(2) Electric drill: A power drill is useful for removing small cores of plaster or brick for moisture determination.

(3) Electronic hygrometer: The humidity and temperature of a building fabric can be measured with an electronic hygrometer. This instrument is essential for investigating condensation dampness (around £300 at 2004 prices) (see Fig. 2.6).

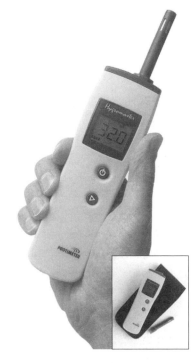

Figure 2.6 Electronic hygrometer (courtesy of Protimeter Ltd).

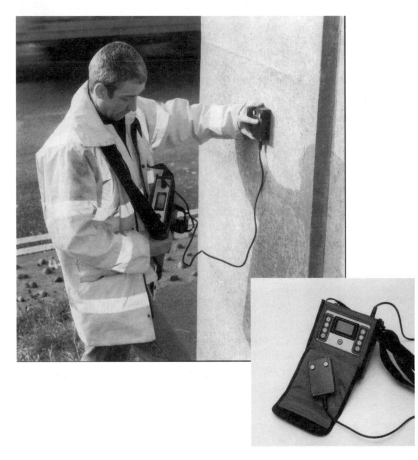

Figure 2.7 Reinforcement cover meter (courtesy of CMT Instruments).

Figure 2.8 'Speedy' calcium carbide moisture meter (courtesy of Protimeter Ltd).

(4) Cover meter: Engineers and surveyors use this inspection tool to detect the presence and depth of steel bars in reinforced concrete (around £1,500 at 2004 prices) (see Fig. 2.7).

(5) Calcium carbide ('Speedy') moisture meter: This instrument is used to ascertain on site the moisture content of particulate building materials such as mortar, brick, stone and concrete (around £700 at 2004 prices) (see Fig. 2.8). These meters have certain advantages over electrical moisture meters. They actually measure the water content of a sample. However, they do not automatically provide the source of that moisture.

(6) Plastic sample bags: These are useful for collecting samples of mortar, brick and other materials for laboratory tests. Ideally they should be clear plastic bags about 100 × 100 mm in size with a seal-

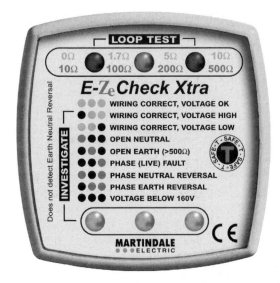

Figure 2.9   Earth loop impedance socket tester (courtesy of Martindale-Electric).

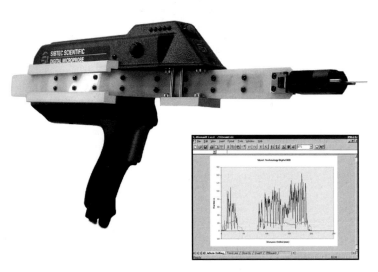

Figure 2.10   Digital microprobe wood-decay-detecting drill (courtesy of Sibert Technology Ltd).

able opening and a front white panel/label for marking or identifying the sample and date of collection.

(7) Socket tester: For use in undertaking earth safety test, voltage and wiring checks (around £50 at 2004 prices) (see Fig. 2.9).

(8) Micro-drilling: This is a very effective tool to determine the presence of decay in structural timbers using a special low-voltage hand-held

drill. It gives digital results which can be printed or viewed in a spreadsheet program such as Microsoft Excel™. (around £3,500 at 2004 prices). Sibert Technology Ltd (http://www.sibtec.com/) is the major supplier of this instrument (see Fig. 2.10).

(9) Impulse radar or ultrasound: This specialist non-destructive technique can be used to determine the extent of any void areas such as flue routes, hidden or blocked up spaces (GB Geotechnics Ltd 2001). It can also be used to help detect leaks in roof and wall elements.

(10) Thermography: An infrared thermal camera can be used as a non-destructive means to determine areas of heat loss, voids or leaks in the external fabric of a building (Hart 1991). Like impulse radar, it is a specialist technique that is normally only used in certain cases – such as for large commercial or residential buildings. Specialist companies such as IRT Surveys Ltd (www.irtsurveys.com) can undertake this work for a fee.

3 Measurement of Existing Buildings

3.1 Preliminaries

If plans of the existing buildings are not available a detailed survey will be necessary (Swallow et al. 2004). The basic principle of measuring existing buildings is easy to learn, but accuracy is essential. A small error in the measurement of an existing building can cause a great deal of trouble when setting out, so care is necessary in taking and checking dimensions in order to give reliable results. For instance, in domestic work the client's proposals may require alterations to kitchens and bathrooms. If this is the case a very accurate survey will be necessary in the areas in which baths or kitchen fittings might be fitted. Even a small error could mean considerable expense if a standard fitting had to be changed or altered.

With practice the art of sketching can easily be acquired. Tracing paper pads are available for survey work. The object of using tracing paper is that once the ground floor plan has been sketched it is easier to sketch the upper floors by sketching over the top of the ground floor plan. Dimensions can also be checked against the upper floor dimensions. A sheet of graph paper placed under the first tracing paper sheet will make it easier to keep the plan rectangular, and easier for the surveyor to maintain a constant scale. The graph paper squares should be to a scale of approximately 1:100 or 1:50 depending on the size and complexity of the building being measured. These approximate scales will allow space for all the small dimensions to be written in legibly. It is important not to allow the survey notes to get too cramped or complicated. If parts of a room contain complicated fittings, then record this on a separate sheet. It cannot be assumed that all rooms are rectangular especially in old buildings. If rooms are irregular in plan they are measured by taking diagonals across the room as shown in Fig. 3.1. It may appear that one diagonal is all that is necessary, but in practice it is wiser to take several as a check.

The majority of measured surveys, especially of large buildings, will be undertaken by two people, but there are occasions where the work must be done single handed. This is often the case in small occupied houses or bungalows which are often so crowded with furniture and fittings that it is not possible to use a tape. In such cases the survey must be carried out using a 2 m rod. Inevitably, there will be some loss of accuracy in this type of survey, but if the overall dimensions are added up and checked before leaving the building, the error need not be significant.

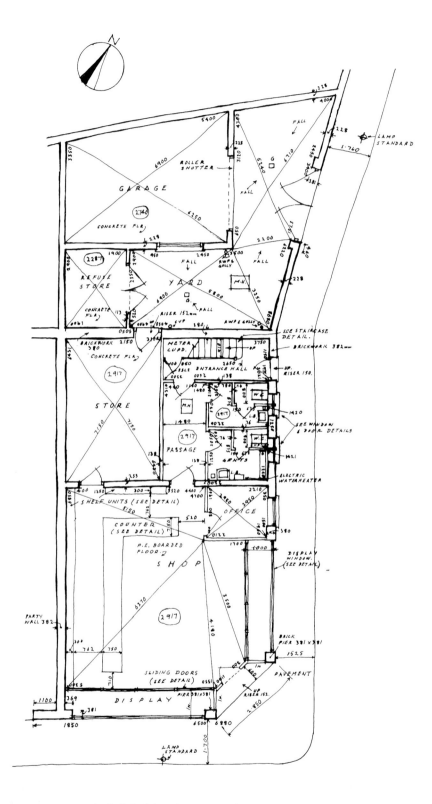

Figure 3.1 Survey of ground floor shop premises.

Internal finishes to rooms facing rear yard

Entrance hall
Floor s.w./p.e. flooring in 152 mm widths. sound condition.
Skirting 152 × 25. s.w. painted. F.G. condition.
Walls plastered & painted. Minor cracks on outer wall. General condition F.G.
Staircase All s.w. painted with polished hardwood handrail – solid panelled balustrade painted. Stair treads & nosing slightly worn. Decorative condition fairly good.
Ceiling plastered & painted with 150-cove cornice. Minor plaster cracks around front entrance. General condition F.G.
Doors Two flush s.w. doors to passage & meter cupboard and two panelled entrance door with glazed top panel & fanlight. All doors, frames & architraves are s.w. painted in good condition.

Store
Floor Concrete screeded. F.G. condition. A few minor cracks in screed.
Skirting 100 × 25. s.w. with rounded top edge painted. Timber sound paintwork in poor condition.
Walls Plastered & painted. Plaster in F.G. condition paintwork poor.
Ceiling Plastered & painted. Plaster sound paintwork poor.
Doors Internal door s.w. flush. s.w. frame & architrave. Timber & paintwork in F.G. condition. External door s.w. two panelled & one glazed top panel. Timber & paintwork F.G. condition.

Refuse store
Floor Concrete. F.G. condition.
Walls Fair face brickwork. Pointing & brickwork in good condition.
Roof Open timber joists (no ceiling) supporting boarded and felt roof. Joists 150 × 50 treated with preservative. All in good condition.
Double doors s.w. framed doors lined with matchboarding. Evidence of movement in tenon joints & minor wet rot in bottom rails. Paintwork poor. Feet of s.w. frame shows evidence of wet rot.

Garage
Floor Concrete. No serious defects apart from oil staining.
Walls Fair face brickwork. Brickwork sound, pointing in poor condition.
Roof R.c. slab. No defects.
Roller shutter Laths & winding gear in F.G. condition. But requires overhaul.
Metal window Metalwork sound. Paintwork poor.

Figure 3.1 Continued

3.2 Internal measuring

On arrival at the premises make a careful sketch of the ground floor of the building followed by a plan of each floor including the basement (if any). When sketches of all floors have been completed the measuring can begin.

Always take dimensions in a clockwise direction and the figures on the tape will be the right way up. It is better to measure the building to a set pattern in order to avoid omitting important dimensions. Measure along walls and partitions on the plaster face including door and window openings and piers etc. 'Running' dimensions should be taken; it is far more accurate and saves time. For example, a wall of a room 5 m long containing a 1 m wide door opening in the centre would not be measured in three separate dimensions 2 m, 1 m and 2 m. By using one length of tape the dimensions are booked 2 m, 3 m and 5 m. The dimension of the door opening should be in the actual door size. When booking running dimensions each figure should be booked near the end of the length to which it refers (see Fig. 3.1).

The thickness of walls and partitions should be obtained at door openings, making an allowance for any rendering or plaster. In old buildings plaster finishes are about 20 mm thick, but in modern buildings they are usually about 13 mm thick.

It sometimes happens that internal partitions cannot be measured by this method and where this is not possible, thicknesses can be obtained by measuring externally between two windows or doors and deducting from this dimension the sum of the distances from the respective window openings to each side of the internal partition or wall. This method is particularly useful when the thickness of a party wall is required as shown in Fig. 3.1. Where it is difficult to obtain the thickness of a partition by either of these two methods it will have to be calculated by deducting the sum of all the internal dimensions from the overall external dimensions. The difference is the thickness of the partition required. However, if the thickness of two partitions is required, the difference will give the total thickness which will have to be divided between the two partitions. The surveyor should not assume that internal walls or partitions on the upper floors will be over those on the floors below; always check by taking measurements inside and outside the rooms.

In old buildings floor to ceiling heights are often found to vary. The heights can be booked in the centre of the room, and enclosed in a circle (see Fig. 3.1). It is often difficult to take the height of a room against the wall where there is a large moulded cornice. In such cases it is easy to obtain the room height by measuring down from the ceiling to the top edge of an open door and adding this dimension to the height of the door. Industrial and warehouse buildings frequently have exposed steel roof trusses supported on stanchions. Measure from centre to centre of all stanchions and at the same time taking a note of their sizes. By using a ladder, details of the truss members and bearing connections etc. can be taken. At the same time the height from floor to the lowest horizontal truss member and the height to the apex can be obtained by using the steel tape.

The floor thickness can be measured at staircase openings or lift wells. Variations in floor thicknesses will be revealed by corresponding variations in floor to ceiling heights below. To obtain the exact size of floor joists and their condition, one or two boards should be removed. This is easily done in unoccupied property, but if the premises are occupied it will be necessary to remove some of the floor coverings. In such cases it is advisable to discuss the problem with the occupier. Usually, an amicable arrangement can be reached and with care the coverings can be lifted up and replaced without undue damage.

Floor levels should be checked with the bricklayer's level and straight edge. All beams and direction of floor joists should be noted and plotted on the sketches. This is particularly important when new openings are to be cut in existing walls. Should the joists run towards the proposed opening, provision will have to be made for the additional load that will be carried on the lintel over the new opening.

The heights of external door and window openings are not always required, but if alterations are contemplated they are essential if sections through the building are to be drawn. The survey sketches should show the plan and elevation of the window or door, and the dimensions must clearly indicate where they are taken; edge of frame, architrave or brickwork. Figure 3.2 suggests methods of measuring window and door openings where future alterations may be necessary. The window and door details are taken from the entrance hall and ladies toilet in Fig. 3.1.

Staircases are usually drawn on both floor plans, the foot being shown on the lower floor and the head on the upper. When measuring the treads and risers it should be noted that timber staircases usually have a nosing

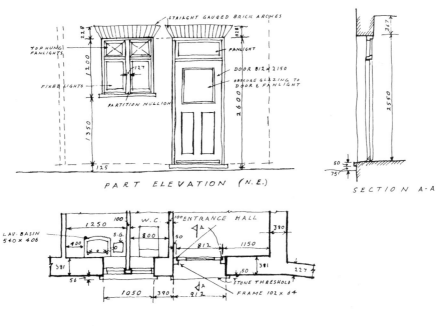

Figure 3.2   Detail of door and window openings.

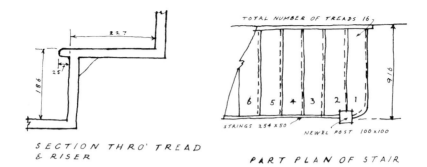

SECTION THRO' TREAD
& RISER

PART PLAN OF STAIR

Figure 3.3   Sketch of stair details.

projecting beyond the face of the riser. As the total of the sums of the width
of the treads will be greater than the total 'going' of the stairs, it is usual to
measure the treads from the face of the lower riser to the face of the top
riser. Enter this dimension with the number of treads on the survey sketch.
The risers are similarly measured from tread to tread (see Fig. 3.3). The plan
and the section shown should be sufficient for a 1:50 or 1:100 scale, but
details of newel posts and balustrade will probably require a larger scale.

It is often difficult to differentiate between the different forms of construction used for internal partitions. Partitions in older buildings often
consist of timber studs lined with lath and plaster. This type of construction can easily be discovered by rapping sharply with the end of a measuring rod which will produce a hollow sound in parts and solid in others.
Partitions in more modern buildings consist of lightweight concrete
blocks, clinker, hollow clay pots or brick. Clinker or concrete blocks 51 mm
to 63 mm thick often produce a slight vibration when struck with a closed
fist. Thicker partitions, however, are often too solid to give an indication
of their construction.

If alterations or additions are contemplated a careful note should be made
of the type and condition of the internal and external finishings adjoining
the area of the proposed alterations. It is quite extraordinary how often
this type of information is wanted. These details can be booked on a separate sheet or written at the side of the sketches as shown in Fig. 3.1 which
describes the internal finishes to the rear rooms of the ground floor plan.

Old wall and ceiling plaster is often suspect and should be carefully
examined; this particularly applies to lath and plaster ceilings which have
lost their 'key'. When plaster repairs are put in hand it usually involves
far more 'cutting away' than was originally anticipated. In such cases it is
often found more economical to replaster a complete ceiling or partition.
Signs of any defects in timber floors and roofs should also be noted at this
stage if they are likely to interfere with the additions or alterations.

It may also be necessary to take measured sketches of cornices, skirtings, window sills and joinery etc. if the proposed work is to match what

already exists. Full size sections of mouldings are taken with the lead strip described under 'Equipment' in Chapter 2. After carefully pressing the strip round the moulding, it is placed on a sheet of tracing paper and drawn by running round the inside of the lead pattern with a grade B pencil. All details should be lettered and this letter noted on the sketch plan of the floor concerned, and in the exact position where the detail was taken. Take note of the main service entry positions, meters, stop valves, heaters, boilers, radiators, lighting points and socket outlets etc. and plot these on the floor plan sketches. This sort of information is useful if alterations or extensions are proposed and service mains or meters have to be repositioned or renewed (Douglas 2002). In fact, the more information you can collect during this type of survey, the easier will be the subsequent work on the drawing board.

3.3   Roof space

The survey of a pitched roof can be divided into two parts, the interior and the exterior. When measurements have been completed on the top floor it is logical to continue into the roof space by way of the trap door, not forgetting to mark the position of the trap door on the top floor plan. Measurements should be taken from ceiling joists to ridge together with sizes, position and direction of all the structural timbers. A detail of the wall plate and joists at the foot of the rafters can also be taken, but the difficulty here is the confined space in the lower portion of the roof especially in low pitched roofs.

Whilst in the roof space ascertain the position and size of any water storage tanks or feed tanks etc. including a description of the cover and any insulation provided. A note of the size and direction of pipe runs and overflows should also be plotted on the sketch.

Care should be exercised when measuring rooms in roof spaces lit by dormer windows. The partitions do not always coincide with the partitions on the floor below and are usually of timber construction. If such is the case, the only point that can be fixed in relation to the floor below is the top of the staircase, and all dimensions should be tied back to this point as shown in the section through a roof space and ground floor plan in Fig. 3.4. Basements also do not always extend over the full area of the ground floor. They are similarly measured back to the staircase serving the ground floor.

3.4   External measuring

When all the internal measuring is completed, the external measurements can be taken. The horizontal dimensions can be booked on the floor plans. Carefully sketch all elevations and measure the widths and heights of all

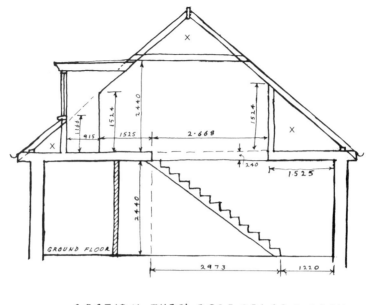

SECTION THRO' ROOF SPACE & GROUND
FLOOR. (X=VOID SPACES, WHICH SHOULD BE VENTILATED).

Figure 3.4   Section through roof space and ground floor.

openings; note the position of rainwater and soil pipes, and their size and material (Swallow et al. 2004).

If difficulties are encountered when measuring external heights, this can be overcome in brick faced buildings by counting the brick courses. To convert these to metric dimensions, four courses should be measured centre to centre of mortar joints in several different parts of the building, and the average rise used as a denominator. This method is particularly useful in obtaining the height of chimney stacks, although binoculars are sometimes necessary when counting the brick courses on a high building. Alternatively, if the building is rendered, the height of the windows, and the heights between the heads of windows and the sills of the windows above can be measured from inside the room through an open window using a 2 m rod. If no levelling has been done along the face of each external wall, the ground line of each elevation should be taken from a constant datum, such as a damp course line or top of a plinth. Measurements can then be taken from this line down to the ground level at appropriate points along each elevation (see Fig. 3.5). Heights of window sills and heads can also be related to this line. This information can then be checked against the internal heights.

Elevations with complicated projections such as cornices and pediments etc. should be measured with the aid of a plumb line, the plumb

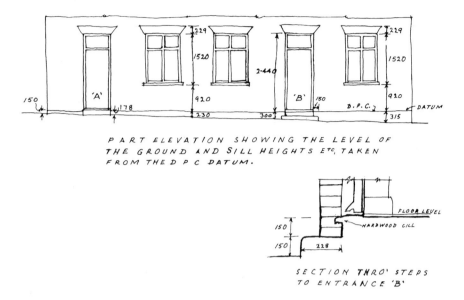

PART ELEVATION SHOWING THE LEVEL OF
THE GROUND AND SILL HEIGHTS ETC, TAKEN
FROM THE D P C DATUM.

SECTION THRO' STEPS
TO ENTRANCE 'B'

Figure 3.5 Part elevation showing the level of the ground and sill heights etc. taken from the DPC datum.

being secured at the foot and offsets taken from the line. The heights from which these offsets are taken must also be measured. This method of measuring is particularly important when walls are out of plumb or bulging.

As previously mentioned in Chapter 2 'Procedure and Equipment' photographs are a great help when surveying a building with complicated elevation details or with extensive defects. The photographs can be enlarged and form valuable records of the structure in its existing condition. They are also of value when preparing the specification.

If alterations or additions are proposed which involve alterations to the drainage system, it will be necessary to plot the positions, sizes and depths of inspection chambers, gullies and drain runs together with the positions of rainwater and waste pipes. The term 'inspection chamber' on all drainage systems is now in general use and is the term used in the Building Regulations.

Inspection chamber positions can be established by taking measurements to suitable points on the external or boundary walls of the building. Figure 3.6 shows an enlarged survey sketch of the drainage system taken from the yard survey in Fig. 3.1. The inspection chamber cover should be removed and the internal measurements of the chamber taken, together with the depth from cover to invert and the position and size of all pipe runs. Do not assume that the inspection chamber is the same size as the cover, as large inspection chambers are often corbelled over at the

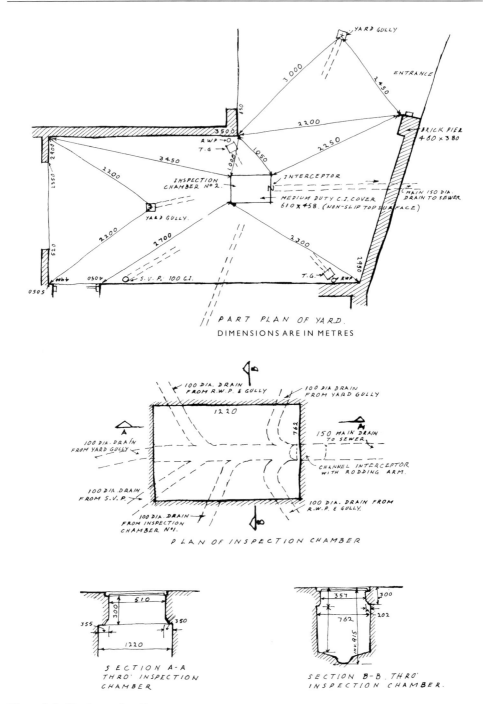

Figure 3.6   Drainage details.

top and fitted with a small cover as shown in Fig. 3.6. Trace all the drain
branches by flushing WCs and running taps in basins and sinks so that
every run is accounted for and at the same time note the condition of the
installation.

At this stage, care should be taken not to overcrowd the sketch sheets.
It is often wise to plot the drainage details on a separate sheet thus avoiding confusion when the survey reaches the plotting stage. This is why the
drainage system has been omitted in the yard of Fig. 3.1. Large irregular
areas are often found around the boundaries of a site as shown in the yard
of Fig. 3.1. These are measured by taking diagonals as explained earlier in
this chapter.

3.5   Levelling

There are several accepted text books on land surveying which deal fully
with the theory and practical side of levelling (e.g. Clancy 1991; Irvine
1995). Here we shall deal briefly with the subject only as far as existing
buildings and their surroundings are concerned. If extensive alterations or
extensions are visualised, it is essential to obtain some information concerning the surface slope of the ground around the building and its relation to the ground floor level of the building. It is desirable to make use
of ordnance survey maps, which are very accurate and can be used as a
check to show whether your angles are correct. If there is no ordnance
survey bench mark near the site, then some suitable fixed point will have
to be selected as a temporary bench mark, such as a door step or inspection chamber cover. This point should be carefully noted on the survey
plan so that it can be easily located at a later date.

Having set up the level the first reading is taken with the staff on the
bench mark followed with further levels on predetermined positions such
as steps, plinths, string courses, openings etc., on the various elevations
which are likely to interfere with the proposed extension. The level of the
existing floors must also be ascertained as well as that of the adjoining
ground. Provided there are no obstructions, it is advisable to use the same
setting up of the level for all readings. In properties of the type shown in
Fig. 3.1 it is possible that this can be easily accomplished. Much depends
on the type of instrument being used, and the eyesight and experience of
the surveyor.

When booking the levels be sure to finish on the point from which you
started and reduce the levels while you are on site so that any error found
can then be checked and adjusted. For small extensions or alteration jobs
it has often been found that only three or four approximate levels are necessary. The levels required are usually between the existing ground floor
and the external paving and can be taken from an entrance using the step
as a datum. In such cases, the simplest method of levelling is to use a

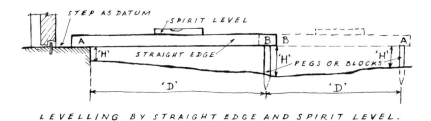

Figure 3.7   Levelling by straight edge and spirit level.

straight edge cut from a board approximately 150 mm × 25 mm × 3 m long. The board is used in conjunction with a spirit level as shown in Fig. 3.7 using a series of pegs or small blocks of timber on a flat base board. The spirit level should be checked for accuracy by reversing it on the straight edge. The bubble should come to rest in the same position. When levelling, the board and level should be reversed at each move as indicated by the letters 'A' and 'B' at each end of the board. By reversing the board and level any error will be minimised. The distance between the pegs or block and their height can then be measured, as indicated by the letters 'D' and 'H' on the sketch.

3.6   Plotting the survey

Plotting the survey is usually done to the scale of 1 : 100 or 1 : 50. If the building has a complicated plan and requires extensive alterations it is better to use the larger scale.

The ground floor plan is drawn first, the dimensions to be plotted being the overall dimensions of the width and length of the building. Then, commencing at one end of the building, set out the walls to each room so as to build up the 'skeleton' of the building. No details of door and window openings, staircase etc. should be drawn in until the partitions and walls have been set out within the overall dimensions.

When this first stage has been completed and checked the details can be filled in. The remaining parts of the plan such as outbuildings, pavings, drainage and soil and rainwater pipes can then be plotted. The upper floors can then be set out in a similar manner together with sections and elevations. It is often difficult to draw a complete section showing structural details. However, if sufficient information has been gathered, an outline section showing the thickness of walls, floors, sizes of roof timbers and beams etc., should be drawn and described.

No doubt the surveyor will set up sections through those parts where structural alterations are proposed, particular attention being paid to floor-to-floor heights and where changes in levels occur. There are several methods of finishing a survey drawing. Existing walls and partitions can be hatched as shown in Fig. 4.1 or filled in in black ink. Alternatively, they can be left in outline and the prints tinted grey or black. Hatching or colouring of plans and sections will tend to make the construction stand out and so simplify the reading of the plans.

4 Surveys of Historic Buildings

4.1 General considerations

The survey of an historic building is perhaps more specialised, but is full of interest (Dallas 2003). The problems which have to be dealt with are of a most intricate nature, but they are often a challenge and a puzzle to be solved. During his examination the surveyor will meet with many forms of construction that are now considered obsolete, but are often the result of native ingenuity and sound practice. It is, therefore, necessary that a surveyor engaged in the examination of an historic building has some knowledge of the traditional methods of construction. From this knowledge he will be able to make reasonable assumptions as to the condition of the structure behind plaster and rendering and decide whether it is necessary to expose certain areas for examination. The surveyor should always bear in mind, that if restoration work is to be sympathetically carried out it can only be achieved by a proper understanding of the original construction and finishings. Many of the most common defects found in an ancient building are described in Chapters 5 to 16 (see also Powys 1995).

Measured surveys of historic buildings are often required to assist the historian or owner of the property who may wish to record the building before it is altered or demolished (Dallas 2003). It is also useful when an application is necessary under the Town and Country Planning Acts, and planning restraints have to be considered or 'listed' building consent obtained. Measured drawings may also be required to identify a particular architectural style prevalent in a given area and would thus make a useful contribution when matters of conservation have to be considered. The type of measured drawings required to deal with these various issues could perhaps be described as 'superficial surveys' and would normally consist of plans and elevations with sufficient notes describing the internal and external treatment, and any special features.

Generally, the site work will follow the methods described earlier in Chapter 3 'Measurement of Existing Buildings'.

In dealing with 'superficial surveys' the following points should be carefully noted:

- Sufficient sketches to enable all elevations and floor plans to be plotted.

- All walling materials – stone, timber framing, brick or rendering – include colour of materials if considered necessary.
- Type and colour of roofing material, including details of chimney stacks.
- Type of doors and windows i.e. leaded lights etc.
- Details of any outbuildings connected with the property.
- Any signs of past alterations and extensions i.e. changes in sizes of door or window openings, roof slopes, eaves, parapet walls, string courses, and cornices.
- Check for wall plaques or inscriptions (if any). These often state the date of erection.
- Roof spaces should be examined for any alterations that may have been carried out.

A surveyor is often called upon to advise on the structural condition of a 'listed building' which is considered to be of special architectural or historic interest. A copy of the listings may be inspected at the offices of the local authority or the Office of the Deputy Prime Minister (ODPM). The Town and Country Planning Act 1990, augmented by the Town and Country Planning (Listed Buildings and Conservation Areas) Act 1990, contains legislation regarding listed buildings and conservation areas which the surveyor should be aware of. The latter Act describes how buildings of special architectural or historic interest are selected for inclusion in the list. Reference should also be made to circulars 14/97 Planning and the Historic Environment and 01/01 Heritage Applications which amend the guidance given in PPG15 Planning and the Historic Environment. If it is proposed that alterations or extensions be carried out, the surveyor should find out exactly what his client intends to do. It is an offence to demolish a listed building or carry out alterations or extensions without first obtaining listed building consent. This consent is necessary even in the case of minor alterations.

The preparation of a satisfactory report cannot be undertaken without considerable experience in dealing with historic buildings. If the surveyor decides to report that repair work is necessary and considers that he has not the appropriate knowledge, he should advise that a person with specialist knowledge of historic buildings be asked to give a report on the works required.

Figures 4.1, 4.2 and 4.3 show the elevations, plans and notes of an 18th-century listed country farmhouse and cottage in Kent. This information is sufficient for a superficial layout as described above. Surveys of historic buildings for repair or restoration purposes will require more detailed information (Dallas 2003). The primary object of this type of survey is to assist accurate diagnosis as a basis for a specification. All available evidence must be collected together and critically examined before conclusions are reached. Measured plans, elevations and sections will be required including all the items mentioned above.

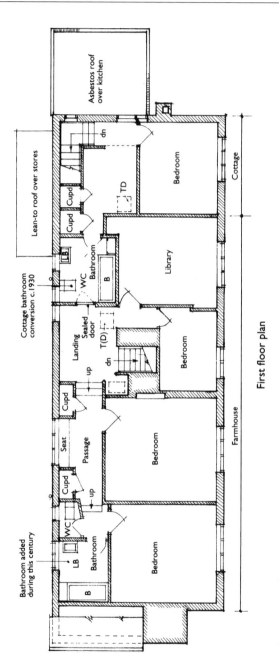

Figure 4.1 Plans.

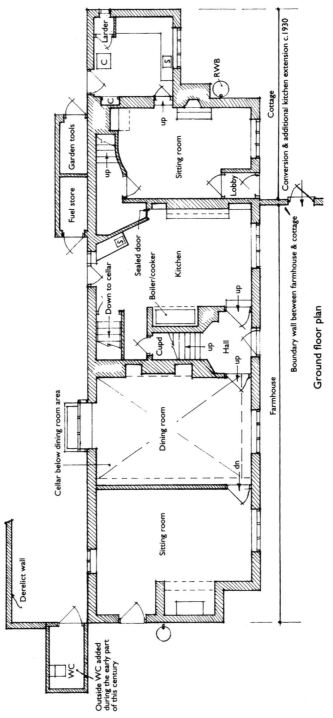

Larder

C

C

S

RWB

up

Cottage

Garden tools

Fuel store

up

up

Sitting room

Lobby

Conversion & additional kitchen extension c.1930

Down to cellar

S

Sealed door

Boiler/cooker

Kitchen

Cupd

up

Hall

up

up

Boundary wall between farmhouse & cottage

Cellar below dining room area

Dining room

dn

Farmhouse

Sitting room

Derelict wall

WC

Outside WC added
during the early part
of this century

Ground floor plan

Figure 4.1 Continued

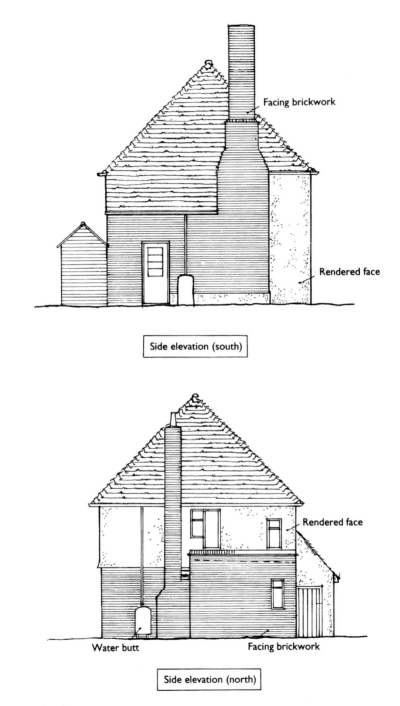

Figure 4.2   Side Elevations.

Special attention should also be paid to the following:

- Carefully note the general structural design and condition of the various materials including the position and extent of wall fractures and leaning or bulging walls.
- Prepare detailed sketches to a fairly large scale of all carved and decorative features i.e. plaster cornices, panelling, timber framing including floor beams and their mouldings.
- Variations in the thickness of walls and partitions and changes in floor levels.
- Prepare list of finishings in each room affected by the restoration work.
- Make a note of any straight joints in blocked windows and doors which may indicate a change of use.
- Plot soil and surface water drainage including position of any cesspit or septic tank.
- Finally, the written report should contain a full description of all the defects and causes of failure. (This matter will be dealt with in Chapter 17 'Report Writing'.)

This type of survey is best drawn to a scale of 1:50 and a larger scale, say 1:20 or full size will be required for timber details or decorative plaster. A problem that is often met with during a survey of an historic building is that the walls and the partitions are not always vertical. Measurements can vary according to the height at which they are taken. In such cases it is advisable to work at a uniform height preferably just about 300 mm above the level of the window board. There is also a tendency for plaster to be thicker near the corners of rooms and around door openings. Another problem is trying to obtain information on the construction of external walls which are often concealed by thick plaster or external rendering. The owner's permission is, of course, needed before cutting away plaster or rendering. In the case of unoccupied property, permission is usually given provided it is carried out with the minimum of disturbance. In occupied property difficulties may sometimes arise with the occupier, who would be unwilling to permit an examination which could result in damage to their property.

Many owners of historic buildings often have documents concerning the building's history, together with old photographs which give valuable information to a surveyor. Owners who possess such information are usually willing to cooperate and supply all that is necessary. The direction of floor boards is often misleading, the original worn flooring having been covered over with later boarding in a crosswise direction. In this event it will be wise to lift one or two boards and check the direction of the boards and joists below.

The photographic survey mentioned earlier should include views of all the elevations and close-up photographs of the various points of interest. Where possible the views should be taken obliquely to show two

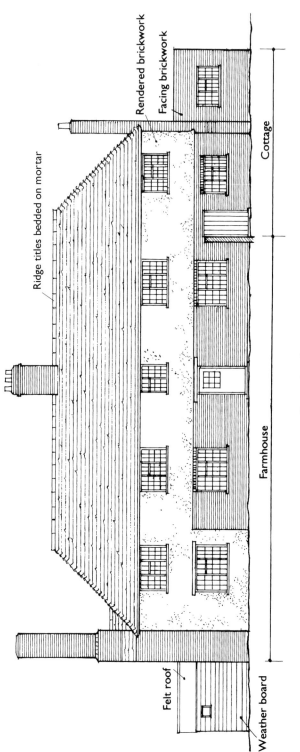

Figure 4.3 Front and back elevations.

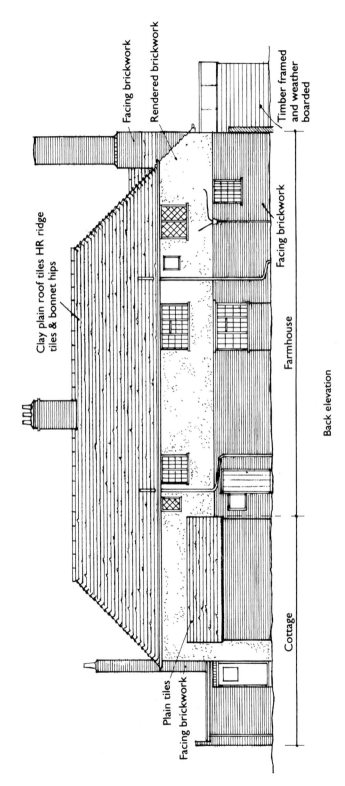

Facing brickwork

Rendered brickwork

Timber framed and weather boarded

Clay plain roof tiles HR ridge tiles & bonnet hips

Facing brickwork

Farmhouse

Plain tiles
Facing brickwork

Cottage

Back elevation

Figure 4.3 Continued

elevations. For interior work a wide angle lens is likely to be needed, complete with flash-gun with adjustable head to match the lens.

4.2   Medieval churches

One of the more specialised branches of 'historic building surveys' is the examination of a medieval church. The system of inspection is of great antiquity and there is a considerable amount of legislation on the subject dating from the thirteenth century. Thus, the church gave a lead in the care and protection of church buildings by pioneering a system of regular inspections by suitably qualified architects and surveyors.

All churches, churchyards and other ecclesiastical buildings are now under the guardianship of the Bishop of the Diocese. Over the past 100 years church law has been regularised by various ecclesiastical measures which have statutory force. The measures are long and complicated, but are worthy of careful study by those interested in the subject. The current measure concerning the fabric of the church is the 'Inspection of churches measure' 1955 which requires that every church must be inspected by a qualified architect (or building surveyor) once every five years. The measure states that a copy of the architect's/surveyor's report with recommendations must be sent to the Archdeacon and to the parochial church council of the parish in which the church is situated. However, quinquennial inspection does not make examination by the church-wardens any less necessary. Many defects may occur in the five years between the architect's/surveyor's visits which can involve the parish council in considerable expenditure if they are not attended to.

The Council for the Care of Churches (1993) recommends that every parochial church council should appoint a member who will keep the church and its fittings under observation and report to the wardens any defects found.

In order to make sure that nothing is missed, it is advisable to follow a definite system of coverage. The checklist shown in Appendix III could be adjusted and used for this purpose. Although quite a number of typical defects will be described in Chapters 5 to 16, the following points should be given careful attention when carrying out an examination of a medieval church:

- Constant dripping from a defective rainwater head or gutter can damage or discolour stone and brickwork and wash out the mortar joints. Damp seeping through the wall will cause internal plaster to deteriorate.
- Many churches have square down pipes which are particularly difficult to maintain as they are normally fixed close to the wall and are therefore difficult to paint behind. The surveyor should bear in mind that lead pipework is usually of historic value.

- Rainwater pipes should discharge into a drain and not into the foundation of the building. Check that pipes and drains are clear of debris.
- Nature of the subsoil and the disposal of ground water especially if the floor is below ground level, and there is no evidence of a damp-proof course.
- Many medieval roofs were designed without tie beams. With no adequate means to resist the outward thrust it is advisable to make sure that the roofs are strengthened so that only a vertical load is carried on the outer walls.
- The various forms of decay that attack old timber. For example, it is important to notice whether the defects are due to dry or wet rot, woodworm or 'shakes'. Examine carefully the bearing ends of beams and roof trusses.
- Whether the disintegration of stone is due to frost, damp penetration or to a chemically charged atmosphere. It may be advisable to obtain a report from an analytical chemist on this matter. Church buildings often took several centuries to complete and the various movements in the structure were spread over a considerable time. Extensions and alterations were carried out over the years and with the contrasting weights of materials and the many forms of construction the pattern of loading becomes very complex.

During the nineteenth and the first half of the twentieth centuries measured elevations including complicated gothic details were measured and drawn by hand. Clients and contractors today can rarely be provided with elevation details drawn by hand. This is partly because of time delays and also because of the expense involved. As mentioned in Section 2.4 photographs are now increasingly used. In this connection the best results are obtained when specification notes are added on the site by hand. This method is superior to hand measured drawings provided care is taken to identify the photograph accurately and that the notes clearly describe the work to be done. With this method quite small defects can be detected which may go unnoticed if reliance was placed solely on measurements and sketches.

It is not within the scope of this book to make available practical information on remedial works. There are several excellent publications on this subject, and these are described in the Bibliography at the end of the book.

4.3   Church towers

One of the problems in the examination of churches is the repair of damage caused by the ringing of bells. Church towers together with bells, cages and beams often require frequent inspections. The surveyor will often find that access is difficult and the belfry very dirty!

If a thorough inspection of the timber is required it is advisable to make arrangements with the Parochial Church Council (PCC) to have the belfry thoroughly cleaned in order that special attention can be given to the beams. In fact, the surveyor could suggest to the PCC that they arrange for an annual cleaning contract which could include the bell frame (Council for the Care of Churches 1993).

The following points should be noted when carrying out an inspection of the tower:

Internally

- Inspect all parts for rubbish and birds' nests etc.
- Window openings in bell towers are usually unglazed. Recommend plastic covered wire netting secured to battens to keep out birds.
- The maintenance of church turret clocks is usually carried out professionally by contract, but it is advisable to check this matter with the PCC.
- Examine all floor boarding and beams including steps and ladders for defects.

Externally

- A careful examination should be carried out to the tower roof coverings, gutters, flashings, trap doors and their fastenings. Leaks can cause all sorts of problems in the timbers and if not repaired the damage can be costly.
- Ensure that rainwater outlets and heads are free of debris and sludge etc.
- Check that the lightning conductors are properly fixed and earthed. If suspect, a test should be recommended and carried out by a qualified engineer.
- The exterior of the tower can be examined with the aid of binoculars during the general examination of the walls and roofs. The most common defects are poor pointing, fractured stonework and loose copings.

4.4   Church bells and fittings

The repair of church bells and the effect of vibrations on the tower masonry caused by the swinging bell is complicated and full of technical problems (Council for the Care of Churches 1993). A set of bells can be extremely heavy often weighing several hundredweight, and when swung in complete circles can cause considerable strain on the tower walls. If in doubt, it is always advisable to consult a firm of bell-hangers. As in most parts of an ancient building, the aim should be to preserve the fine craftsmanship of the bells and bell frames provided all the parts are in sound working order. Some bells are often found to be slightly cracked due to the expansion of the crown staple or corrosion. It is possible to weld a cracked bell, but it is often unsuccessful. Another common defect in an old bell is that the spot where the clapper strikes becomes worn. This problem is easily rectified by giving the bell a 'quarter turn'.

The bell cages or frames are usually frames of heavy oak timbers, braced and cross-braced. These cages usually last for centuries if they receive proper care and attention. The cage should be carried by supporting beams, the ends of which should bear on all four walls of the tower. Thus the forces from the swinging bells may be accommodated in all parts of the tower masonry simultaneously. Usually, if a bell cage is well designed it is unlikely to damage the tower masonry. Problems of vibration and masonry movement arise when the supporting beams are fixed to two walls only. All metal and timber joints should be carefully examined in order to ascertain that all joints are tight. During his examination the surveyor may find it necessary to expose the bearing ends of the timber beams by removing the masonry at one side of each beam end. This operation will no doubt require the services of a local builder who must be carefully instructed as to what is required.

4.5   Measured drawings

Surveyors who care for an ancient church and carry out annual inspections should possess a measured drawing to assist them in their task. The drawings should be to scale say 1:100 and generally follow the details described under 'superficial surveys' at the beginning of this chapter. It is quite remarkable how few surveys exist of church buildings. They are, in fact, of great importance i.e. to assist the historian or the PCC who may wish to record the building before restoration or alterations are carried out.

5 Foundation Failures

5.1 Introductory

Architects, surveyors and structural engineers are all called upon at some time to examine defective foundations and submit reports with recommendations for remedial action. Whilst we can all learn the technicalities that form the basic knowledge of the various building professions, there is always one element where the professional has to rely on skill to diagnose the significance of any symptoms: that is 'experience'. This applies especially to the analysis of foundation problems and their cause.

5.2 Causes of failure

Foundations can move as a result of loads applied causing a downward movement known as settlement. Settlement can be tolerated by the structure provided the loads do not exceed the 'allowable bearing pressures' stated in BS 8004. Other possible causes of foundation movement known as subsidence, brought about by activity in the ground, are:

- Soil erosion caused by flowing water.
- Changes in ground water level.
- Buildings on made up ground.
- Movement associated with mining activities or 'swallow holes' found in chalk.
- Movement due to shrinkage or swelling of clay soils. This is the most common cause of foundation movement.
- Uneven bearing capacities of differing subsoils.

Heave is the upward movement of the ground. It is the result of an increase in moisture content in excess of that which existed when the building was erected. It can occur when trees are removed, but it can be caused by the interruption of a natural water course through climate change. Heave is also caused by the removal of loads on the foundation. A rather uncommon form of heave is when the ground expands when frozen. The problem is usually confined to soils consisting of fine sand and chalk. Where the water table is high and there are prolonged periods of freezing ice layers can cause the foundations to lift, but this only occurs during very severe winter conditions.

Although other possible causes of damage must be considered during the investigation, settlement, subsidence and heave account for damage to most structural elements including floors in low rise buildings (Dickinson and Thornton 2004). Typical symptoms are:

- Cracks in external or internal walls. The cracks may be hairline or much wider.
- Walls bulging or leaning out of vertical.
- Floors slanting out of level.
- Drains or service pipes blocked or malfunctioning.
- Pavings or drives cracking.

When foundation defects are to be investigated a thorough examination is imperative. The primary object of this type of examination is to obtain an accurate diagnosis as a basis for a report. It is therefore extremely important that all available evidence is collected together and carefully examined before decisions are reached as to the method of repair to be adopted. This inspection may take some considerable time, but it is essential that extensive defects are properly investigated. It is always advisable to bear in mind when making a diagnosis that more than one cause may be responsible for a defect, although it is necessary to investigate the primary cause. For example, foundation movements may have been responsible for a fractured external wall, but rainwater could penetrate the fracture causing dampness on the internal face of the wall. Although the two defects will have to be remedied it does not necessarily mean that the crack has caused rainwater to penetrate the wall. The wall may have been damp before the movement took place perhaps due to faulty construction or porous brickwork.

Foundation repairs to existing buildings are generally the most difficult and costly to effect, which is still a good reason why a thorough investigation should be carried out. The object of the investigation will be to determine the nature and strength of the subsoil under load. A visual observation below ground can only be carried out by digging trial holes at intervals along the length of the wall adjacent to the suspected position of the foundation failure. The holes should be of a size to accommodate an adult.

When the underside of the foundation is exposed, the details of the subsoil, together with the condition of the foundation and base of the wall should be recorded. Tests can be carried out by driving an iron bar into the subsoil. A more detailed test for moisture and bearing capacity can be carried out by removing samples of soil with a spade and submitting them to a laboratory for examination.

The surveyor should always bear in mind that the initial examination will only reveal conditions as they are, and they will need to be studied over a period of time before a decision can be made. Prior to carrying out an inspection of the foundation defects it is advisable to have a precise knowledge of the soils present on site. Land used for building varies considerably from hard rock to loose sand. Between these extremes are soft rock, firm earth, firm clay, soft clay, gravel, sand and fill.

Soils may be divided into two categories, non-cohesive and cohesive:

- Non-cohesive soils are the gravels and sands which tend to lack cohesion and have no plasticity.
- Cohesive soils are the various types of clay and silt and possess cohesion and plasticity.

Below are the principal causes of foundation failures that are considered to be most common in the UK.

5.3   Differential movement

Excavating the ground and placing substantial loads on it is sufficient to cause a slight movement as the ground below is compressed to resist the load. Provided the settlement is uniform over the building area the movement does little damage. Alternatively, it may be differential movement where part of the foundation remains stable while the remainder moves (Richardson 2000). A typical example of differential movement is shown in Fig. 5.1 where settlement occurred in the end walls with the centre portion stable. The cracks are usually vertical or diagonal and are often interrupted by window or door openings. In such cases a gap is formed between the frame and brickwork.

Figure 5.1   Example of differential settlement. Settlement in end walls with centre portion stable. Cracks increase in width with height, and appear to be interrupted by the windows, but in fact extend round the openings forming a gap between window frame and brickwork. The movement is at its maximum at roof level and could cause loss of bearing in the roof slab.

5.4   Inadequate foundations

This can be due to the fact that the width of the foundation concrete is not enough to support the building load. In such cases it may be necessary to consider increasing the depth of the foundations by underpinning to a point where the safe bearing capacity of the ground is adequate (IStructE 2000). On the other hand it is not surprising to find that most buildings erected before the 19th century have no foundations and the brick or stone walls were laid directly on the earth or on a bed of consolidated rubble. Occasionally, two or three courses of brick footings were laid on the bed of the excavation to spread the load in lieu of a concrete base.

5.5   Overloading

Every building has its own pattern of loading. Where internal alterations have taken place in the original building or additional loads have been applied due to a change of use, then the loads placed on the subsoils are greater than was originally allowed for. Overloading can also occur where door or window openings have been enlarged which may result in a heavier load being transferred to an adjacent section of brickwork consisting of a narrow pier. Again, the load imposed upon the subsoil is greater than originally allowed for causing the pier to settle (see Fig. 5.2).

Under concentrated loads such an overloaded pier or stanchion on shallow foundations will often show movement cracks between floor and foundation where the foundation has been 'punched' downwards sometimes several centimetres (see Fig. 5.3).

The average domestic building, however, is unlikely to weigh more than 45 kg per metre length of wall with a nominal concrete base width of, say, 680 mm. The naturally occurring subsoils found in the UK are usually able to sustain this type of loading, provided the foundation is deep enough not to be disturbed by the effects of atmospheric action.

5.6   Unequal settlement

Shallow foundations on clay present bearing capacity problems and shrink and expand due to seasonal changes, the effect being felt to a depth of 1.4 m. The bulk of these clays are situated in the SE of England which has the lowest rainfall (Driscoll and Crilly 2000).

The clay soil immediately surrounding a building will shrink and crack during hot weather, but underneath the building the subsoil will be protected from the winter rains and also from the hot sun; thus it will expand and contract much less. Therefore, a differential settlement is set up causing the foundations to the external walls to settle downwards and the

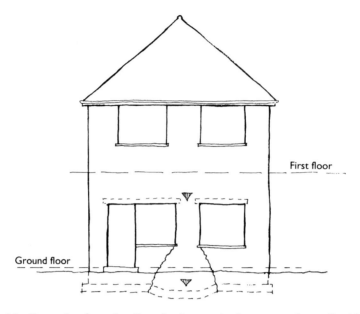

Figure 5.2 Example of overloading. Settlement in the centre of a wall with ends stable. Heavy loads from first floor, roofs and walls above are transferred to a narrow pier of brickwork causing cracks in the wall and foundation settlement. With this type of settlement cracks decrease in width with height.

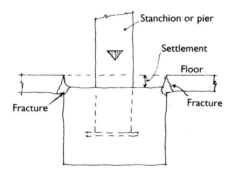

Figure 5.3 Differential settlement under the point load of a stanchion.

walls to lean outward during the hot summer months when the subsoil is dry, and a tendency to lean inwards in the winter months when the exterior wet clay has expanded and the interior has remained stable. During the wet seasons when the clay expands the cracks tend to recover. This movement can also cause diagonal fractures around window and door frames and a drop in the horizontal joints of a string course or bed joints in brickwork. The appearance of cracks in the outer walls can be disturbing so that this seasonal movement can often cause considerable concern out of all proportion to its actual importance (Richardson 1996, 2000).

5.7 Effect of tree roots

Fast growing trees close to buildings can cause unequal settlement when active tree roots dry out the soil causing differential soil shrinkage (BRE Digest 298 1999). Shrinking clays affect the bearing capacity and lead to movement in the building, especially in shallow foundations. Tree roots can extend over a considerable distance and can extract moisture from as deep as 6 m below the surface. It is, therefore, necessary to make an accurate survey of their position and obtain details of the type of tree, and at the same time establish that the tree is the cause of the damage (see Fig. 5.4). Poplars and elms with fast growing root systems can be expected to cause serious seasonal movements.

One way to avoid root problems with tall trees is to maintain a 'safe distance' between the tree and the building. Some species of trees are likely to cause more problems than others. Table 5.1 shows the different types of trees known to have caused damage, ranking in descending order of threat. It also shows their expected maximum height on clay soils. Planting a tree close to a new or existing building will usually entail some risk of damage. It is, therefore, suggested that the recommendations described in Table 5.1 are followed.

Buildings can also be damaged when well-established trees are removed (Bonshor and Bonshor 1996). The resultant pressures due to the removal

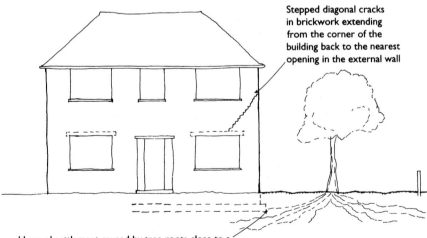

Stepped diagonal cracks in brickwork extending from the corner of the building back to the nearest opening in the external wall

Unequal settlement caused by tree roots close to a building; roots absorb moisture and cause clay to shrink

Figure 5.4 Unequal settlement caused by tree roots.

Table 5.1   Risk of damage by different tree species: the table shows for each tree species the distance between tree and building within which 75% of the cases of damage occurred.

Ranking	Species	Max tree height – H *m*	Max distance for 75 per cent of cases *m*	Min recommended separation in very highly and highly shrinkable clays
1	Oak	16–23	13	1H
2	Poplar	24	15	1H
3	Lime	16–24	8	0.5H
4	Common ash	23	10	0.5H
5	Plane	25–30	7.5	0.5H
6	Willow	15	11	1H
7	Elm	20–25	12	0.5H
8	Hawthorn	10	7	0.5H
9	Maple/Sycamore	17–24	9	0.5H
10	Cherry/Plum	8	6	1H
11	Beech	20	9	0.5H
12	Birch	12–14	7	0.5H
13	White beam/ Rowan	8–12	7	1H
14	Cypress	18–25	3.5	0.5H

Reproduced from the Building Research Establishment Digest 298 by permission of the Controller HMSO: Crown copyright.

of trees and bushes act both vertically and horizontally. In the majority of cases it is the horizontal movement that produces the greatest damage particularly in the upper layer of clay. In such cases there is a danger of the clay expanding over a period of years as it reabsorbs moisture causing the foundation to 'heave' as described in Section 5.2 above. Where window sills crack and rise in the middle this is an indication of soil heave. Differential movements will take place resulting in cracks in walls and partitions. In such cases the removal of a tree may do more harm than good.

5.8   Shallow foundations

The surveyor will often find that foundation movement in older domestic properties can be caused by light strip foundations attached to the main building supporting porches, bay windows and garages. There is a common misconception that these lightweight structures do not need deep foundations as does the main building. If this type of foundation is not carried down sufficiently deep to avoid movement by atmospheric action, the junction between the light and heavy parts of the structure will often show a diagonal crack running down from the lower corner of the rear window as shown in Fig. 5.5.

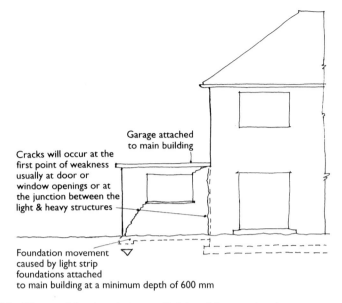

Garage attached
to main building

Cracks will occur at the
first point of weakness
usually at door or
window openings or at
the junction between the
light & heavy structures

Foundation movement ▽
caused by light strip
foundations attached
to main building at a minimum depth of 600 mm

Figure 5.5   Diagonal fracture between light and heavy structure.

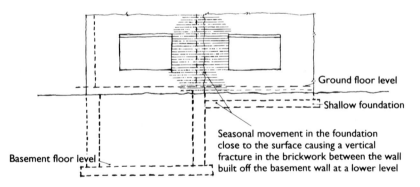

Ground floor level

Shallow foundation

Seasonal movement in the foundation
close to the surface causing a vertical
fracture in the brickwork between the wall
built off the basement wall at a lower level

Basement floor level

Figure 5.6   Unequal settlement between basement and ground floor walls.

Another common example is when part of a building has a basement with foundations set in deep geological beds. Seasonal movement will often take place in the foundations close to the surface causing a vertical fracture at the junction between the walls built off the basement wall and the ground floor. The heavier walls to the basement area will settle relatively more causing movement cracks between the light and heavily loaded walls as shown in Fig. 5.6. This is more noticeable when the ground floor structure is built on shallow foundations in a clay subsoil.

5.9   Building on sloping sites

Buildings on sloping clay sites can often present difficulties (IStructE 2000). The water table on sloping sites tends to follow the topography of the surface and if the natural contours of the site change it does not necessarily alter this line. Where the foundations have been set at a constant depth from the stepped level surface the concentrations of water may affect the foundations at the lowest point causing differential settlement. Figure 5.7 shows such a case where the water table is above the foundation at the highest point and beneath a shallow foundation at the lowest point. In such cases it would have been advisable to form stepped foundations as shown in the sketch in order to be below the water level.

If a building is erected at the lowest point of a sloping site, natural drainage of the water to the lower points can cause saturation of the ground around the base of the wall and foundations thus lowering the bearing capacity (see Fig. 5.8). In both the above cases it is advisable to check that the subsoil is effectively drained at the highest level in order to protect the building against damage by water penetration. The surveyor will often find that surface water drainage has been omitted especially in ancient buildings.

A similar case of water penetration often occurs around the base of the external walls of old buildings where the ground floor level is below the ground level. The surveyor will probably notice that clayware or concrete channels have been inserted around the base of the wall to collect rainwater from the roofs and a certain amount of surface water from paved areas around the building. The channels tend to move with the seasonal changes in the subsoil causing the joints to open, thus allowing water to percolate through to the subsoil and the base of the wall. In such cases

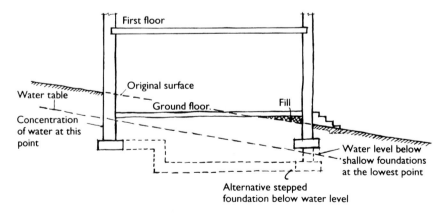

Figure 5.7   Building on sloping clay site.

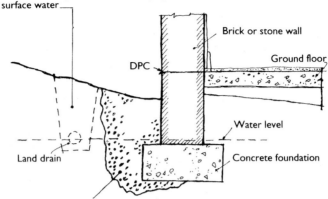

Ensure that the subsoil at the lowest point of the site is effectively drained to intercept ground and surface water

Brick or stone wall

DPC

Ground floor

Water level

Land drain

Concrete foundation

If the ground is not properly drained then the penetrating water will cause saturation of the ground adjacent to the foundations

Figure 5.8 Building at the lowest point of a sloping site.

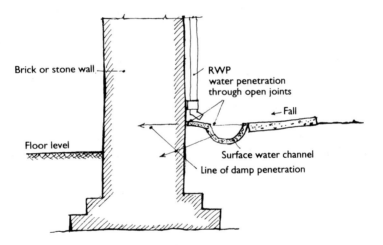

Brick or stone wall

RWP
water penetration
through open joints

Fall

Floor level

Surface water channel

Line of damp penetration

Figure 5.9 Water penetration around base of external walls where the ground floor level is below the ground level.

there is a risk of uneven movement in the subsoil leading to settlement cracks (see Fig. 5.9).

In chalk and limestone areas, cavities in the subsoil can be formed by underground streams or watercourses dissolving the rock. If the sandy overburden falls into the cavity the foundations will drop. These cavities are known as 'swallow holes'. In the three cases described above the foundations need protecting and the ground water must be directed away from the foundations by a system of ground or surface water drainage.

5.10   Building on made up ground

Filled or made up ground is extremely varied in form and should be treated as suspect. Experience has shown that the majority of foundation failures on filled ground have been due to the use of poor fill, and inadequate compaction. Unfortunately, during an inspection, detailed knowledge of the fill is usually lacking. All possible information concerning the site should be obtained by discussion with the local authority and by studying local maps of the area. Apart from digging trial holes the surveyor should observe signs of damage to any adjacent buildings. The trial holes should be deep enough to enable the surveyor to assess the nature of the fill, its depth, composition and degree of compaction. If any remedial work is contemplated such as underpinning or piling, this could well involve the protection and support of the services to the building. A particular note of this matter should be made during the site examination.

5.11   Diagnosis

Crack monitoring may be necessary to see whether the problem is still active (Bonshor and Bonshor 1996). This should be done by the application of 'tell-tales' as described in Section 2.5. 'Tell-tales' should be fixed internally and externally if found necessary. Although 'tell-tales' are most important from the surveyor's point of view, he will sometimes find it difficult to explain the size and direction of the crack to his client by way of notes and sketches. In this respect photographs will be most useful when attached to a report. However, the 'tell-tales' will enable the surveyor to check whether or not the movements are progressive. Movements due to settlement of filled ground usually cause major cracking of external walls and partitions. Bulging can also occur in external walls. Concrete ground floors are also liable to lift and crack.

After all investigations have been completed the surveyor may consider obtaining the services of a specialist in this field to advise on any remedial work required.

6 Defective Walls and Partitions Above Ground

6.1 Type of failure

The failure of brickwork above ground can be divided into the following groups:

- Bulging and leaning walls.
- Overloading.
- Thermal and moisture movement.
- Failure in arches and lintels.
- Defective materials and chemical action.
- Failure in bonding and defects at junctions.
- Frost failure.
- Defects in cavity walls.
- Built-in iron and steel members.
- Tile and slate hanging and weatherboarding.
- Partitions.
- Assessment of cracks.

6.2 Bulging and leaning walls

This usually involves external walls and is more serious than cracking. Bulging or leaning is often due to the lowering of the stability of the wall caused by the following factors:

- Vibrations from traffic and machinery.
- Increasing the floor loads or building on additional floors.
- The original walls being insufficiently thick in relation to the height.
- Lack of restraint between the external walls and floor joists, beams and partitions.

The type of bulging that normally occurs in older buildings built in lime mortar is usually due to lack of restraint. It is a well known principle in the construction of brick walls that some form of restraint should be provided to prevent lateral movement. This is achieved by building in floor joists or by bonding in partitions at right angles to the wall. In Fig. 6.1 the wall is only 228 mm thick at first and second floor levels, and the floor joists run parallel to the flank wall, and do not provide a tie to the external wall. Furthermore, the staircase is positioned against the flank wall so that there is minimal restraint between flank and rear walls. If the brickwork at the junction marked 'X' on the sketch is weak a vertical fracture could develop. The flank wall will also tend to pull away from the timber stud partition as shown on the plan view marked 'Y'. These defects are common in properties built pre-1914-end-of-terrace or semi-detached three-storey houses. Figure 6.2 shows an unrestrained flank gable wall where the shortest span for the first floor and ceiling joists is front to rear. The danger in such cases is the loss of bearing to the ridge.

Figure 6.3 is an example of an untied roof causing the roof to sag. The sagging produces a horizontal thrust on the external walls which may cause them to lean and fracture. With all bulging and leaning defects evidence can also be seen internally by way of cracks in the ceilings, particularly at the junction of wall and ceiling, and where the partitions meet at right angles to the external wall.

In all cases of bulging and leaning walls it is important that the surveyor should accurately examine the defective area taking into consideration the following points:

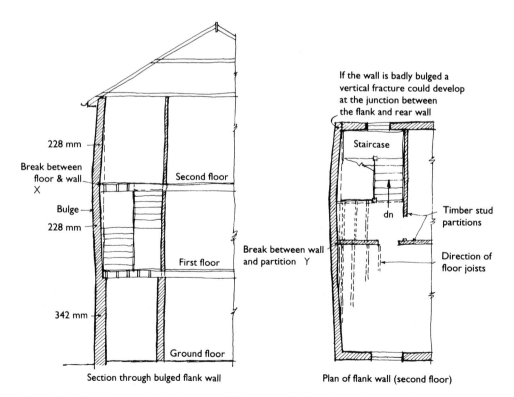

Figure 6.1   Bulging due to inadequate thickness and lack of restraint in walls.

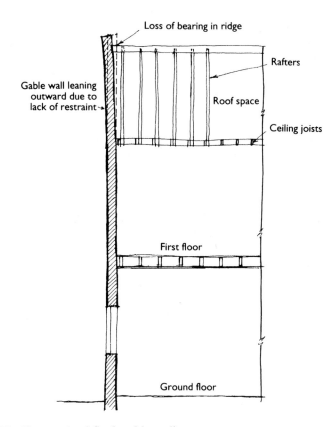

Figure 6.2   Unrestrained flank gable wall.

- The thickness and height of the wall.
- The method of restraint (if any).
- The amount the wall is out of plumb.
- The length, width and position of any fractures, internally and externally.
- General state of repair.
- The number of openings and piers in the defective wall.

A leaning wall which is out of plumb by up to 25 mm or bulges not more than 12 mm in a storey height is not serious on structural grounds alone, but if the leaning wall exceeds this amount then an examination must be carried out, and if necessary be repeated at intervals in order to ascertain whether or not it is still moving. A useful rule to remember is when the amount of lean in the full height of the wall exceeds one third of the thickness of the wall at the base then the wall is dangerous, and remedial measures are no doubt necessary.

Cast-iron bosses attached to the outer face of external walls will often be found on early nineteenth century industrial and warehouse buildings,

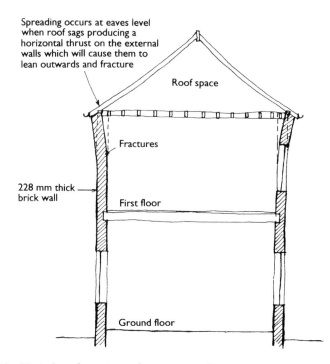

Spreading occurs at eaves level
when roof sags producing a
horizontal thrust on the external
walls which will cause them to
lean outwards and fracture

Roof space

Fractures

228 mm thick
brick wall

First floor

Ground floor

Figure 6.3   Untied roof causing rafters to spread.

and were usually installed as the buildings were erected. The bosses are connected to tie rods to the opposite wall of the building in order to hold the walls in a vertical position. If the rods and bosses are fitted to houses of this period they were probably installed at a later date to restrain bulging walls.

Tie rods and bosses have been used on both brick and stone construction, but were generally abandoned towards the end of the nineteenth century. Corrosion of the iron, which then expands can cause damage to the stonework. It is therefore important for the surveyor to carefully examine the area around the bosses and also check that the ties are doing their job in restraining the walls.

If the plumb of a structure is in doubt then a plumb line should be dropped and offsets taken as described in Section 3.4. However, it is not always possible to plumb the external face of a wall because of obstructions such as cornices, balconies and porches etc. In such cases the only alternative is to set up plumb lines internally. This procedure can only be used satisfactorily in a building constructed with timber joist floors. The operation is more complicated, but essential if a true assessment is to be obtained. It will necessitate removing floor boards and drilling holes in the ceilings in order that plumb lines can be dropped internally. Before measuring the offsets, a small area of the plaster should be removed from the wall face so that offsets can be measured to the brick face since plaster can vary in thickness. A note should also be made of any change in wall thickness between the various storey heights.

Although this method of internal plumb lines is suitable for unoccupied properties, it is often difficult to obtain the owner's permission to carry out such investigation in an occupied property. The majority of owners object to floor coverings and ceiling plaster being disturbed.

6.3   Overloading

Overloading of structural walls will often result in leaning or bulging as previously described, particularly if the original load is in the centre of a beam spanning an opening (Richardson 2000). Excess loads on floors are often due to changes in the use to which the building is put. This change of load pattern imposes an inequality of load on the brickwork causing the walls to show signs of distress. Forming additional openings or enlarging existing openings for doors and windows in older type buildings will transfer loads to smaller areas of brickwork. This type of alteration is easily identified when looking at the difference in the age and condition of the old and fairly recent brickwork.

Excess loads on roofs can also affect brickwork causing bowing or crushing. An example of overloading is shown in Fig. 6.4 and is caused by faulty design, and this is typical of many Victorian domestic properties where about a quarter of the total loading above ground level is carried on the timber beams over the bay window producing a deflection in the beam, and cracked brickwork as shown in the sketch.

In dealing with problems of this nature the surveyor should always remember that the science of structures in past centuries was largely one of trial and error.

When investigating such defects it is quite usual to find that the timber beam consists of two floor joists bolted at intervals and because of their slenderness are readily overloaded. The bearing ends will be embedded into the brickwork with very little protection against moisture penetration. The bearing ends will require close inspection for signs of decay which will entail removing some plasterwork so that observation can be made. Timber may appear to be rotten on the surface, but it is wise to test with a gimlet before assuming that it is so throughout. Woodworm is also found to be particularly active around the bearing ends of beams.

A point to be carefully noted when dealing with bay windows is a tendency for the brickwork to become detached from the main external wall and lean outwards causing tapering cracks between the bay and the main wall. If the wall has a rendered or pebble dash finish then only small cracks will appear at the junction. It is, therefore, important to ascertain the condition of the brickwork behind which will necessitate cutting away some of the rendering each side of the crack. This type of movement can often

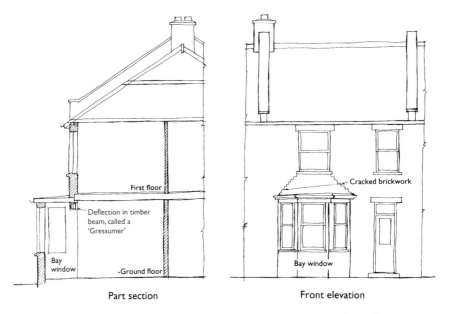

Part section   Front elevation

Figure 6.4   Typical case of overloading in older domestic property where about a quarter of the total load above ground is carried on the timber beam over the bay window.

be the result of shallow foundations which often distort the brickwork around the bay. (This matter is dealt with in Section 5.6.)

6.4   Thermal and moisture movement

Brickwork has a tendency to move due to variations of temperature or moisture content, but the problem is not often encountered in small domestic structure. It is the outer walls of a building and particularly flat roofs that are exposed to the greatest temperature changes and where no provision has been made for thermal movement (Bonshor and Bonshor 1996). Variations of temperature can cause different reactions. There is a tendency for cracks to develop when the temperature drops and the brickwork contracts particularly along a line of separation such as a DPC membrane. Lateral movement can also occur in long parapet walls where expansion forces brickwork outwards and when it contracts it does not usually return to its original position, causing tension cracks to form (see Fig. 6.5). This problem does not often occur in old buildings constructed in lime mortar where the movement tends to be absorbed in the joints (Holmes and Wingate 1997).

Everyone is familiar with drying shrinkage and the changes that take place in building materials due to their moisture content. Bricks vary in their ability to absorb moisture, but very few serious defects arise because

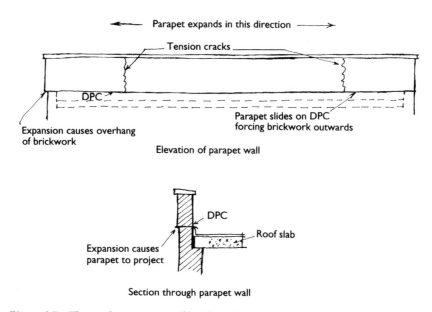

←———— Parapet expands in this direction ————→

Tension cracks

DPC

Parapet slides on DPC
forcing brickwork outwards

Expansion causes overhang
of brickwork

Elevation of parapet wall

DPC

Roof slab

Expansion causes
parapet to project

Section through parapet wall

Figure 6.5   Thermal movement of brickwork.

of this phenomenon. Walls built of porous bricks will often absorb water through the brick and joints whereas in nonporous bricks any water penetration will arise from the jointing material. The external portion of a brick or stone wall in an exposed situation will absorb a considerable amount of water during prolonged wet weather, but will dry out fairly rapidly during the warm summer months. Nevertheless, where moisture is present there is always a possibility of deterioration. Moisture can also rise from the ground or penetrate downwards from roof level. (This matter will be dealt with in Chapter 8.)

Experience has shown that cracks due to thermal and moisture movement tend to act in a vertical direction and are of equal width throughout. They are also localised at their weakest points in the structure, such as around door and window openings. Water will also tend to run downwards from defective window sills to the DPC.

The Building Research Establishment has established another factor which may be the cause of defects in brickwork (Bonshor and Bonshor 1996). It is the irreversible moisture movement, which is often many times greater than the reversible movement. Irreversible movement occurs when newly fired bricks are removed from the kiln and absorb moisture.

This type of expansion can cause several visible defects in brickwork, but difficulties can be avoided by not using bricks fresh from the kiln. The extent of the movement depends upon the type of clay used and the degree of firing. Irreversible moisture expansion of clay is fairly rapid after the brick has cooled, and when built into a wall it can be as much as 4 mm in a 10 m length of brickwork. A typical result of this movement is the oversailing of the DPCs acting as a slip joint, which allows the brickwork above the DPC to expand. The lack of restraint may also lead to cracking of walls near the corner of the building.

In view of the fact that the surveyor will be carrying out an examination several years after the irreversible expansion is effectively over, the diagnosis of the cause is often difficult and cannot be determined from one single examination. As mentioned in Chapters 1 and 2 the early history of the building is important. Local authorities or local builders can often produce useful information.

When recommending remedial measures the surveyor must be extremely cautious. Cracking near the quoins is unlikely to be of structural significance, and if the cracks are fine repointing will be all that is necessary. Where oversailing of the DPC has occurred, there is little that can be done but the movement is unlikely to cause structural instability. Where vertical cracks are wide the surveyor may well recommend replacement bricks. However, the stability of the structure must be checked before such work is carried out. Additional ties may be required. If severe bulging has occurred in a brick panel and the stability of the structure is impaired then the panel would no doubt have to be rebuilt.

6.5   Failure in arches and lintels

Settlement cracks usually arise from the inadequacy in size of the abutments causing the brickwork to thrust outwards as shown in Fig. 6.6 where the opening is too near the corner of the building. If the surveyor considers the fractures to be serious and requiring emergency treatment then the first step is to tie the abutments together by using two or three wire ropes and straining screws connected to timbers through the nearest opening. The arch will also require support by erecting centring similar to that used in the construction of new arches. This will relieve the arch of its load and enable the necessary repairs to be carried out. Alternatively, a raking shore should be erected against the fractured abutment to resist any further movement. A similar problem can arise from the excessive deflection of a lintel, causing converging cracks to run upwards as shown in Fig. 6.7.

A simple rule of thumb method can be used by the surveyor for checking the size of abutments to arches. If the arch is semicircular there should be an abutment on each side equal to three-quarters of the span of the arch. However, this figure can be reduced to half the span for pointed arches. For shallower arches the abutments should equal the span of the arch. Defects of a minor nature often occur in arches in which some Voussoirs have slipped down their beds. These should be carefully examined, and if the arch has settled and cracked it is advisable to examine the brickwork above for signs of settlement and to trace the cause. The brickwork may have been subjected to heavy roof or floor loads in excess of its

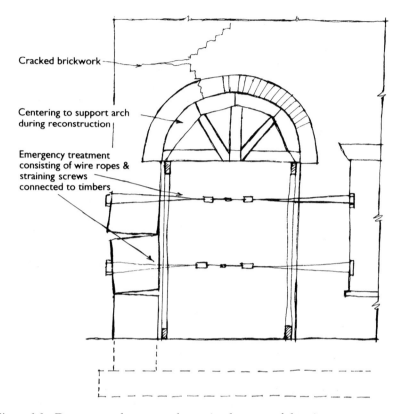

Cracked brickwork

Centering to support arch during reconstruction

Emergency treatment consisting of wire ropes & straining screws connected to timbers

Figure 6.6 Damage to abutments due to inadequacy of the pier.

capabilities. Many failures have occurred in arches where the expansion of long lengths of wall on either side of openings is restricted causing the arch to lift. This defect is usually found in older types of commercial or industrial properties where expansion joints have been omitted.

Many openings in domestic structures built before 1900 were constructed with timber lintels at the rear of the brick arch. The surveyor will no doubt find some difficulty in trying to diagnose the problems in lintels which are hidden. Defection in the lintel and cracked plaster around the bearing ends are invariably due to defective timber which has probably been attacked by woodworm or wet rot. The external arch is often only 119 mm thick and as such it is liable to decay through moisture penetration.

6.6   Defective materials and chemical action

Sulphate attack is one of the principal defects caused by chemical action on materials and is often one of the most difficult to diagnose (Carillion Services 2001).

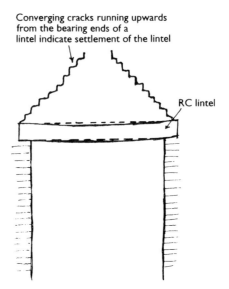

Figure 6.7   Cracked brickwork due to lintel failure.

Firstly, the surveyor must remember that there are three contributory factors that cause sulphate attack: a combination of Portland cement, soluble sulphate salts and water. Sulphate attack cannot occur in the absence of one of these factors.

The majority of clay bricks contain sulphate salts. These salts are soluble in water and when walls are erected the moisture reacts to form a solution which travels by capillary action through the brickwork. The salts form a white deposit on the surface known as efflorescence which is usually harmless and will gradually diminish by the action of rain and wind. This defect is usually found in fairly new buildings (say up to four years old). With older buildings soluble salts will have worked themselves out of the brickwork. A type of salt known as magnesium sulphate may crystallise beneath the surface of the brick causing surface erosion and the outer skin to split away as shown in Fig. 6.8. Portland cement contains tri-calcium aluminate in varying amounts. Wet clay bricks in contact with mortar based on Portland cement will form calcium sulphoaluminate, the crystal of which causes considerable expansion and disintegration of the mortar joints. This expansion produces horizontal cracks in the mortar joints and leaning or bulging sections of brickwork which in time will impair the stability of the wall. This defect is particularly noticeable if the brickwork is restrained at top and bottom or in exposed positions such as parapet walls, retaining walls and chimney stacks. If walls are saturated on one side, then expansion will occur on that side causing the wall to curve. Figures 6.9a and 6.9b show two typical examples of sulphate attack to a parapet wall and retaining wall where continued saturation has

Figure 6.8   Surface decay of brickwork due to crystallisation of salts.

caused the brickwork to bend away towards the dry side. The problem for the surveyor is to decide what remedial work is necessary to eliminate the source of moisture and whether or not the affected brickwork should be rebuilt. He should also check wall surfaces internally and externally between the eaves and the DPC to ascertain that the external walls have been properly designed and to ensure that DPCs and flashings etc. are effective.

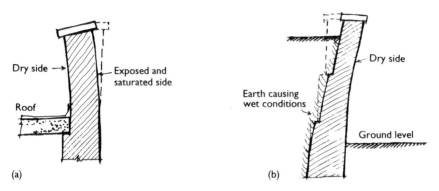

Figure 6.9   Sulphate attack (a) in parapet wall and (b) brick retaining wall.

6.7   Failure in bonding and defects at junctions

The inspection of the external brickwork should follow a set pattern, starting at the highest point of the building and working down to ground level. Brick walls depend on good bonding for their strength particularly at junctions. Poor bonding or fracturing at junctions is a general source of weakness. It is most common in older domestic buildings where fractures have occurred between external and party walls. Examination often shows that the fractures are due to poor bonding or movement of the external walls as previously described in Section 6.2.

When carrying out an inspection of an old wall the surveyor must look back to the bricklaying practices of the eighteenth and nineteenth centuries. What often happened in the past is that external walls were built of two separate skins consisting of an outer facing of 114 mm thick brick and an inner load bearing skin of 228 mm thickness which often consisted of underburnt or broken bricks. The outer skin was tied to the inner skin only infrequently but to give the 'right appearance' on the face of having been properly bonded, snapped headers would be used to save cost. Over the years structural changes due to the thermal and moisture movement can often cause the outer skin to pull away from the inner skin. Bulging will occur between openings and gaps will open between window and door frames and soffits of arches. This defect will not necessarily be apparent on the inside face of the wall.

6.8   Frost failure

Where brick and stone remain saturated in exposed conditions and form ice an expansive force within the crevices of the brick is caused. This expansion is often sufficient to cause crumbling and flaking of the brick surface. This defect is more noticeable in underburnt bricks with low frost resistance. In severe cases it has been known for brickwork to completely disintegrate. The surveyor may well consider it advisable when dealing with an older building to examine the mortar. An old wall that has been repointed using a strong (i.e. cement-based) mortar which constrains the brick may well lead to frost failure.

6.9   Cavity walls

The great majority of buildings are now constructed with cavity walls, either load bearing or in panels within an RC frame. They are constructed with either a brick or concrete block inner leaf. In theory they should prevent the direct penetration of rainwater to the inner leaf. A cavity wall is damp-proof if it is properly designed and built. Unfortunately, faults are rather common. The stability of a cavity wall depends on the proper placing of the metal wall ties particularly when the floor and roof loads are carried on the inner leaf. Accordingly, any defect such as a bulge or lean in the brickwork may cause collapse. This is most likely to occur where there are disturbances in the areas where wall ties can be expected, and where the mortar has deteriorated causing the ties to loosen (Hollis 1990).

The outer leaf of a cavity wall will sometimes move independently of the inner leaf. This may be due to roof thrust or sulphate attack on the mortar. The wall ties are usually of galvanised steel, but due to the presence of moisture and changes in acidity of mortars the zinc coating may have broken down over the course of years. Calcium chloride in cement, lime and sand mortars can also cause corrosion of metal wall ties. Laboratory tests have actually proved that ties manufactured prior to 1981 can corrode at a faster rate than expected. Thus a building erected before 1981 and showing signs of instability in the outer leaf would present a high risk of failure. The thickness of the zinc coating was increased in 1981 following BRE research (BRE IP4/81). Adequate investigation is imperative to ensure the problem is correctly assessed. The usual signs of corrosion are as follows:

- Horizontal cracks in the brickwork every six or eight courses corresponding with the wall tie positions. The cracks are usually several millimetres wide.
- The formation of rust will also cause the external leaf of brickwork to expand vertically.
- Outward bulging of the outer skin. It has been known for the wall facing the prevailing wind to be affected more seriously than other walls.
- It should be noted that zinc corrodes more rapidly when bedded in black ash mortar and affects principally that part of the tie bedded in the outer leaf.

- In severe cases the inner leaf of a cavity wall may show horizontal cracking causing floors and roofs to lift.

Because the defective ties are buried in the masonry their condition is difficult to assess. The following techniques have been used to investigate defective ties:

(1) Metal detectors may be used to locate wall ties along defective bed joints, but give no information concerning their condition. When hiring or purchasing this type of equipment it should be explained to the supplier the sort of information that is required. It is also important to consider the weight and shape of the equipment, since it will no doubt have to be carried on scaffolding or ladders.
(2) Optical probes as described in Section 2.5. These are fitted with light guides which illuminate the cavity and are most useful provided the cavity is not filled with insulation material. The cavity can be viewed through a series of holes drilled in the mortar joints, but only the portion of the tie within the cavity space can be seen. This may be all that is necessary to make a satisfactory diagnosis.

Problems concerning damp penetration will be dealt with in Chapter 8.

6.10   Built-in iron and steel members

Iron and steel members embedded in brickwork can well be affected by corrosion causing the surrounding brickwork to fracture and loosen. In order to diagnose the full extent of the problem, the surveyor will usually find it necessary to remove the surrounding brickwork in order to gain access to the metalwork. Figure 6.10 shows a typical example. Rusting steel stanchions encased in brick cladding may produce vertical cracks.

6.11   Tile and slate hanging and weatherboarding

Tile or slate hanging is very popular in many domestic properties as a feature for bays and upper floors. It provides a cheap weatherproof form of construction and often looks more attractive than rendering. The problem of discharging rainwater down the surface of the tiling is usually solved by providing a brick or two courses of tiles to act as a corbel at the base of the tile hanging to throw the water clear of the wall below. In the construction of tile hung bay windows a timber tilting fillet was often used which over the years would tend to rot by adsorption of damp from the tile. This defect usually occurs when the timber is inadequately protected or not covered by bitumen felt. The surveyor should note the following points as closely as possible.

Figure 6.10   Cracking of brickwork due to corrosion of embedded steel.

- Tile hanging should be checked for missing tiles.
- Check that bottom edges are formed with a double tile course.
- Vertical abutments to be formed with 'tile and half' and weather-proof with lead soakers and stepped flashings inserted into the brick courses.
- External angles should be made with angle tiles or alternatively lead soakers should be fixed under each tile.

The surveyor may well find it difficult to examine all these points from a ladder but will find binoculars most useful.

Weatherboarding is a traditional facing material and was initially restricted to Kent and Essex. In early times the material consisted of rift sawn oak with a feather edge secured to a timber structural frame and protected at the corners by means of angle corner posts. Present day weatherboarding usually consists of deal moulded boards with a form of jointing known as shiplap which provides weatherproof horizontal joints. In cheap work the boards are secured to brick or lightweight concrete blocks without the use of battens. In good class work the wall should be lined with building paper and battens to provide air-space. In recent years, self-coloured PVC panels giving the appearance of white painted shiplap boarding have become available. If properly fixed they require less main-tenance and are probably more durable. The surveyor should note the following points in timber weatherboarding:

- All boards should be level and free from twist or curl.
- Attacks by wet rot or woodworm (these items are dealt with later under 'Timber decay and insect attack' in Chapter 9).
- Where painted, the paint condition should be carefully noted.
- Condition of the abutments of boarding and walls. These should be protected by lead stepped flashings.
- The bottom edge of the boarding should be properly throated.

6.12   Partitions

One of the most common defects the surveyor will encounter is the cracking of partitions or the plaster applied to them. The causes are excessive moisture movement of the materials used in the construction, structural movements caused by settlement, deflection of beams supporting the partitions or thermal movement of the roof.

The cracking of plaster as distinct from cracks which occur in the partitions themselves is discussed under 'Internal finishings'. In order to diagnose the problems the surveyor must first decide if the partition is load bearing or not. This is usually carried out without too much difficulty, but in some properties it is more difficult than it might appear, and under such circumstances the only method of approach is the preparation of accurate plans and sections of the various floors showing the direction of floor joists and beam positions etc. Such complications are often found in eighteenth and nineteenth century buildings where timber trussed partitions were used which were often off-set at the various floor levels rather than directly above each other. Internal partitions being restrained on all sides and bonded into the main walls of the buildings are not likely to show any sign of movement out of the vertical, the movement being limited to settlement.

With the advent of mass produced bricks in the nineteenth century, it became more common for load bearing and non-load bearing partitions to be constructed of 114 mm and 229 mm thick brickwork. It is unusual to require great strength in partitions, and furthermore they are not exposed to damp or extremes of temperature.

During the twentieth century other types of material were introduced, the most common being concrete breeze blocks and hollow clay partition blocks. Since the 1960s, lightweight concrete blocks have been introduced to the exclusion of the former brick partitioning while the introduction of plasterboard has replaced lath and plaster used on the traditional stud partition.

All types of lightweight concrete blocks have relatively high moisture movements and cracking often occurs in partitions built of this material. The greater the length of a partition the more risk there is of the development of vertical shrinkage cracks. However, partitions are interrupted at frequent intervals by door openings. In such circumstances the tendency

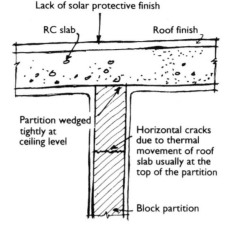

Figure 6.11   Partition failure due to thermal expansion of roof slab.

will be for any shrinkage cracks which develop to occur between the door head and the ceiling.

Another type of failure is associated with the thermal expansion of a concrete roof slab to which the flat roof system is prone. This occurs where the partition has been wedged tightly at ceiling level without allowance in the design for the differential movement between the roof slab and partition. The thermal movements of the slab produce horizontal cracks near the top of the partition as shown in Fig. 6.11. This failure is also due to the lack of solar-protective finish on the roof covering in order to reduce the amount of heat absorbed by the roof slab.

6.13   Assessment of cracks

When assessing the cause of movement in masonry for a 'condition survey' the approximate width, position and direction of the cracks are important factors, although in such cases precise measurements are not essential. When it is necessary to determine that the cracks are caused by progressive movement then repeated observations are required over a period of six months and will require the use of the equipment described in Section 2.5. This procedure is necessary in order to establish whether the movement has ceased or is progressive (Dickinson and Thornton 2004). The following factors should be given careful consideration:

- Whether the crack passes diagonally through the brick joints, or passes through individual bricks as well as the joints and is more or less vertical.

- The width of a crack can vary from a hair line to a large fracture and it is an important factor in determining any remedial measures required. Cracks are usually divided into three groups: fine cracks up to 1.5 mm; medium cracks from 1.5 to 10 mm; and large cracks over 10 mm.
- The nature of the brick must be considered. A simple test to determine whether the brickwork is porous or not is to throw a cupful of water on the brickwork when dry. If the water is not absorbed then there is a risk that rainwater will be drawn through fine cracks by capillary action. On the other hand, if the water is rapidly absorbed the brickwork can be considered porous, and there is little risk in leaving a fine crack in the brickwork.
- Cracks may fluctuate in width with seasonal change sometimes due to movements in clay subsoils or root systems as described in Sections 5.6 and 5.7.
- Pointing should be carefully examined. If the bulk of the pointing is in poor condition then it may be advisable to recommend repointing the whole wall. If the majority of the pointing is in reasonable condition it is best left alone. Much damage can be done to brickwork by raking out sound pointing.

Other factors which must be taken into consideration are the direction and inclination of the cracks:

- Open horizontal joints indicate vertical settlement (see Fig. 6.12a).
- Open vertical joints indicate horizontal movement (see Fig. 6.12b).
- Horizontal and vertical movement indicates transverse failure (see Fig. 6.12c).
- In certain cases one side of a crack may be in advance of the other, which indicates a thrust at right angles to the wall.

6.14   Natural stone masonry

The nature of stone and causes of decay is a complex subject. The following notes are intended to assist the surveyor to identify the more common defects.

Stone was popular in older buildings where it was widely used in natural stone districts where local quarries made this advantageous. The very early stone walls are very thick and were usually constructed of stone throughout, though the inner material may be inferior to the facing material and often consisted of a brick or stone rubble mixed with lime mortar.

To understand the durability of natural stone in old buildings the surveyor must distinguish between the different types and their mineral constituents. Stones are classified as belonging to one of the three following groups according to the manner in which they were formed.

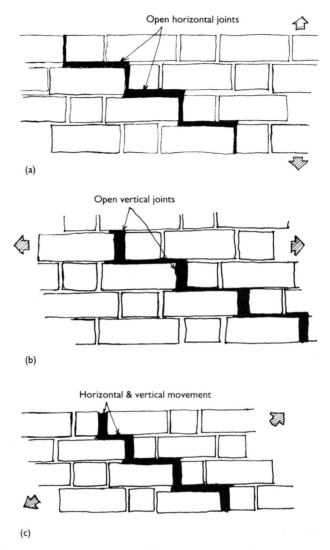

Figure 6.12   Assessment of cracks. (a) Vertical settlement (b) horizontal movement and (c) transverse failure.

6.15   Sedimentary rocks

This is the commonest group consisting of limestone and sandstone which provide most of the building stone in this country. Limestone varies in porosity and density and consists of calcium carbonate cemented together with calcite under pressure. Some limestone contains marine organisms or freshwater molluscs. Sandstone consists of fine grains of quartz cemented together with silica or calcium carbonate under pressure; it

differs in hardness and there are many subdivisions varying from soft calcareous to hard siliceous sandstones.

6.16   Igneous rocks

The building stones in this classification are formed by the slow cooling
and solidification of molten material deep in the earth's crust. They are
commonly known as granites. The stone is extremely hard and is considered the best stone from the point of view of durability especially in damp
conditions.

6.17   Metamorphic rocks

These are derived from sedimentary rocks which have changed from their
original form by heat and pressure arising from movements in the earth's
crust. The two principal rocks of this division are marble and slate. Marble
is very durable and is often used for decorative work such as shop fronts
and interior linings to public buildings. All marbles are vulnerable to
attack by weak atmospheric acids, but resist oils, alkalis and water.

Slate is derived from clays and has a flocculated structure. It is mainly
used for roofing, cladding and DPCs (see Chapter 10).

6.18   Defects in stonework

In carrying out an examination of stone walling it is important to discover
the type of construction and nature of the wall in cross section. As in brick
walls care must be taken in removing material from a load bearing wall.

The causes of structural failure in a stone building are precisely the same
as those for brick walls so that the same procedure must be followed as
described in Chapter 5. However, stone reacts in a slightly different way
and in the following paragraphs these differences will be considered.
Natural stone is found in layers or laminations and in most stones this is
revealed by the fine parallel lines on the faces of the block. It is important
to lay stone walls so that the layers are horizontal or 'natural bedded'
(Davey et al. 1995). If the block is laid with the strata vertical, the surface
will probably spall off. Free bedded stones are more vulnerable than stones
laid on their natural bed. Pointing which is too hard will accelerate the
rate of decay in the stonework. During the investigation it is sometimes
found that a few blocks have decayed while the majority are quite sound.
The defects may be due to the blocks having been laid the wrong way
round as mentioned above.

In the paragraphs following it will be seen that water absorption is the
one common factor associated with practically every form of stone decay.

Atmospheric pollution is the principal cause of deterioration of limestone and sandstones (Richardson 2001). Air usually contains impurities of sulphur and carbon. Sulphuric gases in the atmosphere dissolve in rainwater to form acid solutions and when not freely washed away by rain form a hard skin on the limestone eventually causing surface blistering, scaling and the formation of cracks at arrises (Cook and Gibbs 1993). When the surface blistering breaks off, it pulls away some of the limestone with it. Sulphur gases can also attack calcareous sandstone in a similar manner. Pollution deposits are widely dispersed and travel with the wind and rain, although the effect is more noticeable in urban and industrial districts. Figure 6.13 shows an example of surface blistering of Bath stone and Kentish rag stone due to atmospheric pollution.

Much decay found in porous stonework will be caused by salts which are soluble in water although the majority of natural stones are relatively free from salts. The salt may originate from the atmosphere or from substances applied to the wall surface. These substances then crystallise beneath the surface as the moisture carrying the solutions evaporates, causing pieces of stone to spall off, giving the wall a weather-worn appearance. Another common source of soluble salts is the brick backing to stone walling. Rainwater can penetrate a porous stone face or jointing and into the brick backing. Salts are then carried back to the face of the stone by evaporation. Decay can also arise by rainwater entering through defective copings, sills or other mouldings. Rainwater will tend to wash off soot particles and unsightly stains. However, soot deposits will accumulate on stone surfaces in sheltered or protected areas. Buildings faced with limestone showing these deposits will have a black and white patchy appearance. Buildings in sheltered areas also collect deposits of calcium carbonate and sulphate. The Clean Air Acts 1956, 1968 and 1993 aimed to reduce the source of atmospheric pollution and thus to help alleviate the problem.

Uniform thickness scales separating from standstone and known as 'contour scaling' can cause considerable damage. Although the process is not certain, internal stresses are created in the surface of the sandstone which arise from differences in the moisture or thermal expansion between the sandstone and the crust blocked with calcium sulphate deposited from the atmosphere. The crust breaks away at a depth of 5–20 mm and follows the contours of the surface.

Limestone and sandstone can also be damaged by salts from sea spray; this is usually manifested by a powdering of the stone surface.

Frost does not have any serious effect on sandstone or limestone except in very exposed situations where the stones are lightly saturated. The position and degree of exposure in which the building is located is, therefore, of some importance before a diagnosis can be made. In exposed horizontal surfaces such as copings, sills, cornices and retaining walls where the stone gets saturated then the frost action is likely to occur. Once the face is damaged frost will accelerate the deterioration. If rainwater penetrates

Figure 6.13   Surface blistering of bath stone and Kentish rag stone due to atmospheric pollution.

through fine cracks in the stone or mortar joints it will be held by capillary attraction and the stone will be damaged by expansion or freezing. One major cause of damage is the fractures caused by the corrosion and expansion of iron cramps and dowels particularly in limestone and sandstone. The ironwork is embedded in the stone thus increasing pressure is exerted until the stone immediately surrounding the built-in member splits. This type of defect is readily visible. The close association of limestone with sandstone can result in the decay of the latter. Although sandstone usually weathers well in polluted atmospheres, a chemical reaction is likely to take place in the case of limestone to form calcium sulphate. If this solution is washed into and is absorbed by the sandstone it can cause excessive decay.

Vents and shakes are inherent defects in stone and are not always apparent in newly quarried stone. If a building contains much decorative stonework or carving, vents can be the cause of considerable problems. Vents are small cracks in the stone caused by earth movements during its formation and after the stone leaves the quarry they are exposed to change in climate conditions; these cracks are the first signs of weakness. Shakes are found in most limestone and are not a serious defect. They, too, are minute cracks, but at some stage in their formation calcite has flowed into the cracks resealing them naturally. Vents and shakes can occur in any direction of the stone block.

Various forms of vegetation can cause decay in masonry. Plant life growing in or on stonework or brickwork will tend to keep it permanently damp and therefore contribute to its decay. In some cases much of the charm of old buildings is due to the presence of this growth and the pleasing effect is often encouraged by the owners. In such cases the surveyor may well be advised to investigate the real cause of any damage before recommending the removal of the growth.

On the other hand if any of the defects described in the previous paragraphs should give cause for alarm then it is essential that the stonework is cleaned down and all growth removed in order that a complete examination may be made. Where large areas of growth have to be removed care must be exercised to avoid damage to the stonework (see BRE Digest 418).

Ivy, honeysuckle and wisteria can cause severe damage. They keep the wall damp, and joints are forced apart by the expanding roots which eventually dislodge the stones. Some clinging creepers such as Virginia creeper attach themselves to stonework by suckers, but in many cases are harmless though they do keep the wall damp. In general, most plant growths should be discouraged for if they are allowed to cover a whole wall surface, they are apt to reach the roofs, the most vulnerable part of the building, and lead to blockage of gutters and downpipes. Figure 6.14 shows a fairly common defect caused by shrubs sending their roots into brick joints.

6.19 Cast stone

Cast stone masonry consists of precast concrete units presenting a stonelike appearance and built in a similar way to natural stone masonry. In the British Standard BS 1217: 1997 cast stone is defined as 'A material manufactured from cement and aggregate for use in a manner similar to

Figure 6.14   A fairly common defect caused by shrubs sending their roots into brick joints.

and for the same purpose as natural stone'. Cast stone blocks are usually made to represent a particular stone such as Bath or Portland; the colour being determined by the use of white cement and pigments. Coloured aggregates are sometimes used which also control the texture of the block.

The durability of cast stone is often difficult to define because its properties vary. It may be homogeneous or consist of an inner core of plain concrete with an outer face of a different grade of concrete. Due to its cement content it will behave like concrete and expand slightly on wetting and contract on drying more than natural stone. A proportion of cast stone blocks develop crazing over a period of time; sometimes after exposure to the weather for several months. The cause is often due to the differential moisture shrinkage between the surface layer and the inner core. In the past crazing was difficult to control, but gradually the manufacturers have overcome the problem and have limited its frequency. Some types of block have a porous open texture similar to the softer natural stones, while others have a dense structure and therefore a high resistance to weathering. The open textured blocks are likely to show a definite variation on weathering particularly those made with a semi-dry mix. The appearance of cast stone can be affected by staining and streaking often due to water dripping from a coping or ledge, the projection or flashing being inadequate to throw it clear of the walling.

6.20   Recording defects

The surveyor will often find it impossible at first sight to record how much damage has been caused to natural or cast stone walling. Recording must be carried out systematically so that every defect is included. Projecting features such as cornices and copings will require close investigation. On a small building showing a few minor defects it is a comparatively easy matter to prepare a freehand sketch with a few notes describing the work to be done. When dealing with a large complex structure it may be necessary to cut away the face of some of the stone blocks and test for soundness by tapping. In such cases it is advisable to sketch the elevations, take site measurements and plot them to scale on the drawing board. Photographs of each elevation can be included with the measured drawings as described in Chapters 2 and 3 thus ensuring that everything is included and no defects are omitted.

In most cases the surveyor will be able to ascertain from his examination the cause of the defects in natural or cast stone (Ashurst and Ashurst 1988). However, the problems of stone decay are complex and have been studied in several countries over many years. It is not within the scope of this book to discuss the various methods of treatment. Publications such as Bereton (1995) and Powys (1995) give the available sources of information which are likely to be of use to the architect or surveyor. Alternatively, the surveyor may consider it advisable to consult a specialist in this field.

7 Reinforced Concrete, Cladding Materials and Structural Steelwork

REINFORCED CONCRETE

7.1 Description

The chemistry of reinforced concrete is treated comprehensively in several technical publications (e.g. McDonald 2002) and is, therefore, not dealt with here. Reinforced concrete is strong in both compression and tension and is used extensively in modern buildings in the form of lintels, beams, floor and roof slabs and wall cladding. The success of these various elements depends upon the following factors:

- The use of good quality materials especially aggregates.
- Correct measurement of the ingredients and adequate mixing.
- Correct design in accordance with the codes of practice.
- Correct placing and compaction of the concrete.

Reinforced concrete structures constructed in accordance with the details listed above will be durable and maintenance should not be excessive. However, several problems may occur which often require investigation and extensive repairs. All reinforced concrete is subject to chemical movement, shrinkage of concrete and variable loading. It is the durability properties of the concrete and the effects of external agencies which are the main concern of the surveyor when carrying out an examination of reinforced concrete defects. A lot of patience is necessary to solve some cases, and possible sources of trouble. Some typical faults are described as follows.

7.2 Corrosion and cracking

The cause of cracking in reinforced concrete columns and beams is often difficult to determine. Normally, the reinforcement does not corrode because the concrete provides a protective alkaline environment due to large quantities of calcium hydroxide which are produced as Portland cement hydrates and hardens. Unfortunately the concrete cover can be broken down. This failure is often caused by carbon dioxide from the air which can react with the calcium hydroxide in the concrete to form calcium carbonate. The concrete close to the surface becomes carbonated and this eventually penetrates through to the reinforcement and results in corrosion of the steel. Corrosion of steel reinforcement creates internal expansive forces which will eventually crack and spall off the concrete cover (see BRE Digests 432, 444 and 455).

Many failures are due to the high proportion of calcium chloride present in the admixture to accelerate setting. When chlorides are present in concrete the steel corrodes even though it is in an alkaline environment. This material was widely used during the 1960s and has produced many problems. The current Code of Practice BS 8110:1 1997 has effectively forbidden the use of calcium chloride in reinforced concrete.

When investigating defective concrete structures it should firstly be established whether the corrision in the steel is due to carbonation (BRE Digest 405) or to presence of chlorides (see BRE Digest 444). If the structure is 15–25 years old and there are only a few millimetres of cover to the reinforcement then corrosion is probably due to the presence of chlorides. If the structure is more than 25 years old then the defect is probably due to carbonation. The amount of chloride present can usually be determined by sending samples of the concrete to a laboratory for analysis.

In extremes cases of extensive spalling of concrete or where the reinforcement is badly corroded, causing a reduction in the cross section of the steel, then the structural implications have to be considered (BRE Digest 389). In such cases the surveyor should recommend the services of a structural engineer. A minimum cover of 25 mm of concrete is required for internal work to beams and slabs. For external and internal work in corrosive conditions 50 mm minimum cover for all steel is required except where the concrete is protected by cladding.

It is important that the surveyor checks the depth of the cover, especially in small members such as window sills, string courses and projecting nibs to RC lintels as shown in Fig. 7.1a. If the reinforcing bars are too near the surface the concrete will crack and the bars corrode. Loose portions of concrete can be removed fairly easily and should not present the surveyor with any difficulty. When dealing with cracks it is often necessary to enlarge the crack to gain access to the steelwork before a correct diagnosis can be obtained.

In some cases it may be found that the reinforcing bars are too close together to allow the aggregate to pass freely between the bars and formwork as shown in Fig. 7.1b. Figure 7.2 shows an example of reinforcement corrosion causing cracks to appear in the concrete cover. Cracks can also be caused by structural movement or defects in the foundations; if the cracks continue down below the ground this would suggest some defect in the foundations, possibly due to settlement.

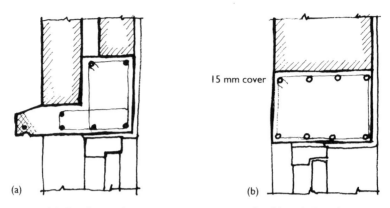

Figure 7.1 (a) Cracking of concrete in projecting nib of lintel. Reinforcement not required to projecting nib. Damp penetration will cause corrosion and spalling of concrete; (b) Reinforcing bars too near the surface causing concrete to crack and bars to corrode. In both cases cracking of the concrete will occur allowing moisture to penetrate and concrete to spall off.

Figure 7.2 Corrosion of reinforcement causing cracking of concrete (courtesy of SGB Group Publicity).

7.3   Aggregates

Experience has shown that durable concrete can only be produced by using good quality aggregates that are clean and free from impurities (Concrete Society 1989). Some impurities have the effect of preventing the cement from setting and hardening. Clay and loam, by coating the surfaces of the aggregate, prevent the cement from coming into actual contact with the aggregate and the resultant concrete is considerably weaker. Aggregates from igneous rocks such as dolerites can expand and contract on wetting and drying causing considerable moisture movement in the concrete mix. This movement can lead to cracking in reinforced concrete units and can also cause deflections in slabs and lintels. Unwashed seashore sand often contains sodium chloride which will increase the risk of corrosion.

The main concrete deterioration risk with aggregates is alkali-aggregate reaction (AAR), of which alkali-silica reaction (ASR) is the most common (BRE Digest 330).

7.4   High alumina cement

This material, in many ways useful, requires great care in application. In recent years its long term durability has become increasingly suspect (BRE Special Digest SD3). During the hardening process a great deal of heat is given off causing excessive expansion together with high humidity; it undergoes 'conversion'. As a result of this change the concrete becomes weaker having a high porosity and lower strength, and also changes colour to light brown. High alumina cement is also vulnerable to chemical attack when in contact with damp gypsum plaster causing the concrete to lose strength. This is an insidious problem which is not apparent until it is in a dangerous condition. It would be advisable, if it is believed that this type of cement has been used, to keep a strict watch for any signs of deterioration.

Specific evidence has to be gathered, analysed and assessed usually by opening up the structure and carrying out tests (BRE Digest 392). Opening up the structural concrete should be carried out by skilled labour using a carborundum saw. The scale of the opening required depends on circumstances, but in all cases the workmen must be properly briefed and instructed in the need for care and protection. The investigation of such a problem will include the need for collecting and recording technical data from several sources. The architect's or specialist's specification used during the construction should provide the surveyor with the data required on the materials used. As mentioned in Section 5.1 the surveyor may consider it advisable to recommend the services of a structural engineer with experience in this field.

7.5   Thermal expansion

Extreme temperature changes are unlikely to occur in the interior of the buildings. Usually the surrounding structure restricts thermal movement and, if the temperature rise is slow, the concrete floors will absorb the expansion without cracking.

Externally, concrete roofs are exposed to rain, snow and wind each having different thermal effects causing the concrete slabs to expand or contract. This movement can result in a fracture between the structural units particularly if there is restraint and the supporting structure does not allow for differential movement. This may occur in large monolithic structures and long parapet walls which have been forced outward. These defects are readily recognised and are often found on large roof areas where expansion joints have been omitted.

7.6   Frost damage

In this country frost failure of concrete is usually confined to new work or to work which because of conditions of exposure remains wet during the winter months. This results in the spalling of the concrete face due to the expansion of ice crystals or the breaking up of weak concrete.

7.7   Electrolytic action

Corrosion can occur by electrolytic action when different metals are in contact or unevenly stressed and moisture is present. Although the latter condition is universal it is unusual for reinforced concrete work to be affected by this type of corrosion.

7.8   Lightweight aggregates

The main object of using lightweight aggregates is to increase the thermal insulation of the concrete; the degree of insulation being dependent upon the low density of the concrete. Concrete made with lightweight aggregates is more porous and therefore offers less resistance to moisture penetration. Thus, the risk of corrosion of the steel and cracking of the concrete face is much greater. Moreover, the bond between lightweight concrete and the steel reinforcement is poor. An essential requirement of all reinforced concrete work is that there must be a good bond between concrete and steel. These two problems can be overcome by coating the bars with cement slurry before placing the concrete. When reinforced lightweight concrete is used externally the concrete cover should not be less than 50 mm with a maximum aggregate size of 13 mm.

7.9   Deflection

Deflection of beams and slabs is rarely encountered in modern large buildings. In small domestic work 'rule of thumb' methods are often used for lintels spanning window and door openings. This sometimes causes problems where little supervision is carried out in that the shuttering is too flimsy for the weight of the concrete or the shuttering is struck too soon. Such movements are relatively slight, and there is usually an absence of surface cracking. This type of defect is easily checked by using a straight edge (see also Section 6.5).

Where steel angles have been used as lintels there is often slight deflection due to corrosion; the metal being insufficiently protected by anticorrosive materials. If there is serious deformation or loosening of the brickwork at the bearing ends of the steelwork then the lintel must be exposed for examination. Rainwater running down the wall surfaces and soaking the steel is often the cause of corrosion in embedded steelwork. Some distortion can be caused in lintels and the surrounding brickwork by differential shrinkage. This type of movement is relatively slight and not likely to be progressive.

7.10   Diagnosis

If difficulties are experienced in arriving at a satisfactory diagnosis then defects are best established by laboratory tests on samples taken from the defective slabs or beams. Such tests can check cement content, cement type, compressive strength, chemical make-up of a sample, etc.

7.11   Brick panel walls in RC frames

In some cases failures have occurred because the design of the concrete structure and the fixings have been insufficient to allow for the thermal movement thus transmitting a compressive force on the surrounding structure. This 'squeezing' effect is often noticed when brick panel walls are used in conjunction with a reinforced concrete frame. In such cases horizontal cracks will show following the line of the floor slabs and frequently accompanied by the displacement of two or three courses of brickwork and spalling of the edges of individual bricks.

Where brickwork is supported on reinforced concrete beams or 'nibs', brick slips are often used to cover the edge of the beam. In recent years there have been a number of failures due to the brick slips buckling. It is usually attributed to the expansion of the brickwork or 'creep' of the concrete frame (see Fig. 7.3). Conventional mortars and adhesive have proved inadequate for fixing the brick slips. A considerable amount of research has been carried out in this field by the Brick Development

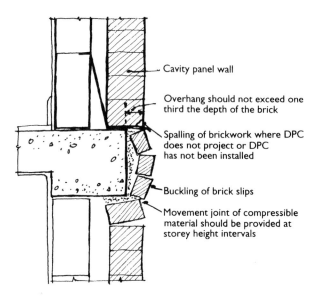

Figure 7.3 Movement in concrete structure due to creep of concrete frame and expansion of brick panel walls.

Association (BDA) and the use of special mortars such as polyester resin or styrene butiadene rubber (SBR) emulsion is now recommended. If the brick slips are to be replaced it is advisable to recommend that they should be isolated by compression joints from the brick panels. One of the important points that should be checked is the overhang of the outer skin of brickwork as shown in Fig. 7.3. The overhang should not exceed one third the depth of the brick. The position of the DPC is also important and should project beyond the outer face of the brickwork to prevent spalling of the lower edge of the brick.

A text of this nature does not lend itself to a complete analysis of all the problems associated with brick panel walls. If extensive remedial works are required then much will depend upon the condition of the building as a whole.

7.12   No-fines concrete housing

Description

No-fines concrete is an unconventional building material used as solid wall construction for dwellings (BRE leaflet BR 160, 1989). It comprises only cement and coarse aggregate (such as clinker), which produces a cellular reinforced concrete with relatively large voids uniformly spread throughout its mass (Scottish Executive Building Directorate 2001). Thus no sand or other 'fine' aggregate is used in the mix (Douglas 2002).

This form of non-traditional construction was employed for building many houses in the late 1940s as a response to the shortage of bricks and bricklayers in the aftermath of World War II. It was developed for dwellings by British contractors such as Weir and Wimpey, and implemented particularly in Scotland during the inter-war period (Scottish Executive Building Directorate 2001). By the 1960s it was also being exploited as an infill-walling component to the in-situ reinforced concrete frame of certain multi-storey flats. Since then it has been mainly used as the main cast in-situ walling element in low-rise housing built throughout many parts of the UK.

According to BRE leaflet BR 160 (1989) the solid walls of no-fines construction are generally between 200 and 225 mm thick. They are rendered externally with a two-coat render and usually finished internally with plasterboard fixed to timber battens attached to the non-fines walling using mild steel cut-nails. Up until the late 1960s this form of walling was considered relatively efficient in thermal terms. With increasing energy efficiency demands following the oil crisis of the early 1970s, however, many of these original no-fines dwellings are now considered thermally inefficient (see typical problems listed below). Some form of 'raincoat' (i.e. insulated render) overcladding to the no-fines blocks would be needed to rectify this deficiency (Douglas 2002).

Problems

Williams and Ward (1991) identified the main problems that surveyors need to be on the lookout for when inspecting no-fines concrete dwellings (see also Reeves and Martin 1989). These can be summarised as follows:

- Lack of energy efficiency.
- Condensation and mould (e.g. due to lack of background ventilation and poor heating).
- Rainwater penetration (particularly in flat roofs).
- Rotting windows and external doors.
- Deterioration of external render.
- Maintenance and adaptation problems (e.g. drilling and fixing, plan modifications, etc.).

7.13   Autoclaved aerated concrete

Description

The term autoclaved aerated concrete (AAC) is a bit of a misnomer in that it is not a true form of concrete. It is neither concrete in its constituent materials nor in its physical properties. 'Siporex' is a common proprietary name for this material.

AAC is made under high-pressure steam-curing conditions by introducing bubbles of gas into a cement or lime mix. The finished product is a uniform cellular material, like a foamed mortar – although it is sometimes described erroneously as foamed concrete.

Siporex can be made into blocks for blockwork walling and planks for wall and roof panels. The blocks are unreinforced and the planks reinforced. The latter range in thickness from 75–150 mm. The wall panels are usually storey height and typically about 1.2 m wide. The roof panels are typically 0.6–1 m wide, and are used as the structural decking on flat roofs and non-pitched roofs.

Advantages

The main advantages that AAC such as Siporex has over equivalent precast concrete units are that:

- It is relatively lightweight.
- It has good thermal resistivity and fire resistance.
- It can be easily nailed and sawn with conventional hand tools.

Drawbacks of unreinforced AAC units

However, Siporex and other AAC blocks are not without their problems. For example:

- They are not as strong as ordinary concrete and have a much higher elastic modulus.
- They are prone to unacceptably high drying shrinkage, which can cause cracking if plastered when wet.
- If plastered when they are too dry the AAC can suck the moisture from the plaster so that a good adherence between plaster and blocks is not achieved.
- They are easily damaged if not handled carefully.
- Minor movements may not be accommodated in the joints without cracking the blocks if too high a mortar strength for building walls is used.
- Cavity walls containing AAC blocks may have insufficient flexural strength to transfer wind loads or be poor at resisting impact loads, all of which would be exacerbated by poor condition of the masonry or the lack of ties between the leaves or the inadequacy of restraint fixings.
- They are prone to premature degradation in damp conditions.

Drawbacks of reinforced AAC panels

- AAC panels are especially vulnerable to moisture – such as that resulting from interstitial condensation and penetrating dampness. They do not have the alkalinity of ordinary concrete to protect the reinforcement from corrosion. As a result, the reinforcement must be protected by the application of a special surface coating.
- AAC does not bond adequately to the steel and must be anchored by hooks or welded crossbars.
- As their modulus of elasticity is low, AAC planks are not as strong as reinforced concrete slabs and are therefore more prone to sagging, which results in ponding when they are used as the structural deck on flat roofs.
- Siporex flat roof panels may have a lower factor of safety against uplift than required by the current British Standard owing to inadequate holding down straps.
- There is a risk of shear failure occurring at the bearing of roof planks on the wallhead.

Surveying properties containing AAC

AAC blocks and panels were used in domestic construction in the 1950s and 1960s. A desk-top survey may reveal that Siporex or other AAC products were used in the construction. Alternatively, problems of deflected roof panels and deteriorated internal blockwork may suggest the presence of this material in the building being surveyed.

In the absence of such evidence, however, it may be difficult for even the experienced building surveyor to immediately identify that Siporex was used in the original construction. Normally an invasive survey would be needed to confirm this, which could be done by:

- Inspecting the cavity and ceiling void areas using a fibre optic borescope (see Chapter 2) to check for cracking, reinforcement corrosion or other deterioration in the AAC units.
- Removing sections of ceiling finish to reveal the soffit of the Siporex flat/mono-pitched planks and check for exposure and corrosion of the reinforcement.
- Removing external infill wall panelling to check reveals of walling containing AAC blockwork for fixing and damp proofing.
- Removing sections of flat roof covering, especially at verges or abutments, to check for deterioration of the planks.

If Siporex or other such AAC products are found in a building, the surveyor should recommend that specialist engineering advice be obtained to ascertain their structural adequacy.

CLADDING MATERIALS

7.14   Description

The term cladding is here taken to mean an infilling to a structural frame and can be considered under the following headings:

- Precast concrete panels.
- Glass fibre reinforced plastic (GRP).
- Glass fibre reinforced cement (GRC) (see BRE Digest 331).
- Profiled metal and asbestos cement cladding.
- Curtain walling which can be defined as a cladding enclosing the entire structure.

Cladding panels are often a feature of proprietary methods of construction, which may include forms of fixing especially adapted to the type of panel and frame employed. Many forms of reinforced concrete panels have appeared over the past forty years and various aggregates have been used to give a textured coloured surface externally.

The weight of the concrete panels is reduced to a minimum by casting the panel as thin as possible, rarely more than 75 mm thick, rigidity being achieved by casting ribs at the edges and at intermediate positions. The panels are usually reinforced with welded galvanised mesh, the reinforcement being kept back approximately 30 mm from the external face of the panel. The panels are secured to the structural frame in various ways by non-ferrous or galvanised steel plates, clips and bolts. The edges of the panels may be designed to give a self drained anti-capillary joint or the joints may be sealed.

Glass fibre reinforced plastic (GRP) is a popular cladding material, and consists of a glass fibre reinforcement impregnated with resin, incorporating fillers and a hardener. Curtain walling consists essentially of covering the steel or reinforced concrete frame externally with a light cladding material consisting of an outer weather resisting layer and an inner layer of insulating material with or without an air space. The cladding depends on the high insulation of the inner layer to provide the necessary protection against heat and cold. The external layer consists of any of the materials described above that will resist weather conditions. The internal insulation is usually a mat material such as fibre glass, or expanded polystyrene with an inner lining of plasterboard or fibreboard. It is important that a vapour barrier such as aluminium foil, building paper or polythene sheet be provided between the insulation layer and the lining board (i.e. on the warm side of construction) in order to avoid condensation problems within the cavities of these composite materials.

7.15   Cladding defects

Concrete cladding defects, particularly on high-rise buildings, have caused serious problems in many post-war buildings. Failures have occurred due to inadequate provision for thermal expansion and construction processes. In the past architects relied on the technical literature produced by the various manufacturers, which although generally comprehensive did not always show the material in relation to others. It was not until the 1970s that technical guidance on profiled cladding was available from the (now defunct) Property Services Agency. Many of the profiled panels and exposed structural grids all pose new methods of jointing and fixing and are not necessarily covered by the established text books on building technology. The surveyor will often find that defects are repeated in a consistent pattern which could supply useful clues during an examination (BRE Digest 217).

When carrying out an examination the following items are the more common problems to look for:

- Cracks in concrete units.
- Staining and spalling on concrete panels (this may indicate that the reinforcement or fixings have corroded).
- Defective throatings and drips in projecting members.
- Defective flashings, usually at corners or steps where the flashing has been cut or shaped.
- Check for signs of displacement or distortion.
- Where dissimilar metals have been used check that isolating sleeves have been correctly used.
- Over-tight fixing points particularly in GRP cladding.

Substantial differential movements can occur between cladding and background, and this unfortunately has often been overlooked during the design stage. The cost of rectifying such defects can be extremely high, partly due to the problems of access. Large areas of scaffolding and hoisting equipment can often cost nearly as much as the remedial work. The examination can also be a costly matter due to the fact that a satisfactory diagnosis can only be carried out from cradles or scaffolding. In this type of examination the defects are often hidden and difficult to trace. In many cases it may be impossible to gain access to the inner parts of the cladding units to examine baffles, gaskets and sealants etc.

In such cases the surveyor may well be advised to consult the specialist firms or their architects before undertaking diagnostic or remedial work. The detailed drawings and specifications, including those supplied by specialist firms used during the construction of the cladding should provide useful information. Manufacturing and erection tolerances are also important so that in case of any problems the surveyor is capable of checking the dimensions of a particular component under examination.

However, such information can sometimes be misleading. Drawings do not necessarily include revisions made during construction or inaccuracies during construction on site. The architect's site instructions may well provide useful information concerning any modifications made during construction. Thus, the information obtained in drawings and specifications can only provide indications; the work actually executed will have to be determined by site investigation, and will no doubt necessitate removing some of the cladding units.

Many curtain wall systems are constructed of impervious materials. Thus, a considerable amount of water runs down the surfaces of a curtain wall structure. To resist this 'run off' of water the system must be carefully designed, joints must be waterproof and sills and transoms must have good projection and generous drips. Gaskets and baffles must be put together with the greatest care.

7.16   Joint problems

Flexible sealing compounds used in cladding are based on a variety of materials including drying and non-drying oils, bitumen or butyl rubbers. They are extremely good from the point of view of elasticity and resistance to weathering. The different sealants have properties which allow the various movements to be accommodated in relation to the width of the sealant used. Maximum joint width, minimum joint width, and movement tolerance are usually stated in the manufacturer's literature. Hardening and splitting of the sealant often indicates ageing of the material.

Accurate assembly of the cladding units is essential and the movement expected between the sides of the joint must be properly calculated. Unfortunately, there have been many failures of sealants some of which are caused by butt joints which are too deep and too narrow causing loss of adhesion at the sides of the joint or loss of cohesion in the sealant (see Fig. 7.4a butt joint). If the sealant material is unduly stressed it will tend to break down. The lap joint shown in Fig. 7.4b is often preferred because it is better protected from the rain, snow and wind and therefore imposes less strain on the sealant. In order to limit the depth of the sealant and distribute the stress and strain over the width of the joint, a back-up material is often used. The back-up materials usually consist of expanded plastics or fibre building board, but if butyl mastics are used as a sealant then it is usual to use a polyurethane sponge (see Fig. 7.4c).

Self curing sealants based on polysulphide rubbers, silicones or polyurethanes require back-up materials with a parting agent in order that the movement of the sealant is not restricted. Polyethylene or polyurethane sponge are the most suitable materials for this purpose. When investigating sealant problems the surveyor would be well advised to check this backing material.

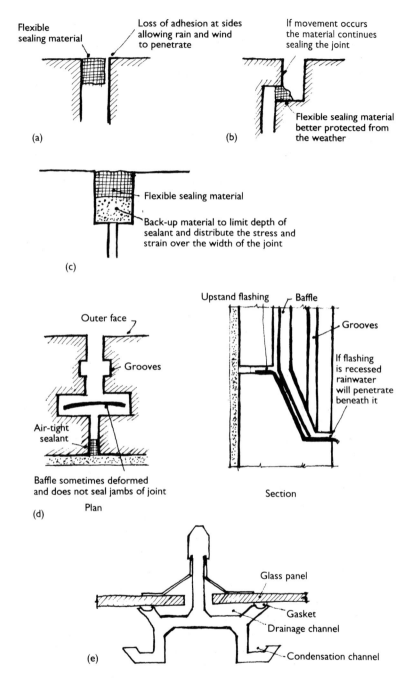

Figure 7.4 Joint problems. (a) Butt joint, (b) Lap joint, (c) back-up material, (d) drained joint and (e) vertical gasket joint in patent glazing.

One of the most successful concrete cladding joints is the open drained joint which reduces the risk of rainwater penetration. The vertical joints between the panels shown in Fig. 7.4d usually consist of an outer groove to trap rainwater which discharges over an upstand flashing at the base of the panel as shown in the section of Fig. 7.4d. The centre portion consists of a baffle which if deformed does not form a satisfactory seal at the jambs of the recessed joint. If rainwater penetrates the grooves in the outer face and the defective baffle there is only the inner barrier consisting of a sealant to prevent damage to the inner face of the cladding. The inner sealant must be air and weather-tight. Flashings at the lower edges direct any rainwater to the outside face which may have penetrated the baffle. One common failure is that the flashing may be recessed behind the panel face allowing rainwater to penetrate beneath it. Gaskets are pre-formed materials extruded from plastics or rubber, and widely used for securing glass panels and for waterproofing a joint (see Fig. 7.4e). They do not depend on adhesion, but on pressure against the joint surfaces to provide a seal. Gaskets cannot adjust well to rough surfaces, and are, therefore, best used between smooth surfaces. One of the difficulties in forming a gasket joint on site is the mitre at the corners. This is a joint that should be carefully checked during the site investigation. Points of weakness are also likely to be found where there is a change in the construction of the cladding or where window or door openings occur. Sills should be carefully checked to ensure that they throw rainwater clear of the cladding below.

7.17   Metallic fasteners

In most buildings, particularly those in very exposed situations, corrosion is a potential risk. Defects associated with metallic fasteners have often been due to corrosion. If the metal fixings are suspect the first point to ascertain is the type of fixings installed. All fixings should be of non-ferrous metal e.g. phosphor bronze, gunmetal, copper or stainless steel. Iron or steel fixings should not be used even if protected with a corrosion resisting coating. Damage to the coating will result in corrosion and ultimate failure of the metal fixing.

Moisture penetration and atmospheric pollution sometimes attack manganese bronze cramps used for fixing stone cladding. Corrosion of fixings can also cause staining on the face of the concrete or stone including cracking or spalling of the cladding material usually close to the fixing points.

Other faults are lack of proper engagement of dowels in the holes or slots meant to receive them; holes not properly formed makes fixing difficult to achieve. Fixings and anchor slots for stone or concrete cladding should preferably be of the same metal thus obviating electrolytic attack.

STRUCTURAL STEELWORK

7.18 Description

Steelwork designs have usually been carried out by a structural engineer and the frames erected by a specialist firm, thus eliminating any serious problems such as faulty design.

Mild steel used for structural purposes can have superficial defects resulting from poor jointing during erection and corrosion (British Steel 1996). Mild steel is seldom directly exposed, but has a protective coating consisting of paint or zinc, or is enclosed by a concrete casing or other fire-resisting material. These various coatings reduce the rate of corrosion (Hudson 1984). Weathering steels have been developed for external use. The exposed surfaces build up a stable oxide coating which protects the metal beneath and provides a maintenance free finish. This finish can vary from an attractive purple to a dark brown hue depending on the length of exposure. Exposed steelwork is usually found in the older type of industrial buildings supporting open steel roof trusses.

7.19 Diagnosis

The two chief points of interest to the surveyor are the rigidity of the joints and the protection provided against corrosion (see British Steel 1996). Care should be taken to see that no bolts are omitted and that all nuts are properly tightened. Protection against corrosion should be in accordance with the latest recommendations of the Codes of Practice (e.g. BS 5493: 1997).

8 Damp Penetration and Condensation

8.1 Description

Damp penetration is one of the most damaging failures that can occur whether the building is old or of a modern type of construction (Massari and Massari 1993). It can damage brickwork by saturating it; cause decay and breaking up of mortar joints; dry and wet rot in timber and corrosion in iron and steel and stained wall surfaces internally and externally (see Trotman et al. 2004).

A considerable amount of water is used in the construction of a new building for mixing concrete, mortar and plaster and for wetting bricks. It is estimated that the average house will contain a tonne of water in the brickwork alone. Some of this water evaporates before the building is occupied and some will be immobilised in the hydration of plaster and cement. However, much of the water used in the constructional process is retained and will dry out slowly. Not much can be done about this type of moisture, but good ventilation and low heating during the first twelve months will greatly assist the drying out.

The majority of building materials are porous and therefore soak up a considerable amount of water. Ground water will be drawn up through the base of the walls by capillary action. This penetration can easily happen even on a fairly dry site unless stopped by a damp-proof course. The amount of rainwater blowing onto a wall can be considerable. It is not unusual for rain or snow to penetrate a 342 mm or 457 mm thick solid brick or stone wall if it has the force of the wind behind it, and if the brickwork is of a porous nature. The south west aspect of a building is the most likely position for this type of dampness in view of the fact that the prevailing wind blows upon this face.

The common causes of rising damp from the ground and penetrating through walls are all considered in Sections 8.4 to 8.14.

8.2 Damp courses

The most common damp course that the surveyor will encounter will be a layer of bitumen impregnated felt or felt with a lead core. They are completely impervious and are obtainable in long lengths. Metallic damp courses are occasionally found consisting of sheet lead or copper. Both are pliable and easily laid; due to their smooth surfaces they tend to encourage 'slip' but at the same time will accommodate a great deal of movement. Mastic asphalt has been used in older buildings, but is rarely used in modern construction. The work was usually carried out by specialist firms and can be expensive. The material is durable but in warm weather tends to extrude under load and bleed through on to the external face of the wall. This is also a problem with bituminous felt damp courses.

Before the Public Health Act of 1875 it was not compulsory to provide a damp course at the base of a wall. The majority of buildings erected before this date will usually have damp problems. During the early years of the new act damp-proof courses were often badly laid or have become defective with age. The early type of damp course usually consisted of two courses of slate laid breaking joint or two courses of hard engineering bricks. Slate provides excellent protection but tends to fracture if stressed by the unequal settlement of the wall. The weakness in using engineering bricks lies in the number of joints, and any settlement can cause cracking in the mortar joints, providing a path for damp to penetrate by capillary action.

8.3   Diagnosis

When investigating a damp problem it is most essential for the surveyor to understand the construction of the building and be certain as to the cause before attempting a remedy (Burkinshaw and Parrett 2003). Some damp problems are straightforward and fairly easy to diagnose, whereas others are more complex and may consist of several damp areas in a more complicated pattern, often at a point some distance from the original source of the trouble. It is well to remember that moisture can travel a considerable distance before revealing itself inside the building.

Damp problems in a building are not always new and are often due to long-standing defects, and in some cases several temporary repairs may have been carried out over many years. The occupier of the building will no doubt point out certain defects caused by damp penetration, but on some occasions the surveyor will consider it necessary to look at the structure as a whole especially if he is dealing with an old building.

It is well to remember that old buildings (built before 1880) were constructed of solid brick or stone walls, and no damp course or roofing felt were provided although a few had boarding beneath the slates or tiles.

One of the most common forms of damp penetration occurs through defective roof coverings or worn flashings and damp courses to chimney stacks. The surveyor will usually notice damp stains on the top floor ceiling or top of the chimney breasts no doubt due to defects in the roof. (These defects will be considered in Chapters 10 and 11.)

As already mentioned in Chapter 2 a moisture meter can be of great assistance in determining the moisture content of building materials or tracing the extent of the damp penetration. However, care must be taken where the moisture meter is being used internally. If the wall plaster has been lined with an impervious damp-proof lining under the wallpaper or other covering it will be necessary to remove a small portion of the surface covering, possibly in some inconspicuous part of the room, to enable the meter probes to be pressed into the damp area of the plaster. In the following paragraphs the identification of the causes of damp penetration will be dealt with. They can be divided into three categories as follows:

- Rising damp from the ground.
- Penetrating damp through walls.
- Extraneous causes.

RISING DAMP FROM THE GROUND

8.4   Solid walls with DPC absent or defective

As mentioned earlier in Section 8.1 the ground around the foundations is always damp, moisture rises into the brick or stonework by capillary attraction and unless a damp-proof course is provided above ground level the moisture will rise causing considerable dampness to walls and floors. It is, therefore, important that the surveyor makes a careful examination of the base of the wall to ascertain the existence or absence of a damp-proof course. If there is no damp course, dampness will show along the bottom of both external and internal faces causing discoloration, mildew and peeling decorations. The damp will rise in a wall or partition to a height at which there is a balance between the rate of evaporation and the rate which the damp can be drawn up by capillary attraction. The line of dampness due to the absence of DPC is usually continuous and fairly horizontal and in severe cases can reach a height of approximately 1.5 m (see BRE Digest 245).

Figure 8.1a shows a typical early nineteenth century external wall construction with no DPC and no foundation concrete. Care must be exercised when checking for rising damp due to a defective DPC. The cracks usually pass through the horizontal joint with a semicircular damp patch internally spreading up from the line of the DPC. The damp course line should be checked for continuity especially where it is stepped due to changes in floor levels or external pavings. At the same time the condition of the pointing to the damp course should be noted.

Another factor the surveyor must consider is that the ground water usually contains dissolved salts which form a fine deposit on the wall surface during evaporation of the water. Some of these salts are

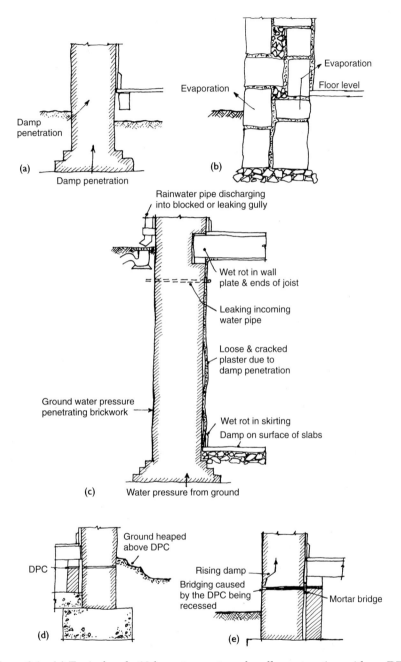

Figure 8.1   (a) Typical early 19th century external wall construction with no DPC (b) Traditional solid stone wall. No DPC. (c) Typical damp problems in old basement walls. (d) Damp course covered by earth. (e) Defects caused by bridging DPC.

hygroscopic and absorb moisture from the air. In such cases the plaster-work becomes contaminated with salts and usually requires replacement.

8.5   Stone walls in older buildings

Older buildings with solid masonry walls built before the nineteenth century have no damp course and are usually of stone throughout, though the interior may be inferior to the facing material. The type of construction is not necessarily unsound and, in fact, is often quite effective in repelling rising damp. The water drawn up from the ground will be dissipated by evaporation through the porous masonry walls and lime mortar joints. In most cases the damp does not rise more than about 300 mm above ground level. The masonry also relies on its thickness to prevent damp reaching the inner face of the wall before evaporating.

The important point for the surveyor to remember is that any treatment which prevents evaporation such as waterproof rendering internally or externally, will only cause the damp to rise higher before it can evaporate causing more problems! Any treatment recommended is a matter requiring careful judgement (see Fig. 8.1b).

When undertaking an examination the surveyor must closely inspect the walling material internally and externally making full use of the moisture meter. Water is drawn through fine fissures by capillary attraction particularly in stone with a coarse grain which contains open pores into which rainwater may be drawn. Fine fissures are often found between mortar and stone forming paths for moisture. Examine the joints cutting away some of the pointing if necessary.

8.6   Basement walls and floors

The conditions under which walls and floors below ground level are exposed to damp penetration are very different from those occurring above ground level. Rooms which are partially or fully below ground level can be affected by penetrating damp, where there is a high water table. For every 300 mm below the water table the water pressure increases and the outside faces of basement walls may remain permanently damp. Ground water penetration can be seasonal depending on the relative height of the water table.

In modern buildings it has become common practice to protect basements by external or internal waterproof coatings. By this means the basements can be as dry as the rooms in the superstructure. Provided this tanking is efficient and the rooms are adequately ventilated the surveyor should have no problems. However, the exact cause of a 'damp patch' may be difficult to discover. Damp patches on walls or floors could emanate from an intermittently leaking water pipe or may be caused by condensation given off by a washing machine or drying appliance in an unventilated basement.

A common defect is the jointing of the damp-proof membrane at the internal angle of wall and floor. The slightest leakage at this junction will result in gradual percolation of moisture and a head of water may build up between the two leaves of brickwork or concrete walling. If the water pressure is strong below the floor it can cause actual flooding of the basement.

As far as the surveyor is concerned, older buildings (those built before the twentieth century) are usually easier to diagnose. They are usually damp throughout and no doubt have been prone to damp problems from the day they were built. The walls are usually constructed of fair-faced solid brickwork or stone set in lime based mortar and it is unlikely that any damp proofing will be located.

Flag stones or brick floors were usually laid on the bare earth and moisture rising from the ground is common. The surveyor will often find an impervious finish has been laid over the solid floor which has become very wet underneath, and probably affected by rot. Figure 8.1c shows typical problems in an old basement wall.

8.7   Heaped earth or paving against walls and bridging of rendering

A common defect is where the horizontal DPC is placed nearer the ground than the 150 mm clearance required by the Building Regulations, and becomes ineffective through being bridged by earth or paving slabs laid at a higher level. In wet weather moisture may be drawn into the wall from the soil above the DPC. Where dampness appears on the interior face near the floor the ground level in relation to the DPC should be examined (see Fig. 8.1d).

Rendered plinths were common during the early part of the twentieth century, often in speculative built houses. In many cases the plinths were applied to buildings with solid walls and they invariably bridged the DPC causing rising damp to penetrate through to the wall plaster and skirtings. Bridging can also be caused by the DPC being slightly recessed and the joint filled with pointing mortar as shown in Fig. 8.1e.

Defects due to 'bridging' are relatively easy to diagnose, but where the elevations have been rendered, it must definitely be established that the wall is solid. Damp patches internally at skirting level cannot be properly diagnosed until this is known.

8.8   Internal partitions

Another common defect found in domestic property is a partition wall with a solid floor on one side and a joist floor on the other. If the DPC is a few millimeters below the concrete floor then damp can rise through the unprotected gap (see Fig. 8.2a)

8.9   Rising damp in ground floors

The surveyor will often find that the ground floors of properties built prior to about 1860 usually consisted of boarding secured to timber joists either laid directly on the earth or separated from it by a few millimetres, the joists being secured to timber plates on brick sleeper walls. No DPC or ventilation was provided. These conditions beneath timber floor joists are the main cause of dry rot outbreaks and will be dealt with in detail in Chapter 9. After about 1880 the principles of construction improved and the floor joists were laid on honeycomb pattern sleeper walls with ventilators at the base of the external wall.

The ventilators were often small and insufficient in number, but they did allow a certain amount of air to circulate around the joists. Damp-proof courses consisting of slates are likely to be found below the wall plates of sleeper walls during this period. In both the above categories very few will be found to have been provided with site concrete (see Fig. 8.2b).

During his examination the surveyor will usually find the floor boarding to be in sound condition, but after removing a few floor boards the joists may well be found to be rotted or saturated with wet rot on their undersides.

After the First World War solid ground floors appeared both in domestic and commercial properties. However, it was soon realised that a concrete slab was not sufficiently impermeable to prevent moisture penetration. Many failures were reported in the timber fillets and boarding as a result of moisture penetration and thus a clause was added to the Model By-laws of 1937 requiring that timber in contact with a concrete base must be protected by a damp-proof membrane consisting of 3 mm thick bitumen or coal tar pitch. The floors of a building erected before this date are therefore suspect and the surveyor will be wise to remove a few floorboards and examine the base concrete. Rising damp in solid floors is easily seen on the surface of the floor finishing. Timber strip flooring or wood block will often have a water mark on the surface and in some cases will be loose and easily lifted.

A number of damp-proofing materials have been developed over the past sixty years providing a completely impervious membrane over the solid floors. Any failures that may occur in the damp-proof membrane are usually due to inadequate jointing with the external wall damp course. It is essential that the damp-proof membrane is continuous with the DPC in the external wall. It is under such circumstances as these that the moisture meter is of great assistance, and in order to determine accurately the cause of the defect it will probably be necessary for the surveyor to obtain the owner's permission to remove a portion of the skirting and expose the

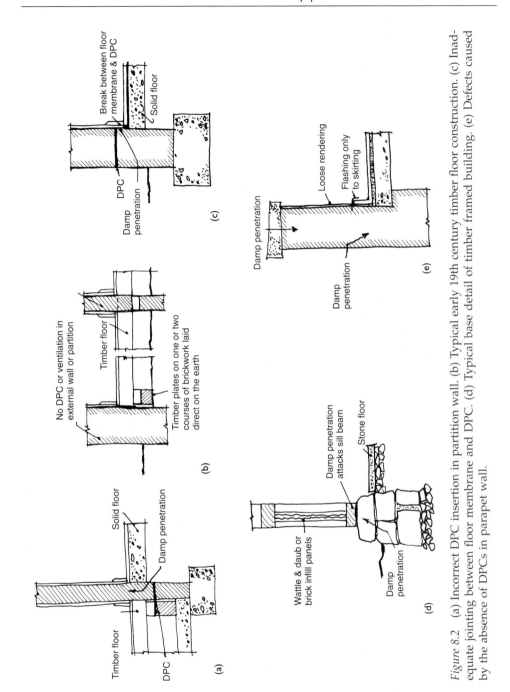

Figure 8.2 (a) Incorrect DPC insertion in partition wall. (b) Typical early 19th century timber floor construction. (c) Inadequate jointing between floor membrane and DPC. (d) Typical base detail of timber framed building. (e) Defects caused by the absence of DPCs in parapet wall.

internal face of the wall. If moisture can be found to have affected the walling behind the skirting the cause can only be due to inadequate jointing between the floor membrane and the DPC in the external wall (see Fig. 8.2c). An important point to check is the air space and ventilation to the ground floor. This space must be properly ventilated to protect the joists and wall plates against dry rot. Air is induced from outside by the insertion of air vents at frequent intervals immediately under the damp course level (giving at least $1500\,mm^2$ fresh air inlet per lineal metre of external wall). The air vents are usually constructed of cast-iron or terracotta (size $230\,mm \times 75\,mm$) although tile slips are often used in good quality work. It is not unusual to find that the air vents have been covered with perforated zinc or even solid metal plates in order to discourage insects or mice. When carrying out the examination it is important to ensure that the air vents are free from obstruction. If close to garden planting or shrubs the apertures are often found to be filled with earth.

8.10   Rising damp in old timber framed buildings

Early timber framed buildings usually had their main oak posts dug into the ground, but those that have survived were supported on a horizontal member or sill beam into which the vertical posts were jointed (Swindells and Hutchings 1991). The sill beam was laid on a brick or stone dwarf wall to keep the timber away from the damp ground. With no damp course reliance was placed on the dwarf wall to keep rising damp to a reasonable level. However, this wall was not sufficient protection, and rainwater and rising damp from the ground gradually attacked the timbers by a combination of wind pressure and capillary action producing wet rot in the sill and the base of the vertical posts (see Fig. 8.2d). This defect can often lead to undue stress being thrown upon other members of the frame. It is therefore advisable to examine all the framing thoroughly, however sound the timbers may appear externally.

Damp penetration in framed buildings also occurs at the joints between the panel infilling and the timber framing. The systems usually consist of cracks between the frame and the panel infilling allowing rainwater to penetrate. Distortion of the frame is also a common defect resulting from twisting and warping of the timber over many years. Figure 8.3 shows the effects of rising damp and woodworm at the foot of a timber post in a timber framed building.

PENETRATING DAMP THROUGH WALLS

8.11   Locating damp penetration

It is quite possible for rainwater to penetrate a 343 mm solid brick wall if the bricks are of a porous nature and the force of the wind is behind it. In

Figure 8.3 Effects of rising damp and woodworm in a timber framed building.

the past various forms of overhang have been used to protect walls from
rain penetration, but during the past sixty years their use has declined.
The overhangs mainly consisted of canopies, projecting eaves, string
courses and cornices. On modern properties it is quite probable that the
external walls are much wetter for long periods now that these overhangs
are absent. If overhangs do exist they should have sufficient fall to throw
rainwater off and preferably be covered with some sort of protective flashing. In old buildings many cornices and string courses have been covered
with cement fillets. These fillets will often be found to have shrunk and
cracked at the junction with the wall and thus form a trap for water.

In most parts of England the prevailing wind blows from the south west so this aspect of a building is the most likely position for this type of penetrating damp. Water is drawn through fine fissures by capillary attraction, and it is often assisted by internal warmth which tends to draw the moisture inside where it can evaporate easily. Fine cracks and fissures are evidence of poor quality bricks, and stonework with a coarse grain also contains open pores into which rainwater may be drawn.

If the walling material appears to be sound and not excessively absorbent, the joints should be examined and if necessary some of the pointing should be removed. If the mortar is of poor quality the risk of damp penetration is increased. Dampness is often caused by using strong cement mortars which tend to shrink. In such cases water is drawn in through the fine fissures between mortar and brick. Window and door openings in solid walls are often a source of trouble and should be carefully examined. In old or modern buildings the wood frames are often flush with the external face of the wall. If the joint between the frame and the masonry is defective allowing moisture to penetrate then the timber is prone to wet rot.

8.12   Parapet walls

Parapets are undoubtedly the most vulnerable part of a building as far as damp is concerned especially as they are exposed to the weather on three sides (Oliver 1997). The majority of parapet walls are of one brick thickness, and it is common to find that the inner face has been rendered to prevent damp penetration. Whether the parapet is solid or of cavity construction can easily be checked by measurement at coping level. When the type of construction has been established the surveyor should examine the coping. Copings of natural or artificial stone should always have sufficient fall to throw the water off the upper surfaces and a good throating should always be provided on the underside so that water draining from the top will not run round the underside to the wall beneath. Copings should also prevent the downward penetration of rainwater, but being broken by joints they may be defective causing the parapet beneath to become very damp. Brick copings are particularly susceptible to penetration by driving rain through the mortar joints and do not provide a good weathering surface.

The external faces of the parapet should be carefully examined. If the bricks show signs of decay or surface flaking this will indicate that the wall is saturated and frost damage or sulphate attack has occurred. Cracks or bulging of any rendering either on the internal or external faces will also indicate severe damp penetration. These defects are usually caused by the absence of a DPC below the coping (see Fig. 8.2e).

The next point to investigate is the provision of a DPC under the coping and in the parapet wall at the top of the skirting to the roof. If these DPCs are missing or ineffective then damp will penetrate through the main fabric of the structure. A common defect found in a number of parapet DPCs has been to stop the DPC short of the internal brick face and not turn it down. On a flat roof the DPC should be taken right through to the internal face of the parapet to form a cover flashing over the skirting as shown in Fig. 8.4a. On the external face, the DPC should extend to form a small projecting drip on the face of the brickwork. If the wall is rendered externally the DPC should pass through the thickness of the rendering and be turned down to prevent 'bridging' of the damp course by the rendered face.

If the precautions described above are not taken then failure is almost certain. Cavity parapets are the best solution and if designed in accordance with the current Code of Practice no problems should ensue.

8.13   Cavity walls

A common problem in cavity wall construction is bridging of the cavity wall by excessive mortar droppings at the bottom of the wall or on the wall ties (Hollis 1990). Moisture will then be absorbed by the inner wall and irregular damp patches will appear adjacent to the mortar droppings or near the floor. Inclined wall ties which slope downwards from outer to inner leaf also allow water to penetrate. This is also a common fault with DPCs over window and door openings as shown in Figs. 8.4b and 8.4c where bridging of the cavity is caused by excessive mortar droppings building up off the DPC or due to a faulty or missing damp-proof tray. Damp patches on the plaster about 300 mm above the head of the window usually indicate mortar droppings collecting at the head of the damp-proof tray. The weep holes externally should also be examined to ensure that they are not blocked.

The reveal to a window opening seems the obvious place to expect the entry of water particularly if the jointing between wall and frame is in poor condition. Rainwater penetration at this point is often due to poor detailing of the DPC. Semicircular damp patches at the junction of the window frame and plaster indicate a defective or missing vertical DPC which can also cause wet rot in the frame. The vertical DPC should project into the cavity to prevent 'bridging' by mortar droppings (see Fig. 8.4d). Another common danger point is the sill. All sills should be provided with a DPC for their whole length and width, and turned up at the ends.

Where a flexible DPC crosses the cavity the underside should be supported on an asbestos or slate sheet. Figure 8.4e shows the line of damp penetration due to omission of the DPC under the sill. In recent years the cavities of cavity walls have often been filled with plastic foam to increase the thermal insulation value of the wall. The foam usually consists of urea-formaldehyde and is injected into the cavity by drilling holes through the mortar joints. In view of the fact that the main purpose of a cavity is to prevent damp penetration, a filling material, in principle, increases the risk.

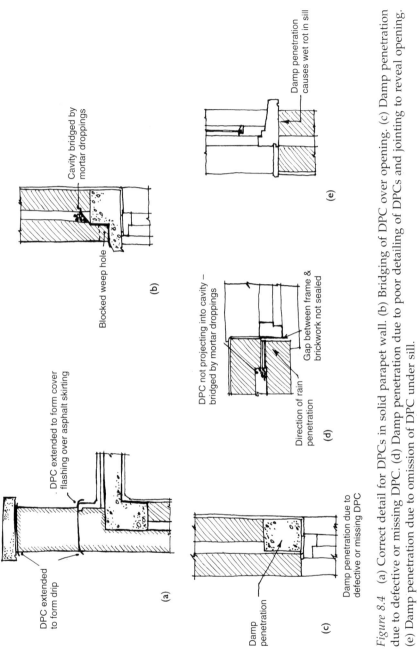
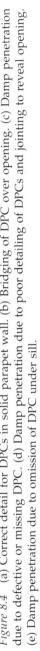

Figure 8.4 (a) Correct detail for DPCs in solid parapet wall. (b) Bridging of DPC over opening. (c) Damp penetration due to defective or missing DPC. (d) Damp penetration due to poor detailing of DPCs and jointing to reveal opening. (e) Damp penetration due to omission of DPC under sill.

There have been a few cases in the past where urea-formaldehyde foam has shrunk causing voids to form, allowing moisture to cross the cavity and penetrate the inner leaf. However, where the technique is correctly applied the original function of the wall is not greatly impaired.

Defects can also be caused by large amounts of driving rain against brickwork that has a high porosity. The visible signs of damp on the inner leaf take the form of damp patches which are often difficult to distinguish from damp penetration caused by mortar droppings on wall ties.

BS 5618 is a code of practice for the use of urea-formaldehyde and contains a local driving rain index value. This document may well give guidance to the surveyor when engaged on a particularly difficult problem concerning filled cavity walls.

Unfortunately, in cases of doubt with cavity wall problems it will often be necessary to cut away several bricks in the outer leaf in order to ascertain the extent of the defect.

EXTRANEOUS CAUSES

8.14   Leaks in plumbing systems

An extraneous cause of dampness which is sometimes overlooked is water that has become entrapped under impervious floor coverings laid on boarded floors. Water passes through the joints or cracks and spreads out underneath making evaporation almost impossible. This type of dampness is usually caused by leaks from plumbing systems or weeping joints in sanitary fittings.

Defective pipes buried in an external wall can sometimes give rise to damp patches on the wall surface similar to rainwater penetration. Prolonged dampness from this source has been known to cause wet rot in floorboards particularly in rooms containing sanitary fittings. Care must be taken not to confuse this problem with condensation which is similar in appearance.

Many cold water pipes collect condensation on their surface and in severe cases water will run down the exterior of the pipe and on to the floor covering or boarding. This problem will be dealt with in more detail in the following section.

CONDENSATION

8.15   Description

Condensation and its effects has been one of the worst post-war building problems particularly in new housing and blocks of flats (Garratt and Nowak 1991). Dampness together with mould growth caused by surface condensation or high humidity is not only difficult for the surveyor to diagnose, but can be very distressing for the occupiers of the building.

8.16   Causes

In order that the surveyor can discuss possible remedies for condensation it is necessary for him to understand how it occurs, and be able to identify the defects. Condensation must not be confused with dampness due to water penetration from the outside nor with dampness caused by the normal drying out of a new building (Burkinshaw and Parrett 2003). Moisture which condenses on internal surfaces is derived from the internal air and is generally produced by the occupant's activities. Air at all temperatures absorbs moisture, but the higher its temperature the more moisture it can retain. However, air at any temperature will ultimately reach a state when it cannot absorb any more and it will therefore have reached saturation point. Condensation will occur when the warm air is cooled to a temperature known as its 'dewpoint' temperature, either by being brought into contact with the cold surfaces of the structure or by passage into a cooler part of the building. Condensation will also occur on absorbent surfaces, but will not always show until the surface is very damp. In such cases mould growth will appear consisting of green or black patches which will cause deterioration of decorative finishings. This type of growth can also form on clothing etc. stored in unventilated and unheated built-in cupboards.

8.17   Diagnosis

Generally speaking, this type of dampness applies to all types and ages of buildings. However, a lot of present-day condensation problems have been aggravated by the use of modern materials which have a low thermal capacity (Trotman et al. 2004).

When carrying out an examination of older type property the surveyor will probably not have to consider the various factors that have created condensation (Massari and Massari 1993). The features built into these properties created the correct conditions to avoid condensation problems although they were probably not realised by the early builders. The materials they used had a high thermal capacity and included such features as air bricks, box windows and lime plaster on the walls and ceilings. The most important feature was the open fire which gave good ventilation to the room and chimney breasts while their warm flues passing through the building heated the internal wall surfaces. This method of heating the principal rooms of a house has ceased and has thus also done away with the natural ventilation from the flue. All these features played an important part in the removal of moisture laden air. Gypsum plaster in lieu of lime plaster has been used considerably during the past sixty years. The gypsum plaster being denser and colder has not the power to absorb moisture. However, this problem has largely been overcome by the introduction of vapour check thermal boards and lightweight retarded hemihydrate plasters.

'Cold bridges' formed by cavity wall ties can produce small discoloured patches of mould on wall plaster. Openings in cavity walls sealed with slates in cement mortar can also cause a similar condition on the wall plaster. The use of steel or concrete lintels over door or window openings which completely bridge the cavity transfers a completely cold area from the outer face to the internal face of a building. Metal windows have a high thermal conductivity which can cause similar conditions when built direct to brick openings that provide a cold bridge between the exterior and interior. The appliances in domestic properties should be carefully noted. Washing machines, gas cookers and tumble driers are often the cause of condensation problems, particularly where some form of ventilation to the outside has not been provided.

A considerable amount of double glazing has been carried out during the past twenty-five years which no doubt reduces the risk of condensation, but does not entirely prevent it on the glass and frame. This will depend on the humidity and temperature in the building. Condensation can occur in pitched and flat roofs of modern buildings unless precautions are taken. One cause is due to the increased use of thermal insulation. Insulants are used up to 100 mm in thickness immediately above the ceiling level in order to prevent heat loss and pattern staining. Pitched roofs usually have an underlay of felt and the ceiling below lined with aluminium foil backed plasterboard. Both these methods tend to make the space above the insulation colder and the dewpoint is thus more easily reached. As previously mentioned chimneys are now either removed or not used. In such cases there will be no warmth from the flues to help raise the temperature in the roof space and so help reduce the risk of condensation.

Where condensation is noticeable on the roof timbers and metal fasteners it is no doubt due to the lack of ventilation especially if the insulation is packed tightly up to the eaves. The surveyor should carefully note this point and if necessary recommend that a gap of approximately 20 mm should be provided along the eaves and a few small holes drilled into the soffit board on two opposite sides of the building to provide adequate ventilation to the roof space. Natural ventilation is the cheapest form of remedial work and is nearly always beneficial in reducing condensation. The important factor is that moisture laden air is extracted and a current of air promoted. Some severe cases of condensation require knowledge of many complex factors and the reader is recommended to study BS 5250 and Garratt and Nowak (1991) for further information.

8.18   Problems with flues

During his examination of older type buildings the surveyor will occasionally find that the fireplace openings and flues have been sealed off, and that damp patches have appeared on the chimney breast. This defect is often mistaken for damp penetration or condensation, when the basic problem is due to the lack of ventilation. If this stagnant air is not released, a chemical corrosive process known as soot damp will commence which gradually eats into the brickwork and through to the plaster. The problem is easily overcome by installing a louvre vent at the bottom of the chimney breast thus ensuring a flow of air through the flue.

9 Timber Decay and Insect Attack

9.1 Introduction

Every year enormous sums of money are spent on repairing the damage caused by timber decay and insect attack (Ridout 1999). It can cost several thousand pounds just to reinstate timber attacked by dry rot in a single house. A considerable amount of this damage could be prevented if the conditions which favour this type of fungal decay were more generally understood, and all necessary precautions taken to check any outbreak. Occupiers of buildings do not always appreciate the fact that dry rot can exist with no visible evidence on the surface of the wood. If timber in a building is properly seasoned before it is used and is subsequently protected against the penetration of damp it will not decay. Most outbreaks of dry rot are due either to faulty construction or to the lack of proper maintenance, particularly defects in the rainwater disposal system.

The surveyor who undertakes to examine a building in order to assess the damage caused by dry rot, wet rot or beetle attack should have sufficient knowledge of timber pests and their control to be able to identify the fungi or insects. There are several technical publications on this subject obtainable from the Building Research Establishment and Timber Research and Development Association (see list in Bibliography). He should also have a sound knowledge of the ventilation and insulation of buildings, surface water drainage and damp problems in the structure as described in Chapter 8.

DRY ROT

9.2 Description

Dry rot is the most common form of fungal decay in timber and one of the most serious as far as the surveyor is concerned (BRE Digest 299). It is known technically as *Serpula lacrymans*. The fungus feeds on softwood in moist and poorly ventilated underfloor timbers and in conditions favourable to its growth it will spread and affect large areas of timber. It produces very light and minute spores which float about in the atmosphere; alight on timber and destroy the fibres, finally making it brittle and decayed both along and across the grain, and the cube effect thus produced is characteristic. The growth thrives best at a temperature around

23°C and therefore spreads more rapidly during the summer months. While the fungus is growing it forms white cotton-wool-like cushions with patches of bright yellow, and its feelers can creep across the brickwork, stone or plaster to attack timber in other positions which may be comparatively dry. The fungus is likely to attack timber having a moisture content above 20%. A great quantity of spores are produced by the fruit body of the fungus, and are often widespread as a red dust throughout a room.

9.3   Diagnosis

Below are listed some of the most likely places where dry rot is a possibility in a neglected building:

- In timber around floor construction where there is no DPC or the DPC is defective, and the ventilation is inadequate.
- Flat roof construction which is usually unventilated. Dry rot attacks are not only caused by defects in the roof coverings but also through penetration from parapet walls.
- Where timbers are built into brickwork or otherwise concealed, for example, built-in floor joists, timber lintels (called 'safe lintels') in old properties, and fixing blocks.
- Timbers supporting parapet gutters or valley gutters in pitched roofs (often poorly ventilated). Boarded roofs are also prone to dry rot attack.
- In skirtings and panelling to walls. Also around door and window openings. Internal window shutter boxes in old houses are particularly prone to dry rot attack. A slight waviness on the painted face of these materials can indicate that the back of the member has been attacked by dry rot.
- In basement construction where there is no damp-proof lining and poor ventilation.

In the first case described above the ground floor construction is the most likely place for dry rot to appear. The construction consists of timber joists supported on sleeper walls built of the site concrete. If the sleeper walls are well provided with honeycomb openings, and the external walls have enough air vents, the circulation of air may be sufficient to carry away the moisture and so prevent the fungus starting. However, it often happens in course of time the air vents become clogged or covered with earth (see Chapter 8). Pockets of stagnant air and rising damp through poor quality site concrete or brickwork are the primary contributory causes of fungal decay. The timber will absorb the moisture and so produce the fungi which decompose the wood. Figure 9.1 shows ground floor joists badly affected by dry rot attack.

Figure 9.1   Dry rot spore dust in floor joists and partition (courtesy of SGB Group Publicity).

Usually, the first indication of dry rot is the appearance of fruit bodies or spore dust. The fruit bodies have a fleshy consistency in the shape of a plate about 300 mm in diameter growing out of cracks in a skirting or wall panelling. If the underfloor joists are badly affected by dry rot it will not necessarily show on the boarding especially if it has been covered with a permanent finish such as linoleum or tiling, but a useful indication of an attack is the spongy depression in the floor covering which is characteristic. A further indication of the presence of dry rot is a musty smell, but the decay is usually well advanced before this is apparent.

Dry rot may also occur in floorboards that are laid on timber fillets either resting on top of the concrete slab or embedded in. If the underside of the concrete is exposed to damp this will creep into the fillets or attack the underside of the floorboards.

Another likely place is parapet gutters or valley gutters in pitched roofs. Trouble usually arises when damp penetrates the roof coverings which may be due to cracked or missing tiles, or parapet walls due to rain penetrating the brickwork. An examination of a pitched roof can be complicated, and if it is large it is advisable to divide the roof area into sections giving each rafter, collar or purlin a number and use corresponding numbers in the report. Wall plates and lower ends of rafters are often hidden particularly if they are under parapets or valley gutters. In such cases it may be necessary to remove some of the tiles or slates to the eaves in order to expose the feet of the rafters and wall plates. These areas are the most vulnerable parts of the roof structure. Never be satisfied until these areas have been examined. Fascia and soffit boards are likely to be suspect. The backs of these timbers are unlikely to be protected other than by a coat of primer, and in most cases are in contact with the brickwork, thus the conditions for dry rot attack are produced.

The interiors of many domestic roofs are unlit so always carry a powerful torch or handlamp. This piece of equipment is very necessary when trying to locate timber decay or small holes made by the furniture beetle. To probe the wood use a bradawl or small screwdriver. Decayed timber can be pierced without difficulty while the fibres of sound wood grip the point. For testing large sections of timber for fungal attack use a large gimlet. If the timber is sound when struck a ringing note will be heard in contrast to a dull sound if the timber is decayed.

The dry rot fungus produces hyphae which can grow over brickwork and penetrate the wall or ceiling plaster. Their presence is indicated by raised 'blisters' or cracks in the plaster. The hyphae can carry water to dry timber, and where they come in contact with door frames and skirtings etc., it is likely that dry rot will be found.

WET ROT

9.4   Description

Wet rot is the most common of the fungi and requires timber with a high moisture content (usually above 25%) to thrive, but it does not have the ability to penetrate into masonry, and because of this is usually localised (BRE Digest 345). The most common of the wet rot fungi found attacking building timbers are the cellar fungus (*Coniophora puteana*) and the pore fungus (*Poria vaillartii*). The wet rot fungus requires very damp conditions to germinate and a continuous supply of moisture for its existence. Once the source of moisture is removed the fungi will die.

9.5   Diagnosis

Much wet rot fungus in external joinery has been reported in recent years (BRE Digest 345). A high percentage of this type of rot has been due to rainwater penetration on the external faces and condensation on the internal faces. A contributory cause has been rain penetration through defective tenon joints in windows and doors. A close inspection of these areas is essential particularly at the bottom of each vertical member. Sills and jambs which abut the structural walls should be carefully examined. External door panels must be considered suspect because internal grade plywood is often used for the panels. Owing to damp penetration the ply becomes wrinkled and split at the edges. Glazing bars are particularly susceptible where the putty or glazing beads have broken away.

Cellar rot requires very wet conditions in which to grow and is found in cellars and similar situations below ground. The cracks run longitudinally with the grain and the fruit bodies, which are rarely found in buildings, consist of olive green strands when fresh but soon darken to a dull olive brown. Unlike dry rot it does not extend its activity beyond the area of dampness.

Roof timbers are sometimes affected by wet rot resulting from defects in the roof coverings, and this usually involves the replacement of the defective timbers. The most vulnerable parts of a timber roof structure are those situated under gutters or buried in masonry. Early slate or tiled roofs with no underlay provide good ventilation to the roof timbers, so although the rafters may be damp stained they may not have rotted. Floor joists are also affected, particularly joists built into outer walls which are often inaccessible. External timbers such as fences and gates, even if treated with a preservative, are liable to deteriorate when exposed to ground moisture.

The surveyor should always remember that the wet rot fungus is of microscopic size and is invisible to the naked eye. Consequently, the moisture meter must be used to ensure that the full extent of the rot has been detected.

BEETLE ATTACK

9.6 Description

In recent years beetle attack has had much prominence and is probably the prime reason for a property owner seeking a structural survey. Beetles of various types breed and live in the cellular structure of the timber and the four most common in this country are described below.

Furniture beetle

In the majority of cases of infestation the furniture beetle is the most common. This is a small beetle about 3 to 5 mm long, reddish to brown in colour and has rows of fine pits longitudinally along the wing cases. It is clothed with a fine covering of short yellow hairs. The life cycle is one to three years. The flight holes are 2 to 3 mm in diameter. The eggs are laid in cracks and joints and hatch out three or four weeks after laying. The grub leaves the eggs, boring straight into the timber and tunnelling along the grain. The bore dust consists of cylindrical pellets, gritty to the touch. They attack seasoned softwood, but also attack the sapwood of hardwoods and plywood.

Powder Post beetle

The beetles are about 5 mm long, elongated and reddish brown to black in colour. They have no hood over the head. The life cycle is approximately one year. The flight holes are circular in shape and are about 3 to 4 mm in diameter. The female lays 30 to 50 eggs during a season. The grub is about 6 mm long and yellowish white in colour with brown jaws. The grubs begin boring along the grain and later branch out in all directions. The bore dust consists of a very fine flour-like powder. The beetle only attacks hardwoods and usually the sapwood. Softwoods are never affected.

Death Watch beetle

The Death Watch beetle derives its name from the tapping sound both sexes make during the mating season. This is a larger beetle, about 6 to 8 mm long, chocolate brown in colour, spotted with thick patches of short yellowish grey hairs. The life cycle is one to eleven years. The flight holes are circular, about 3 mm in diameter. They lay their eggs in crevices or knots or old flight holes and the eggs hatch in two to eight weeks. The grubs wander over the surface before tunnelling. The bore dust contains coarse bun-shaped pellets. The beetle is often found in old hardwoods, usually oak and chestnut, and most damage occurs in built-in parts of a structure such as end beams, wall plates and poorly ventilated places (see Fig. 9.2).

House Longhorn beetle

This beetle was originally confined to the northern parts of Surrey, but has now been reported in parts of Hampshire and Berkshire. It is a large beetle around 10 to 20 mm long. The life cycle can last up to ten years. The flight holes are oval, 10 by 5 mm and are widely spaced. The bore dust consists of finely divided wood dust and the excrement consists of short cylinders almost as broad as they are long. The beetle is usually found in the sapwoods of softwood and can cause serious damage particularly in roof timbers. The Building Regulations require that all timbers in the above mentioned counties should be treated with timber preservatives.

Termites

Climate change in the UK has increased the risk of termite attack to timber in buildings, especially in the south of England. The potential damage that termites can make to timber is extensive (see BRE Digest 443).

9.7   Diagnosis

In the first instance, it is necessary to identify the type of insect and the extent of the outbreak before considering what remedial measures are necessary. Also a knowledge of the life cycle of the beetle is helpful which can be summarised in the following four stages:

Figure 9.2   Effects of Death Watch beetle (courtesy of SGB Group Publicity).

- The egg laid by the beetle.
- The grub which develops from the egg.
- The pupa where the grub works into the timber surface.
- The pupa changes into an adult beetle and emerges from the timber to lay further eggs.

It is in the second stage that most of the damage is done to the timber as it is eaten for food.

The examination of timber floors involves a close inspection of all visible timbers with the aid of a torch and bradawl. This can be a very tedious operation requiring much patience. When dealing with floors it is often only possible to remove occasional floorboards for checking the undersides where exit holes are usually found. Evidence of a recent beetle attack is usually identified by clearly formed exit holes and piles of dust on the horizontal members. The beetles themselves can also be seen during the following periods:

- Furniture beetle – June to August.
- Powder Post beetle – April to October.
- Death Watch beetle – April to June.
- House Longhorn beetle – June to August.

When dealing with roof structures the surveyor must remember that the most vulnerable parts are those timbers which are under parapet gutters or partly buried in masonry, such as ends of tie beams and wallplates. The Death Watch beetle will often attack these inaccessible parts of the structure especially under leaky gutters which may be invisibly infested with fungus as the fungus grows as a result of damp conditions. Roofs of old farm buildings are particularly prone to beetle attack, usually the Furniture beetle.

Timber joists supporting the flat roofs or valley gutters may be impossible to examine and if so, this must be clearly stated in the report.

Roofs constructed of softwood and particularly if they are over twenty years old are often attacked by the Furniture beetle. Old buildings with hardwood roof structures are often neglected and when examined will often be found to be riddled with Furniture or Death Watch beetle. Another serious timber pest found in roof structures is the House Longhorn beetle. This type of infestation can cause far more damage in a much shorter time than other types of woodboring beetle. The insect often works below the surface of the timber for many years before emerging and is not easily detected until considerable damage has occurred.

9.8   Conclusion

It should be remembered that treatment of large outbreaks of dry rot or wood boring beetle is generally beyond the capabilities of the average builder. During the past thirty years there have been many changes in the way that most outbreaks of these timber pests are being tackled. Very few surveyors at the present time are asked to examine property and report on dry rot and insect attack and appoint a builder to carry out the work. Most property owners now engage a specialist contractor who usually employs his own surveyors who have been trained in this type of investigation. Names and addresses of such firms are usually obtained from

'Yellow Pages' or from advertisements in various magazines. The advantage of employing a specialist is the fact that he will guarantee the work and will no doubt use proprietary solutions obtained from recognised manufacturers which are guaranteed to be effective. The object of the treatment is to destroy the beetle infestation in all its stages. This operation is best carried out during the period of greatest activity, namely spring or summer.

10 Roof Structures and Coverings

ROOF STRUCTURES

10.1 Introduction

The structural framework of a flat or pitched roof is usually of timber or steel framing which supports the roof coverings or in the case of a rigid or portal frame consists of a continuous member which may have a solid or lattice web.

The timber lean-to and collar roofs are the simplest type of roofs for short spans and are often found on back additions or on outbuildings to older domestic properties (see Figs. 10.1a and 10.1b).

Figure 10.1c shows a typical purlin roof found in many domestic properties erected during the first half of the twentieth century and consists of a ridge piece, rafters, purlins, collars and struts. The purlins are fixed under the rafters to provide intermediate support and are secured to struts which bear on an internal load bearing partition. It was also common practice to stiffen the roof by providing collars at every fourth pair of rafters. The size of the various members is important as it is essential to use economical sizes governed by the span of the roof and the load it supports.

During the past 30 years many experiments have been made in the construction of both flat and pitched roofs with the object of effecting economies in the use of material. A very economical system is the trussed rafter which is now used in over 90% of new domestic work. It consists of lightweight truss units formed into a number of strongly built triangles that span wide areas, generally spaced at about 600 mm centre to centre. The trusses are available for pitches of 40° for plain tiles and 35° for interlocking tiles or slates. No purlins or common rafters are used. The advantages are that there is no need for a load bearing partition on the first floor and the trusses are suitable for prefabrication. The trusses are made from timbers of uniform thickness secured at the main joints with double sided timber connectors and bolts (see Fig. 10.2a).

The room in the roof space has become popular during the past 60 years. Buildings of this type are economical to erect with the minimum of brickwork and a high pitched roof containing the bedrooms. Ceilings are secured to collars which are also the ceiling joists and timber studs clad in plasterboard with skim coat plaster finish form the walls. Dormer windows and sloping ceilings make attractive but sometimes cold rooms, but in the more modern dwellings glass fibre quilt is often used to insu-

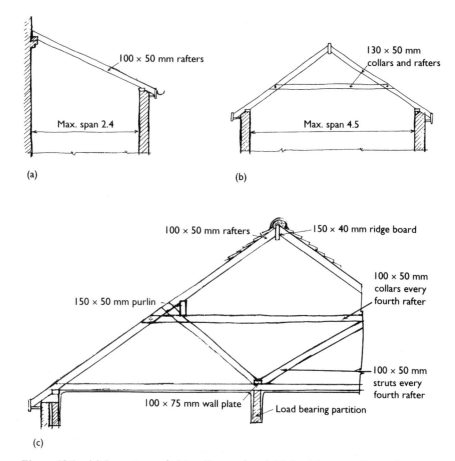

Figure 10.1   (a) Lean-to roof, (b) collar roof and (c) double or purlin roof.

late the triangular space and ceiling (see Fig. 10.2b). The mansard roof is of similar construction, but without the sloping ceilings.

Flat roofs for medium spans say from 6 to 7.5 m are usually constructed of timber joists which are freely adaptable and are particularly suitable for domestic roofing (see Fig. 10.3a).

Reinforced concrete slabs cast *in situ* or reinforced concrete beams supporting a proprietary deck construction have become very popular. The decking consists of troughed galvanised steel, asbestos cement or aluminium. The depth of the troughs varies to suit the span required. The manufacturers often provide a complete roofing system consisting of a metal deck, a vapour barrier, thermal insulation and built-up felt roofing.

During the eighteenth and nineteenth centuries larger span roof trusses followed the pattern shown in Fig. 10.3b for the majority of the larger domestic dwellings, light industrial and commercial buildings. This type of roof has been included because surveyors are often called upon to

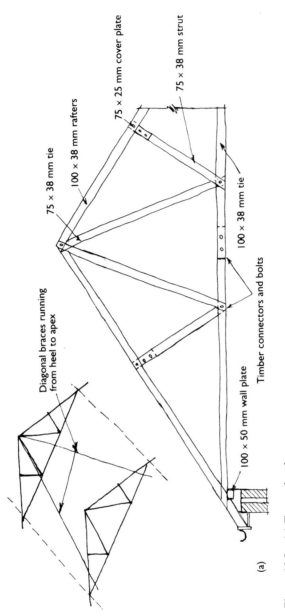

(a)

Figure 10.2 (a) Trussed rafters,

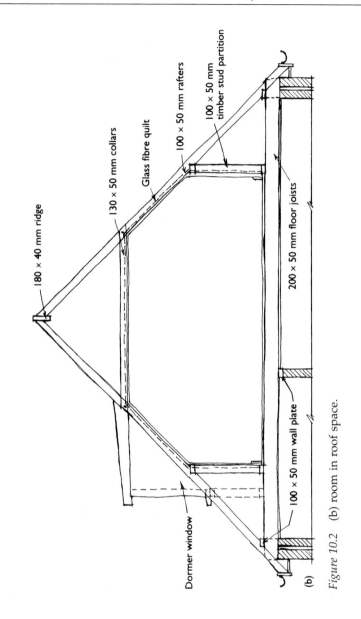

180 × 40 mm ridge

130 × 50 mm collars

Glass fibre quilt

100 × 50 mm rafters

100 × 50 mm
timber stud partition

200 × 50 mm floor joists

100 × 50 mm wall plate

Dormer window

(b)

Figure 10.2 (b) room in roof space.

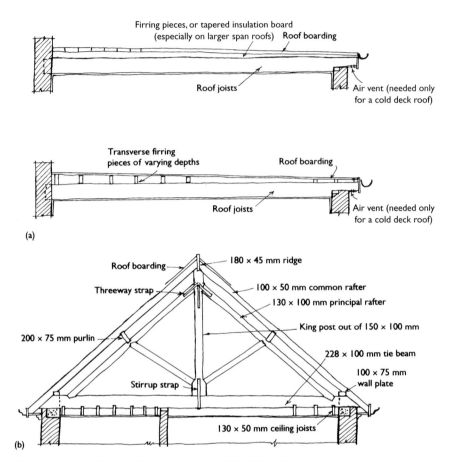

Figure 10.3　(a) Flat roof construction and (b) King Post roof.

examine and old building where this type of construction is met with. The example shown is a King Post truss, but there were many variations and this type together with the Queen Post reached their ultimate design during the mid nineteenth century. The truss supports the purlins which in turn support the common rafters. The roofs were usually close boarded and covered with slates making a sound insulated loft. In all cases the joints were strengthened with iron straps and bolts.

Medium and long span roofs cover a large variety of single and multi-storey buildings including industrial, commercial, warehouses and assembly halls. Many of these roofs are constructed of steel trusses, lattice girders or portal frames which are fabricated in structural steel sections and assembled by bolts or welding. The variations in design are unlimited. Figures 10.4a to 10.4e show the outlines of the various types in common use.

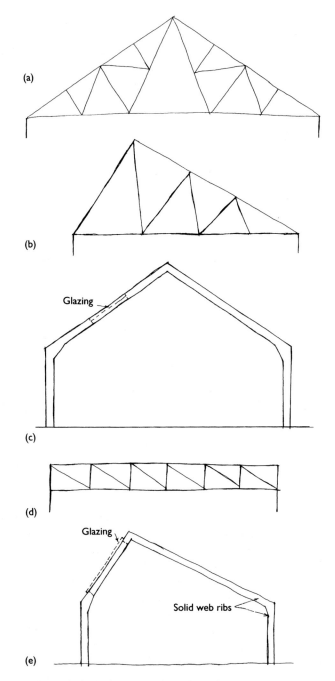

Figure 10.4   (a) Typical steel truss arrangement for a 25 m span. (b) Typical steel north light truss. (c) Solid web portal frame. (d) Lattice girder for flat roof. (e) North light frame.

10.2   General investigations

The survey of a pitched or flat roof is best carried out during adverse weather conditions which makes any defects obvious. As this is unlikely to take place careful inspection is required of all relevant parts of the structure. It can also be a long and costly job, for the condition of the timbers cannot be assessed unless a close examination is made. Inspection of the roof space should begin by making a note of the details of the access i.e. folding ladders or other means of access. Check the extent of any boarded walkways and any parts of the roof space where access is difficult. The next important point is the strength and stability of the timbers and joints which in a fairly new building should conform to the requirements of the Building Regulations and Codes of Practice.

The timber used in the construction of pitched or flat roofs is usually softwood. The most common form of defect found in structural timbers are fungi, wood boring insects and moisture penetration. (These defects have been dealt with in Chapter 9.)

10.3   Defects from natural causes

The principal defects arising from natural causes are described. They should be carefully noted and if necessary recommendations for their treatment should be made in the surveyor's report.

- *Knots.* Large dead knots should be examined; they affect the strength of the timbers particularly if they are situated on edges or arrises.
- *Shakes.* Shakes or splints usually appear in timber due to stress developing in the standing tree or during seasoning where they are exposed to the weather, although they are often caused by felling. They are not always visible except when they occur at the end of a timber.
- *Wane.* This is the rugged arris produced by the timber being cut too close to the outside of the tree during conversion of the log.

10.4   Timber pitched roofs

The stability of the roof timbers is most important. The failure of any one member of the frame is liable to place undue stress on the remainder. Any failure in the roof frame due to inadequate jointing or the corrosion of screws, bolts or nails will also result in movement of the roof. Rafters, collars and struts should be securely nailed; purlins should be strutted at appropriate intervals; and the struts notched over a timber plate and to the underside of the purlin as shown in Fig. 10.1c.

Rafters and ceiling joists should be in one piece, but if they are in two pieces the joint should be made by lapping the timbers for approximately

600 mm and securely bolting one to the other, the joint being placed centrally in the span. If rafters are not properly birdsmouthed to the wall plates and adequately nailed to the ceiling joists, movement may occur at the eaves.

In older type buildings the purlins will be found built into gable end walls, or extended through the wall to provide support for barge boards. Due to lack of protective coating, the purlin ends are often found to be affected with wet rot. Alternatively, they will be found to be supported on brick corbels or concrete blocks built into the wall. Occasionally, the purlins will be found to simply butt against the gable and wall with additional struts to support the end of the purlin.

Roofs of terraced or semi-detached properties often show a hump over the party wall which may be the result of any one or more of the following defects:

- Failure of purlins or insufficient number of purlin struts.
- Rafters being undersized.
- Raising the upper surface of the party wall above the upper surface of the tile or slate battens bent over the wall.

A check should be made to ensure that all trimming to ceiling joists and rafters has been properly carried out. Trimmers are normally 25 mm thicker than the adjacent joists or rafters. Rafters are often found to rest against a chimney stack without a trimmer or brick corbel support. This defect can cause settlement in the rafters and a depression in the roof covering. A careful examination should be made to ensure that no roof timbers are built into the brickwork or flue. The portion of the stack within the roof space should be examined for movement. If the stack is rendered movement cracks will show through the rendered surfaces.

When carrying out a survey of a pitched roof which has a hipped end it is necessary to check the hip rafter where it is fastened to the wall plate. As there is a considerable thrust at the foot of the hip the roof should be strengthened with an angle tie (size approximately 100×75 mm) across the corner. The tie should be dovetailed, housed and securely screwed to the wall plates.

During the examination of a pitched roof a note should be made of the internal finish of the roof covering and its condition. Modern pitched roof coverings have a layer of underfelt, but in many houses built before 1940 felt is rarely provided so consequently there is a tendency for rain and snow to penetrate between the tiles or slates into the roof space. These conditions are conducive to wet rot in the battens and rafters and will generally be visible. The better class of buildings usually had boarded roofs which were either butt jointed or feather edged.

There are very few problems with the trussed rafter and the majority of roofs where this method has been adopted have performed successfully. One or two weaknesses have occurred over the years which are usually due to badly made joints or the omission of diagonal bracing, causing the trusses to move laterally. If not corrected, these defects can cause the roof tiles to lift and rainwater to penetrate the roof space. Care should be taken to ensure that the fasteners and diagonal braces have been properly installed and if they are not provided or are defective, a definite recommendation should be made for their provision (see Fig. 10.2a).

It is advisable to clearly mark any defective timbers with a felt pen or wax crayon using a number or a letter. This reference can be mentioned in the client's report and will also assist the builder when carrying out repairs.

10.5   Timber flat roofs

There are two main types of timber flat roofs: cold deck and warm deck. Cold deck flat roofs, where the insulation is below the deck (i.e. between the joists), were common up until about the 1970s. As they have a high risk of causing interstitial condensation within their structure, cold deck flat roofs are no longer recommended. Warm deck roofs (where the insulation is placed above the deck) should be used (BRE Digests 312 and 324).

Where flat roofs are provided it is usually impossible to inspect the timbers for defects. If the defects are due to water penetration a damp patch or water mark will be seen on the ceiling. If the problem is serious it is likely to be indicated by deflection in the roof timbers, coupled with depressions in the roof covering and cracked or loose ceiling plaster. If the roof covering is seriously defective and will obviously require renewing then it is often possible for the surveyor to arrange for a portion of the roof covering to be removed allowing access to the boarding and joists. As mentioned in Chapter 2, fibre-optic probes and boroscope systems have come to the surveyor's rescue and it is now possible to introduce a 600 mm long boroscope into the roof structure via a small drill hole in order to view the condition of the joists etc. A polaroid camera and attachment can be fitted to this instrument to record the information. Flat roofs need to be laid to adequate falls and to incorporate a vapour barrier. The fall is achieved by either using tapered firring pieces secured to the joists or tapered insulation board on a vapour control layer on the warm side of the timber deck. Various types of boarding are used to support the roof covering e.g. rough boarding, T and G boarding, blockboard and chipboard may be used. Care should be taken to examine the method of ventilation of the timbers if it is a cold deck flat roof, as shown in Fig. 10.3a.

10.6   Steel trusses and lattice girders

The design of steel framed roofs is usually entrusted to specialist firms of structural engineers and in consequence serious defects are uncommon.

The chief points of interest to the surveyor are the rigidity of the joints and the protection provided against corrosion. Care should be taken to see that no bolts or rivets are omitted; that all nuts are properly tightened and at least one thread of the bolt projects through the nut. Where tapered washers are necessary the connections should be examined to ensure that an even bearing and tight joint is formed.

The corrosion of steel members may be due to one or both of the following defects:

- Water entering the building through defective roof coverings.
- Condensation from high moisture producing activities.

All exposed structural steelwork requires protective treatment either by painting or galvanising and special care should be taken to examine the steel sections for rust. The surveyor's report should include a description of the areas affected and method of preparation and treatment. Painting over rust is undesirable as the paint is likely to deteriorate (see Figs. 10.4a to 10.4e).

10.7 Older type roofs

The internal examination of a large roof of an eighteenth or nineteenth century building (see Fig. 10.3b) can be a very long and costly operation. If the roof is of the open type with no ceiling then a long extension ladder or tower scaffold is usually required. An inspection of this type of roof will reveal that the greatest number of failures are due to decay of the timbers, often resulting from the deterioration of the roof covering. It is, therefore, necessary to check carefully for evidence of wet rot and wood boring insects. (All these defects have been dealt with in Chapter 9.)

Many older type buildings have roofs without tie beams which exert an outward thrust on the walls. Many medieval church roofs and halls were designed on this principle particularly those known as hammer beam collar roofs. This pressure may distort the roof causing fractures or other defects in the timbers. In such cases the answer to the problem may well lie in strengthening the walls rather than the roof trusses which will probably only require repairing. Also, it is advisable to check that the cross-walls have been properly bonded to the outer walls. The following parts of a roof are particularly vulnerable to decay and if necessary the masonry around them should be opened out and the timber exposed for examination:

- The feet of common rafters.
- Wall plates.
- The concealed ends of tie or hammer beams which are embedded in walls and are often affected by damp penetration.

- Gutter bearers behind parapet walls.
- Bearing ends of purlins embedded in gable end walls.
- As mentioned in Section 10.3 the timbers may have inherent problems if they are suffering from 'shakes' or longitudinal cracks. Depending on where the shakes occur the timbers may require strengthening. Another disadvantage of cracks is that they provide crevices that are ideal hiding places for wood boring insects.

When carrying out an examination of an old roof it must be remembered that movement of the timbers is a natural function. Many old roofs were constructed of oak and considerable movement takes place in this timber for many years after felling. If the timber has adopted an eccentric posture and provided no decay is found it is inadvisable to attempt to straighten any deformation.

10.8   Services and other fittings in the roof space

Whilst in the roof space the surveyor will no doubt wish to check the various services. These usually consist of cold water storage tanks and pipe runs. In modern buildings the material used for storage tanks is usually PVC or glass fibre. These materials are immune to corrosion and are light, thus having the advantage of being easy to move into position and they also reduce the dead weight on the structure.

The surveyor will no doubt see many different types of water storage tanks in old buildings. These will vary from slate slabs with bevelled joints to lead tanks some of which are contained inside a timber casing. Most of the old tanks will be found to be defective and therefore there is a risk of contamination. It is, therefore, wise to recommend that they are replaced with a PVC or glass fibre tank. Galvanised steel tanks are strong, but are liable to corrosion by oxidation or electrolytic action.

Many failures have occurred when galvanised steel tanks have been used in association with copper pipes. Electrolytic action between copper and zinc coating causes the zinc to deteriorate and ultimately perforate. If a galvanised steel tank is found it should be drained down and checked for traces of corrosion. Corrosion usually takes the form of orange coloured rust spots around the inlet and outlet connections of the tank. The bottom of the tank may also be affected but these spots are usually masked by sediment. If there is any doubt regarding the tank's suitability it should be replaced. This is particularly important when it is proposed to replace the pipework throughout the building. The economics of this are obvious; it is not worth risking a defective tank when most of the pipe runs in the building are being replaced. It is, therefore, essential that clear advice should be given in the report on the condition of the tank. However, for minor rust problems there are a number of proprietary linings which can be applied to the inside of galvanised tanks to give protection against corrosion. A recommendation to this effect should be included in the report.

Tanks should be insulated on all sides except their base and a suitable cover provided. If the service pipes are secured to the top of the ceiling joists they should be lagged and the insulation material connected with the tank insulation. No difficulty should be experienced in removing some of the insulation in order to examine the pipework or joists. A description of the insulation and condition of the pipework and valves should be noted. Tanks should be supported on stout timber bearers and in the case of PVC or glass fibre tanks a boarded platform should be provided. Ball valves should be closely examined especially in old buildings. They may have one or more of the following defects:

- Many old copper floats can corrode and become perforated and partly filled with water.
- High velocity discharge from a ball valve may erode the seating.
- Worn washers will require replacement.
- Ball valves may stick in the open position caused by corrosion coupled with lime deposit. This is merely a sign of the need for periodical cleaning and the piston to be greased.

Plastic floats are now used in the majority of new buildings. They do not corrode and are more durable. An overflow or warning pipe should also be checked. They are often found to be too small and sagging. In order to prevent the entry of cold external air the overflow pipe should be turned down within the cistern and terminate 50 mm below the normal water level. However, it should be noted that this is rarely found in older buildings. The size of the pipe must be larger than that of the inlet and should discharge externally in a conspicuous position where it can be easily seen. The capacity of the tank and the depth of water below the top should be ascertained. The water authority's requirement for domestic storage is at least 250 litres for a tank serving both hot and cold supplies. Larger domestic properties or a building containing three or four flats should be provided with proportionately larger storage tanks.

The cold water storage tanks for a large building such as a block of flats or a commercial building are usually situated in a tank room with access from a flat roof. In such cases consumption is calculated either from the number of fittings served or from the number of persons in the building. A typical example of a cold water storage installation for a large building is shown in Fig. 10.5. If two rising mains have been installed they are usually coupled together below the ball valves. The two tanks are connected at the base by a larger header pipe. The header accommodates all the cold and hot water connections. A valve should be installed where the header is connected to each tank to permit shutting off any tank for replacement or cleaning without interrupting the supply.

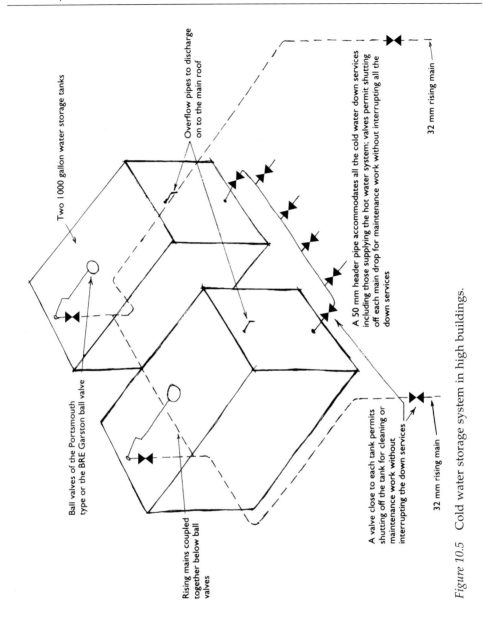

Two 1 000 gallon water storage tanks

Overflow pipes to discharge on to the main roof

Ball valves of the Portsmouth type or the BRE Garston ball valve

Rising mains coupled together below ball valves

A 50 mm header pipe accommodates all the cold water down services including those supplying the hot water system; valves permit shutting off each main drop for maintenance work without interrupting all the down services

A valve close to each tank permits shutting off the tank for cleaning or maintenance work without interrupting the down services

32 mm rising main

32 mm rising main

Figure 10.5   Cold water storage system in high buildings.

Valves should also be installed to each drop close to the header as shown in the sketch. The pipe layout, valves and joints should be carefully checked as many failures in this type of installation could have serious consequences. For buildings over ten storeys the rising mains may well be supplied with a booster pump at the bottom. The pump would be controlled by a hydraulic type float in the storage cistern. In most high blocks between fourteen and twenty storeys there is often a separate drinking water system with its own pumps.

When dealing with large installations of this type the need for a specialist engineer to test the mechanical equipment and water pressure may well be necessary. The specialist report should be integrated with other engineering services, particularly sanitary plumbing and hot water supply.

10.9   Electrical installation

In the roof space the electrical cables serving pendant drops will be visible. If the cables run with the joists they should be fixed to the sides and not on the top of the joints where they are liable to be damaged. If the cables run at right angles to the joists, they should be supported on battens. In both cases the cables should be properly secured with cable clips. Cable joints should always be made in junction boxes screwed to the joists or battens. Examine the junction boxes for faulty joists in elbows and tees and at the same time check the type of cable used and its condition. Cables deteriorate due to ageing of the insulation material and mechanical damage. In older properties where the cable is over forty years old it is extremely likely that the remainder of the lighting and power points will require replacement. This matter will be dealt with in more detail in Chapter 14.

10.10   Roof insulation

An insulation layer in the roof space is now required in modern structures, but it may be absent in older buildings. The required thermal conditions are usually met by the installation of a layer of glass fibre quilt or loose material fixed between the ceiling joists. The glass fibre should be 200 mm thick and all pipes and tank covers should be covered. Where insulation material is not provided a definite recommendation should be made in the report for its provision (i.e. polyethelene pre-formed insulation for the pipes, and polyurethene for the sides and top of the tanks). However, care should be taken to ensure that there is adequate ventilation in the roof space to counteract the probable effects of condensation. It is very necessary in roof spaces to keep the metal fixings from becoming damp and the moisture content of the timbers at a low level. Condensation is normally due to the lack of ventilation when the insulation material is packed tightly into the eaves at plate level. If it appears that adequate ventilation is lacking then it is advisable to recommend the provision of a series of holes in the soffit boards.

10.11   Party walls in roof space

Party walls are usually found in the roof space of terrace properties and should be constructed to provide a degree of fire protection between the properties consisting of brick or concrete block work 225 mm thick. The joint between the roof covering and the top of the party wall should be checked to ensure that the area has been properly sealed against water penetration. Some party walls are carried up above the roof slope but this matter will be dealt with in Section 10.15. In many cases the surveyor will find that the walls are roughly pointed, but are generally sound. If the wall is rendered it should be carefully examined for defects. In older buildings the structural backing is often constructed of poor quality materials which have deteriorated, allowing the rendering to pull away. Gentle tapping on the face will indicate whether the rendering has lost its 'key'. In some cases it may be necessary to remove parts of the rendering in order to examine the quality of the substrate or structural backing.

ROOF COVERINGS

10.12   Introduction

The repair and maintenance of roof coverings is of fundamental importance to all buildings. The life of roofing materials is dependent upon a number of factors including climatic conditions, degree of pollution, methods of fixing and manufacturing techniques.

The first indication of a roofing defect is a damp patch on the ceiling or at the internal angle of wall and ceiling. These problems usually occur as a result of failure at the junctions to chimneys and parapet walls. Defective flashings, cracked lead parapet gutters, soakers wind-lifted and cement fillets on old roofs that have broken away are all typical defects. Junctions with roof lights and dormers are also potential trouble spots and should be closely examined. However, the surveyor must not only concentrate upon these defective areas, but should carefully assess the whole condition of the roof.

Rain and snow may penetrate pitched roofs because the roof coverings are laid to too flat a pitch without increasing the lap of the tile or slate although much depends on the degree of exposure. A common problem concerning the pitch of a roof is when sprocketed eaves are used for the sake of appearance. This frequently gives rise to rainwater penetration. If the pitch of the main roof is too low then the eaves will be too flat to prevent rain or snow being driven in.

Where parapet walls have been erected, an adequate inspection of the lower portion of the roof cannot be carried out with binoculars. If an extension ladder is not available then access to flat or pitched roofs is usually gained by means of a roof light or hatch from the interior of the building.

When carrying out an examination of a pitched or flat roof the surveyor will have to decide whether the existing coverings can be patched or should be completely renewed. There inevitably comes a time in the life of a roof when refixing and patching becomes uneconomic. One of the factors which will affect this decision is the question of matching the existing tiles or slates as closely as possible in colour and size. This can often be done with materials taken from demolished buildings or buildings where alterations have been carried out. When dealing with older buildings it will often be found that a particular tile or slate is no longer produced or the cost is prohibitive. In such cases it is possible to gather old roofing materials together and fix them to the most important slopes of the roof and use new materials for slopes that are concealed from view.

Old tile or slate roofs will often be found to be in a deplorable condition with over half the covering loose or damaged (i.e. the roof is termed 'nailsick'). Close examination generally reveals that this is due to corrosion of the nails or decayed pegs. This is a continuous process over a period of years and in such cases it is more economical to strip and renew the whole of the roof covering. Before recommending total replacement it is advisable to examine the timber structure. For example the weight of concrete tiles is much greater than that of slates and may require the purlins, collars and rafters to be strengthened.

As mentioned in Section 10.4 roofing felt is seldom found in older domestic properties so if total replacement of the covering is required then the problem is easily solved by the addition of a layer of underfelt. However, when making a decision to repair or renew it is wise to consider all the circumstances carefully. The decision to carry out a considerable number of repairs when it would be advisable to recommend a complete renewal of the covering may be regretted in a few years' time.

It is unusual to find a metal flat roof supported on a concrete slab. Commercial and industrial buildings erected during the past seventy years are usually covered with asphalt. Timber joist constructions with lead, copper or zinc coverings are usually found on small domestic properties or the older type of small commercial building. It was always considered to be more economical to provide timber structures for this type of roof.

One of the disadvantages of metal flat roof coverings is their liability to 'creep' or 'buckle' by thermal expansion and a considerable number of failures are caused by the materials remaining deformed after thermal movement has taken place rather than reverting to their original form. These items will be considered in the following paragraphs.

10.13   Types of slate

There are several types of natural slate most of which are found on the roofs of older buildings.

- *Cornish slate*. This type of slate is brownish-grey in colour and from 6 to 10 mm thick. It can be used in random or diminishing sizes.
- *Welsh slate*. This type has been widely used throughout the UK and is generally bluish in colour. It is strong and durable and from 4 to 5 mm thick.
- *Westmorland slate*. These slates are of excellent quality, are produced in random or diminishing sizes and are fairly thick (8 to 10 mm). They have a pleasing appearance and are greenish-grey in colour.
- *Swithland slate*. This type of slate is used extensively throughout the Midlands and slates are 10 to 15 mm thick. They can be used in random or diminishing sizes.
- *Scottish slate*. This type of slate is used throughout Scotland and is about 10 mm thick (CCUS 2000).

10.14   Ridges, hips and valleys

In older buildings the ridges were often formed with slate wings and rolls, wood rolls with lead dressings or half-round clay or concrete tiles. Of these the half-round tiles appear to be the most popular.

The methods enumerated above for ridges are equally applicable to hips, but an alternative to hips is to close mitre with lead soakers under. Where half-round tiles are used hip irons must be fixed at the foot of the hip to prevent the hip tiles slipping.

Valleys can be formed with open gutters, or with close mitred slates with soakers or a secret gutter under. Open gutters appear to be the most popular means of forming a valley with a minimum width of 200 mm to allow foot room to carry out repairs and cleaning.

10.15   Examination of a slate roof

The various points to be noted during an examination of a slate roof may be summarised as follows:

(1)   Slates lie close to the roof with very fine spaces between them thus causing capillary action. Defects usually arise through the slates slipping or cracking, the former generally being caused by the use of inferior nails which have rusted due to water penetration.

(2)   Rainwater will often penetrate through the wide joints. This is due to the side lap being less than the minimum required for the length of

the slate. However, this type of penetration is less likely to occur on a steeply pitched roof than one of lower pitch.

(3) Poor quality slates will often soften as a result of atmospheric pollution causing a breakdown of the bond between the laminated layers (termed 'delamination').

(4) Cracking can be caused by high winds lifting the slates without dislodging the nails or by single centre nailing which also tends to allow the slates to lift.

(5) A further point for the surveyor to note is the presence of a considerable number of lead clips or 'tingles' about 13 mm wide bent over the bottom edge of the slate. This indicates 'nail sickness' and means that a large number of fixing nails have failed and the slates are moving down the roof slope.

(6) Where abutments occur the soakers and cover flashings should be carefully checked. The flashings should be properly secured with wedges and pointed. In the older domestic work it was common practice to substitute cement fillets for lead flashings. Much trouble is caused by moisture penetration which arises when cracks form in the fillets as a result of drying shrinkage. If the fillets are found to be loose and cracked it is advisable to recommend replacement with a lead flashing.

(7) Hips and ridges are situated in inaccessible positions and are very exposed to the weather. They should be carefully examined particularly the leadwork which could be defective due to age. The hip irons to the half-round hips should be securely fixed to the hip rafter to prevent slipping and the hip joints properly pointed.

(8) Lead valley gutters should be given close attention. They are often a source of trouble where the lead has perished due to 'creep'. If the lead is split water will penetrate through to the valley board causing wet rot in the timber. Secret valley gutters have always been troublesome due to the fact that they are inaccessible. They are usually found choked with dead leaves and other debris. This type of valley is not recommended and the surveyor may well consider it advisable to recommend that the valley be reconstructed as an open valley gutter where the lead sheeting is laid on boarding giving a clear width of not less than 200 mm.

10.16   Tiled roofs

Roofing tiles are principally manufactured with clay, made either by hand or by machine. Machine made concrete tiles are now used more frequently but lack the character of a clay tile and are easily recognised by their mechanical appearance. A tiled roof, although satisfactory in use, is never completely watertight. Water penetration is only prevented because the tiles usually dry out before they produce moisture on the inner surface.

Valleys and roof gutters can be made of various materials such as copper, lead and zinc and can be formed as open gutters, close mitred with soakers or secret gutters as described for slating. Special valley tiles are also made for this purpose and provide a quick and cheap method of finishing a valley. An alternative to these are laced valleys which are formed by using valley boards and a tile-and-a-half tiles laid diagonally across the valley boards while the last few tiles in each course are cut splayed and given an upward lift.

Nevertheless, the durability of a tiled roof can be affected in the following ways:

- If moisture penetration occurs it is usually the result of rainwater or snow being driven by wind underneath the lower edges of the tiles. Roofs of modern buildings are usually lined with a layer of underfelt below the roof covering which acts as a second line of defence.
- A great number of domestic buildings built before 1940 have roofs without either underfelt or boarding. To prevent rain or snow penetration 'torching' has been used on the underside of the tiles and consists of a mixture of lime mortar and cow hair. The mixture was spread behind each batten to fill the space between the tiles. The disadvantage with 'torching' is that it tends to shrink and fall away. The torching should be carefully examined and commented on (see Fig. 10.6).
- Apart from making a note of all missing or broken tiles it is advisable to check that the tiles are properly cambered as this ensures that the tile has a tight seating and at the same time allowing air space between the tiles which facilitates drying out.
- Plain machine made tiles can deteriorate as a result of frost action or by crystallisation of salts. Poor quality tiles may have a laminated structure with air pockets between the laminations which tend to draw

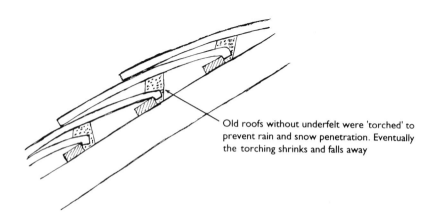

Old roofs without underfelt were 'torched' to prevent rain and snow penetration. Eventually the torching shrinks and falls away

Figure 10.6   Method of 'torching' to underside of roof tiling.

moisture into the voids. Following the freezing of the entrapped moisture the tile eventually falls apart.

- The tile ribs may be defective or nails corroded causing the tile to slide down the roof slope.
- Flashings to abutments should be examined. They are liable to become dislodged and the free edge against the rising brickwork requires refixing and pointing. If cement fillets are used they should be examined for defects as described in Section 10.15.
- The remarks made in Section 10.15 concerning hips and ridges equally apply to tiles, but an alternative finish to hips largely used in better class work is to provide bonnet tiles. Bonnet hip tiles must be bedded at the tail and should be examined for shrinkage. The fixing nails sometimes corrode causing the bedding mortar to fall away. When bonnet hips are used a hip iron is not required, but tile slips are usually inserted in the bedding material of the bonnet tile at the foot of the hip.
- When dealing with old buildings erected during the seventeenth century the clay tiles were usually hand made and secured by means of oak pegs hung over riven oak battens or by iron nails. The fixing pegs and battens will often be found to have deteriorated and will require complete replacement.

Other types of tile frequently met with are pantiles, single and double Roman tiles, Spanish, Italian and various proprietary interlocking tiles. All these tiles are single lap having head and side lap only. The roof pitches of such coverings are usually recommended by the manufacturers and if possible the surveyor should check the pitch and fixing methods against their technical details. A careful examination of the eaves and verges should be carried out to ensure that they are properly constructed. Current practice when laying pantiles is that a course of plain tiles is laid under the lower course of pantiles at the eaves and should project 50 mm over the fascia. In some cases it may be found that the plain tiling has been omitted and the roll of the pantile filled with mortar which has a tendency to shrink and fall away. This defect is often caused by slight lifting due to the action of the wind.

10.17　Bituminous felt and polymeric sheet roofing

Built-up felt roofs are usually laid by specialist firms who are prepared to give a guarantee over a reasonable period (see BRE Digest 372). The use of this material has become widespread during the past forty years particularly for domestic properties, mainly because it is cheap compared with other roof coverings.

The covering consists of built-up layers of felt based on organic, asbestos or glass fibre bonded together with a bitumen compound. In recent years a range of polymeric materials have become available. For the purpose of this text they are defined as factory made flexible polymeric sheet materials, of which one layer is used to achieve a waterproof barrier. This material has become increasingly important to surveyors as they are often called upon to examine this type of roof covering and advise on remedial measures. Polymeric materials are designed to be fixed to the main roof structure with special adhesives, heat welding or solvent welding. They have good flexibility, but there have been problems due to water penetration through inadequately bonded joints and mechanical damage (see 'common defects' below, item 8).

The upper surfaces of the older type roofs were often finished with plain felt and the surface painted white to provide a reflecting surface. The current Building Regulations now require the surfaces to be covered with chippings secured to the felt with a dressing of bitumen or mineral self-finish, for solar protection.

Common defects

The following items are typical of the defects which affect felt roofing and should be carefully considered when making an examination:

(1)   The felt is liable to crack mainly through differential movement between the felt and the substructure and the shrinkage of the base due to thermal changes.

(2)   As mentioned earlier in Chapter 8, condensation can present serious problems in roofs of this type. Moisture can be prevented from entering the roof by the provision of a vapour barrier such as plaster board and ventilation in the roof void to prevent fungal growth. The entrapped air or moisture between the layers of felt or between the felt and the substructure will cause the felt to lift in the form of 'blisters'. The blisters are not necessarily a sign of failure, but should be carefully examined to see whether any foot pressure has been applied to cause the blister to crack and form a passage for water penetration. One of the most common defects is to find that no provision has been made for ventilation and a vapour barrier. It is, therefore, important that the surveyor endeavours to ascertain the nature of the original construction (see Fig. 10.7).

(3)   Insufficient fall may cause water to gather in pools which can cause frost damage during the cold months and blistering during the summer months. The fall should not be less than 38 mm in 3 m. Deterioration of the felt is often difficult to assess, but is accelerated by standing puddles and by the choice of the wrong felt for the top layer.

(4)   The loss of protective chippings through wind and rain can also be accelerated by the use of bitumen emulsion instead of bitumen in a volatile solvent. Loose chippings can lead to choking of gutters and rainwater heads. Chippings can also penetrate the felt and cause

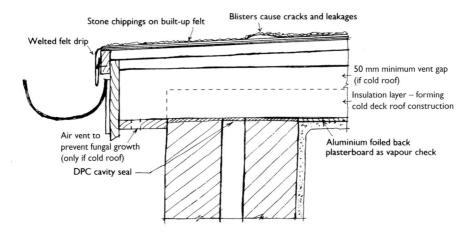

Figure 10.7   Built-up roofing – eaves detail.

leaks. Reflective paints mentioned earlier have been used in the past and often result in crazing.

(5) Another common cause of failure results from differential movement between skirtings and parapets. Skirtings should be integral with the surface felt and should be formed by turning up the two top layers of felt against the abutments to a height of 150 mm. Failures have often been caused by the omission of a separate flashing wedged into a groove and pointed. An examination will often reveal that felt skirtings against parapet walls have no timber or concrete angle fillets at the corner so that the felt may have cracked.

(6) It is advisable for the surveyor to try and find out the number of layers provided and assess the quality of the felt. This operation may be difficult, but can often be done at a corner of the roof where the chippings are loose or missing.

(7) Where eaves and verges occur, check that the welted felt drips have been properly formed in accordance with the Code of Practice. Projections through the roof covering such as a vent pipe should be properly formed with a metal hood.

(8) The greatest concern in the use of single layer systems is ensuring that the laps are sealed. Welds give a higher strength lap than adhesives. It is easy to puncture a single ply roof. Stones and grit trapped beneath the membrane during laying can cause holes to form. If the surveyor is instructed to recommend repairs then these must be executed by specialists using the appropriate materials. Under no circumstances should temporary 'Flashband repair strips' be recommended.

Unfortunately experience has shown that built-up felt roofs have not enjoyed a good reputation and tend to deteriorate fairly quickly, their life being limited to about sixteen years and maintenance costs are high. The surveyor carrying out an examination of this type of roof must, therefore, be extremely careful to detect all imperfections before recommending repairs or replacement.

10.18   Asphalt

Mastic asphalt has one great advantage over metal roof coverings in that it provides a jointless covering, thus avoiding the weaknesses inherent in seams, rolls, drips and welts. It also gives an unobstructive smooth finish which is eminently suitable for flat roofs especially where frequent traffic is anticipated. Mastic asphalt should be laid in two coats each 10 mm thick and to keep the roof free from standing water it should have a minimum fall of 1:40. An isolating membrane should be provided beneath the asphalt to allow for differential movement (BRE Digest 144).

The recognition during the past fifty years of the value of flat roofs as a means of providing car parking, roof gardens and playgrounds etc., has created a demand for roof coverings which combine water resisting qualities with an attractive finish such as precast paving slabs preferably with solar reflective properties. The pavings should be laid in bays with expansion joints where necessary and a separating membrane should be provided beneath the paving as shown in Fig. 10.8. The durability of asphalt can be affected in several ways and the most common defects the surveyor will have to investigate are described below:

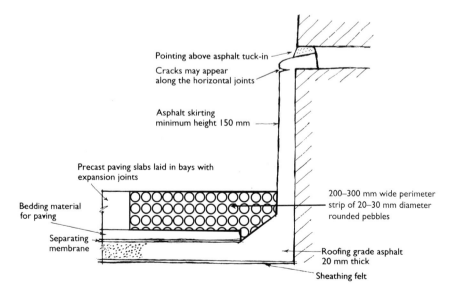

Figure 10.8   Mastic asphalt to roof garden or car park roof.

(1) Surface crazing is fairly common and is generally due to the lack of reflective treatment, but can also be due to 'ponding' of rainwater on the roof. Stresses imposed on the asphalt during hot weather will tend to cause the material to expand and consequently the top surface will contract during the cold winter months. Thus the surface of the asphalt is liable to form a 'map-pattern' cracking. Certain bitumenised aluminium coatings including emulsion paint are often used as a reflective treatment to reduce the absorption of solar heat. Such paints can cause crazing by shrinkage and even pronounced cracks.

(2) Blisters may appear on asphalt roofs, especially when the roof is constructed of concrete with a lightweight concrete screed. The blisters are usually caused by pressure of air or water vapour. Concrete roofs are frequently asphalted before they are completely dry. The heat from the sun causes expansion of any moisture vapour or air which is present between the concrete and asphalt and tends to lift the asphalt locally from the concrete surface. Rainwater may also be trapped under the asphalt during wet weather which may ultimately give rise to blistering. Blisters may be found in many sizes up to about 150 mm and if not split they may be left, but a note should be made stating that they should be inspected periodically. Those which have split will need to be cut out and patched.

(3) Cracking is fairly common in old asphalt roofs and is caused by movement of the substructure and the asphalt often due to the absence of an isolating membrane. Careful inspection should be made of old asphalt roofs which may have become brittle and cracked, although this is sometimes due to excessive wear.

(4) Many failures in asphalt roofing have been caused by differential movements taking place at the joints between the horizontal surfaces and the vertical abutments of parapet walls, particularly where a timber substructure has been installed. In many cases it will be found that rainwater penetration has occurred behind the asphalt skirting, either because the skirting has not been properly turned into the brick joint or the cover flashing has been omitted.

In old asphalt roofs a crack may have been formed at the junction between the skirting and the asphalt roof due to the absence of an angle fillet. Angle fillets should be provided at the base of all skirtings to avoid sharp angles which have a tendency to crack. The eaves should also be checked to ensure that the rainwater is properly discharged into the gutter. In the best practice a non-ferrous metal apron or drip is fixed under the asphalt with metal cleats and dressed down into the gutter.

(5) Railings and pipes should be checked to ensure that the asphalt has been properly dressed up to the pipes to a height of approximately 150 mm and that the top is sealed to the metal or tucked under the purpose made flange.

(6) Cracks sometimes occur on sloped surfaces, usually at right angles to the slope and are caused by changes in temperature giving rise to alternate expansion and contraction. This movement causes the asphalt to creep towards the lower part of the roof. In severe cases it may be wise to recommend stripping the defective material which will also provide an opportunity to examine the base and if necessary to provide a suitable key such as metal lathing for the new asphalt.

(7) Asphalt roofing is easily damaged by the careless use of builders' plant and materials thrown down on new asphalt. Contact with spilt paint and oil can also cause damage by dissolving some of the asphalt.

10.19   Copper

The surveyor will not encounter copper roofs as frequently as other metal roof coverings. Copper is usually found in large Victorian houses or commercial buildings not only as a roof covering but as an architectural feature such as a dome or turret. The material is durable, readily cut and bent and very resistant to corrosion. On exposure to the atmosphere, a protective green patina is formed on the surface which is generally considered to enhance the appearance. An underlay of felt is necessary to provide a satisfactory base for the copper sheet and provide a degree of thermal insulation and also to prevent chafing. Joints in the direction of the fall are either formed with a standing seam or dressed to conical timber rolls screwed to the boarding. The minimum fall should be 1 in 60. Joints running across the flow should be formed with drips spaced not more than 3 m apart and 65 mm deep. However, copper sheeting is usually laid in 1.8 m lengths and therefore requires joints between drips. The joints are dressed flat in the direction of the flow and formed with flattened welted seams. Single lock welts are suitable for pitches greater than 45° and for pitches below this a double lock cross welt is essential. Where sheeting abuts upstands such as parapets and chimneys the metal should be turned up a minimum of 150 mm and a cover flashing fixed into the wall wedged and pointed.

Common defects

Defects occur in some situations and are usually due to one or more of the following items:

- The majority of defects to copper roofs are mainly concerned with redressing seams, rolls, welts or loose aprons and pointing to upstands.

- Copper must not be restrained as there is a risk of not only fatigue but of thermal expansion lifting the sheets from the roof. It is, therefore, important to examine the sheets for cracks and splits.
- If several sheets have failed and stripping is recommended, the substructure should be carefully examined before the new copper sheeting is laid. If necessary a new felt underlay should be provided which should be non-bituminous.
- It is important to remember that jointing or patching copper sheets with solder is bad practice. The two materials have a different expansion rate and will break away. This method of repair is often found in old copper roofs.
- A further point to investigate is the run-off from a copper roof on to other metals such as galvanized iron. This run-off has been known to have a corrosive effect on these metals.

10.20 Lead

In the past lead was by far the most widely used metal roof covering. The material is durable and one of its virtues is that it is easily cut and dressed. Lead is generally laid on boarding with the grain running in the direction of the fall. The minimum fall for lead flats is 1:80 but 1:40 is often preferred. A felt underlay should be laid over the boarding to allow the lead to move freely and to minimise irregularities on the boarding. Many of the old lead roofs were laid direct on the boarding with no provision for thermal expansion which causes the lead to 'creep'. Vertical joints parallel to the roof fall are provided for by the use of wood rolls.

Horizontal joints at right angles to the roof fall are formed by the adoption of drips. As for copper, where the lead abuts upstands a minimum skirting height of 150 mm should be provided together with a covering flashing.

The following items should be noted and carefully examined by the surveyor:

(1) If the joints of the boarding show on the upper surface of the lead it usually means that the underlay is missing.
(2) If thermal expansion occurs the sheets will tend to lift in folds and not return to their natural position. The lead will eventually become brittle and crack allowing rainwater to penetrate through to the substructure. The sheets should be carefully examined for such defects.
(3) One of the most common faults is due to inadequate drips. They are normally 60 mm deep as shown in Fig. 10.9a but a minimum depth of 40 mm is permissible if an anti-capillary groove is provided in the vertical face of the drip. This groove prevents rainwater rising and penetrating through to the timber as shown in Fig. 10.9.

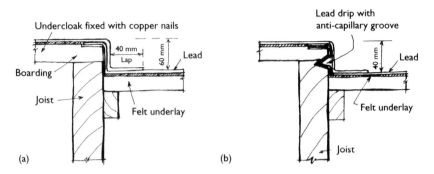

Figure 10.9   (a) Lead drip. (b) Lead drip with anti-capillary groove.

(4)   Slight corrosion by acidic or alkaline conditions can occur on lead flat roofs. Acidic conditions are usually due to acid charged rainwater or the discharge from an adjoining roof containing moss or algae. Deterioration of lead can also be caused by organic acids in contact with western red cedar or oak, or as a result of underside corrosion (Douglas 2002). Alkaline from cement and lime mortars used for pointing can also cause corrosion. Slight problems can occur in the fixing clips and nails of another metal causing electrolytic action.

(5)   Depressions in the lead are easily recognisable and indicate defects in the boarding below. The cause may well be due to damp penetration through cracks in the lead or dislodged coverings to the wood rolls. If there are large areas of depression it would be advisable to arrange for some of the lead and boarding to be removed and the structural timbers examined for defects.

(6)   Particular attention should be paid to vertical or steeply pitched lead surfaces particularly where the material is used on the cheeks of dormer windows and fixed with solder dots. In such situations the lead tends to creep causing the solder to become dislodged.

(7)   Old buildings covered with lead derive much of their character and beauty from this material and it should, therefore, be repaired or replaced by new lead. There are very few roofs on medieval buildings that are covered with the original lead. Many of the lead coverings on medieval church roofs were recast and relaid in the eighteenth and nineteenth centuries and many of these have since been renewed. Lead roofs constructed centuries ago were often laid in long and wide sheets and have therefore suffered from thermal movements as a result of changes in climatic conditions. Fixings were often poor and drips too shallow.

When confronted with an old building the surveyor may have difficulty in deciding whether the roof should be condemned or repaired. If the lead shows signs of stress or has worn thin and is covered with pin holes it should be taken up. Old cast or milled lead can be melted down and recast but soldered patches and other impurities should be removed from the molten metal.

10.21   Zinc

Zinc has a shorter life than copper or lead and for this reason its popularity has waned during the past eighty years. Zinc is much less malleable than lead and was often regarded as a cheap substitute. Its principal defect is that it is liable to corrode in both alkaline and acid conditions and is also susceptible to electrochemical corrosion when in contact with copper. With the exception of western red cedar, oak and chestnut, seasoned timbers do not affect zinc. To prevent condensation on the underside of the metal and to allow the zinc to expand and contract, felt or building paper should be laid over the roof boarding. However, it is often found with a zinc flat roof that the underlays are omitted. The recommended minimum thickness should be English Zinc Gauge 14. The joints should run parallel to the flow, being formed with tapered wood rolls similar to those described for copper. The zinc is turned up the roll both sides stopping short of the upper surface and secured with metal fixing clips, which are fitted round the rolls and bent over the free edges of the sheets. A separate zinc capping piece is placed over the roll to overlap the sheets either side. The capping is secured by holding down clips nailed to the battens. Drips are provided as in lead work for joints across the flow and bent outwards. The upper sheet projects over the drip and is bent under the angle formed by the lower sheet.

Common defects

The following items should be noted and carefully examined by the surveyor:

- Zinc does not readily show depressions indicating defects in the substructure. A metallic noise is produced when the suspected area is tapped, but it is often difficult to assess whether or not it is due to some defect in the boarding or if the zinc has lifted. Unlike lead covering, zinc is not easily bent back and the surveyor may find difficulty in making a thorough examination.
- The fall of the roof should be checked. The fall should not be less than 50 mm in 3 m.
- The capping pieces, stop ends, drips and upstands should be closely examined. They are often found to be distorted or damaged by foot traffic. It is often found that drips have been omitted. With steeper pitches at 15° or more this is permissible, but lower pitches should have a welted or beaded drip.

- Apron flashings against parapets will often be found to be loose or curled upwards. Zinc embedded in a mortar joint will deteriorate if not protected with a coating of bitumen.
- Old zinc roofs will usually be found to be corroded and pitted and some may have been coated with bituminous paint. Under such circumstances the surveyor may consider that its life is limited and a patch and repair job is not practicable. In such cases it is advisable to recommend a complete renewal of the metal.

10.22   Aluminium

Aluminium as a roofing material is comparatively new and it is therefore unlikely that the surveyor will encounter many roofs that are defective. The metal is used as a flat roof covering in the form of sheets or as a pitched roof covering consisting of corrugated or troughed sheeting. The jointing, fixing and falls to flat roofs are similar to copper using rolls, drips and welts with a felt underlay. Metal fixings should be made of aluminium or galvanised steel and non-conducting sleeves and washers should be used to insulate the aluminium sheets from other dissimilar materials. The thickness of the sheets whether preformed or flat is usually 20 swg to 24 swg, but where there are severe conditions of exposure thicker sheets are often used. In the past there have been problems with condensation on the underside of the metal and dripping on to the structure below. This problem was often found in poorly ventilated buildings with uninsulated single skin roofs. However, this has largely been overcome by various proprietary systems using insulated panels having a high degree of thermal insulation. For pitched roofs a wide selection of composite insulated panels are made which conform to BS 6868:1972. They are ideally suited for industrial and commercial buildings where a high standard of insulation is required.

As far as making an examination is concerned the following points should be checked:

- As with copper and lead, defects in aluminium are mainly concerned with redressing seams, rolls, welts or loose aprons and pointing to upstands.
- As mentioned above, the underside of any single skin roof should be carefully examined for evidence of condensation. In certain instances the aluminium will be seen to have suffered severe corrosion and have a chalky appearance.
- Under certain wet conditions, western red cedar, oak and sweet chestnut can cause corrosion. Timber that has been treated with copper preservatives can also attack aluminium in contact with it.
- Aluminium is often more electronegative than other materials and must, therefore, have no contact with copper, brass or bronze including the run off of rainwater flowing from copper to aluminium.

- Aluminium can be degraded by alkaline materials such as mortar and concrete.

10.23   Stone slates

In the following 'stone slating' has a completely different meaning to the term slate roofing mentioned earlier. Stone slates are formed from sandstone and limestone and are quarried from stratified layers which are then split along the natural bedding planes. They vary enormously in weight, quality and thickness.

There are virtually no new stone slates produced, but one or two advisory bodies are attempting to revitalise the industry and encourage production. However, some types of stone are obtainable from salvage yards. Surveyors requiring stone slates for repair work may be able to find suppliers by contacting the local Conservation Officer.

Stone slates provide an attractive roof covering, but are seldom used on new work because of the very heavy roof timbers required to take the loading. They are, therefore, usually encountered in old buildings where alterations or repairs are required.

Thicknesses vary between 16 or 19 and 25 mm and the colours are natural and differ in accordance with the locality of their quarry. Laying is usually carried out in random widths or random sizes laid in diminishing courses. The common method of fixing is by turned oak pegs in a drilled hole secured between two battens as shown in Fig. 10.10b. The extra batten is used to prevent the peg from twisting out of position due to the weight of the slate. Alternatively, battens are laid on open rafters, plasterer's laths being laid between them parallel to the battens and rendered with lime mortar solid to the top face of the batten. The stone slates are then bedded, pointed and head nailed to the battens as shown in Fig. 10.10a. Ridges and hips are usually finished with sawn stone angular cappings or half-round clay tiles bedded in mortar. Valleys are generally swept as open gutters. Close mitring is not particularly easy with this type of roof. Figure 10.11 shows a timber roof structure after stone slates have been removed.

The town or city surveyor unaccustomed to inspecting stone slate roofs in the stone districts must not be misled into thinking that the roof is about to disintegrate. The natural settlement of the roof, the uneven shape of the stone slates and the damage by frost, causing the top surface of the slates to spall, will obviously contribute to all the various undulations found in the covering.

With a suitable ladder the following items affecting the durability of stone slates may now be considered:

- If there are signs of settlement of the roof timbers the surveyor may wish to satisfy himself that the roof construction is sufficiently strong

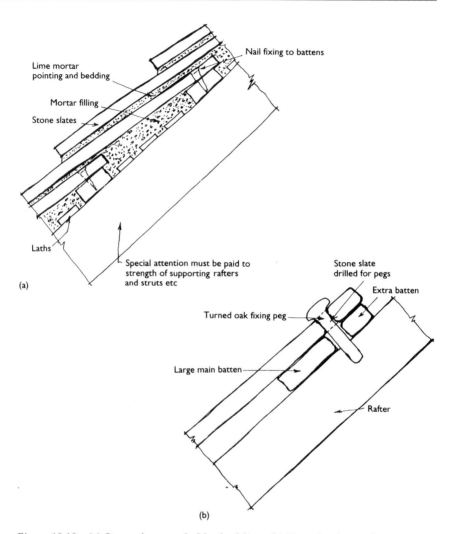

Figure 10.10   (a) Stone slates sealed by bedding. (b) Tarred oak peg fixing.

to support the roof covering. A note should be made of any defective timbers or weaknesses in the rafters and collars etc. Wood pegs were usually of oak, but many will inevitably have rotted or have become loose as a result of movement in the structure.

- Slates laid on a bed of lime mortar should be carefully checked. Moisture may have penetrated through to the laths and battens causing wet rot and the mortar bedding to become damp and loose. If these defects are far advanced then the surveyor may well consider stripping and renewing all the fixings.

- Externally, the very old slates may well be seriously damaged by frost or have become worn and fractured and require complete renewal.

Figure 10.11   Timber roof structure after worn stone slates have been removed.

As methods of fixing vary considerably in different parts of the country, the best course is to check the roofs of similar material found in the district and follow the local tradition.

10.24   Asbestos cement and translucent roofing sheets

Asbestos sheeting is composed of asbestos fibre and cement and is used extensively on industrial buildings, outbuildings, garages or as a temporary covering on other structures. In addition to the usual corrugated sheets there are various profiles available together with a variety of mastic waterproof sealers. The sheets are non-combustible and durable and are secured to timber or steel purlins by galvanised driving screws or hook bolts fitted with lead or plastic washers.

One of the great advantages of asbestos sheets is that they are light in weight and therefore easy to handle and fix. However, asbestos is now classed as a deleterious material and as a result its use is no longer allowed in new construction or refurbishment schemes (Douglas 2002).

Asbestos sheets have very little thermal insulation value. In the more permanent structures an insulation board lining is usually fitted either under or over the purlins. Widths range from 710 to 1142 mm according to the type and make of sheets and lengths from 1225 mm rising in increases of 150 mm to 3050 mm. The roof pitch is normally at a minimum of 22.5° and end laps of 150 mm. However, when laps are sealed with a waterproof mastic the pitch may be reduced. A comprehensive range of accessories is available to meet a variety of situations such as ridge cappings, ventilators, eaves closures and filler pieces. The material has a high resistance to atmospheric pollution and is resistant to alkalis and acids although rainwater made acidic by atmospheric pollution may soften the surface of the asbestos. The sheets are obtainable with a coloured factory applied process which has good wearing qualities. Where the face of the sheets is treated, the internal face should be primed and painted in order to reduce differential moisture movement and carboration shrinkage between faces.

This precaution will also prevent the sheets from cracking. Small cracks often develop into large cracks causing moisture penetration. Asbestos cement sheets tend to harden with age through hydration of the cement matrix and with no natural weathering the sheets become more brittle, but this does not necessarily affect the watertightness of the roof. However, the sheets tend to crack easily when subjected to pressure. Asbestos sheets are not susceptible to attack by insects, but fungal growths are common on the unpainted sheets. The growth can cause surface deterioration and together with acid waste products carried down the roof slopes by rainwater they will slowly eat into the flashings and gutter linings.

Translucent sheets are obtainable having corrugated profiles to match the asbestos roofing sheets. They are used as dead lights or as a complete roofing system laid to the same pitch as the asbestos sheeting. They are also used for covered walk-ways, car ports and canopies. Translucent sheets are composed of glass fibre polyester resin, acrylic resin or corrugated glass with wire reinforcement. Sudden impact loads may result in cracking. Surface deterioration is also a problem caused by weathering and the accumulation of dirt which often reduces light transmission. With this type of material some condensation is inevitable and usually occurs in buildings where industrial processes involve high temperature and humid conditions.

If the building is extensive, a builder's 'attendance' will no doubt be necessary to assist with ladders, crawling boards and walkways. Special platforms which sit astride the ridge and are fitted with guard rails are available and are adjustable to any roof pitch. The platforms are very useful when a close examination of ridges or chimney stacks is necessary.

The following items should be checked when carrying out an examination of asbestos cement or translucent roofing sheets:

- When over-purlin insulation is used it is advisable to check that the insulation material has not been affected by rainwater penetration.
- If sheets are cracked or otherwise damaged they cannot be repaired and replacement or overcladding with a new non-asbestos sheeting system (e.g. GRP or powder coated steel) should be recommended.
- Check that fixings are through the crown of the corrugation. Screws or bolts may have corroded.
- Moss and lichen growths may cause a slight softening of the asbestos surface.

- Check end laps which should be consistent at 150 mm, but side laps are dependent on the type of sheet used.
- Joints between asbestos sheets and translucent sheets should be checked for watertightness. The joints should be sealed with a mastic or foam strip.

10.25 Asbestos cement slates

Asbestos cement slates are manufactured from asbestos fibre, Portland cement and water. They are compressed and coloured by the addition of pigments to the cement matrix. Asbestos slates resemble natural slates and are resin coated which enhances their appearance and their resistance to algae. British Standard 690 Part 4 governs quality and size. Being light in weight, durable, impervious to climatic conditions, rot and most acids, the demand for this material is increasing rapidly. The slates are non-combustible in accordance with BS 476 Part 3 and are, therefore, suitable for all types of buildings.

Manufacturers usually give a 30 year guarantee but experience has shown that in normal conditions a 50 year life span can be expected. The slates are suitable for roof pitches of 20° and above, and are available in a limited range of colours including natural, brown and blue. The slates are supplied in varying sizes up to $600 \times 300 \, mm^2$, and standard ridge and hip tiles are available in colours to match the slates. Valleys may be formed by either an open gutter or by close mitring using individual soakers. The preparation and the method of laying is similar to natural slate except that copper disc rivets are used to hold down the tail of the slate. This is necessary because of the tendency of the slate to lift in a high wind.

Asbestos slates have not been used for such long periods as other roofing materials and it is, therefore, unlikely that the surveyor will encounter many defective roofs covered with this material. Nevertheless, the durability of an asbestos slate roof can be affected in the following ways:

- A note should be made of broken or cracked slates. This is usually the result of pressure on the surface such as that from a falling object.
- All the abutments should be checked for loose or defective soakers or cover flashings. The flashings should be properly secured and pointed. Soakers should be equal in length to the gauge of the slate, plus lap, plus 25 mm laid to the slate coursing.

The same precautions described above for asbestos sheets should be applied to asbestos tiles.

10.26 Corrugated iron

This type of covering is usually found on outbuildings and buildings of a temporary nature where appearance is of no consequence. It provides a cheap covering but is gradually being superseded by corrugated asbestos sheets. However, corrugated iron sheets are resilient and will usually stand up to sudden impact loads, whereas asbestos sheets are brittle and tend to fracture easily. The galvanised zinc coating applied to corrugated iron has a limited life and the sheets tend to deteriorate as a result of corrosion after the zinc coating has failed. The fixing screws and bolts also require checking for corrosion.

10.27   Thatch

Thatch has been a roof covering in Britain for many centuries and its application is a highly skilled operation (Hall 1988). Straw and reeds drawn from local cornfields or reed beds provided one of the earliest coverings for small dwellings and at one period even churches and large manor houses were thatched. The great advantage of this form of roofing lay in the fact that the materials were procured locally (Cox and Letts 2000). Other advantages are as follows:

- It is a lightweight material and thus saves on structural timbers.
- It insulates the building against heat, cold and noise.
- It saves on gutters and rainwater pipes which are not necessary with a thatched roof.

There are four types of thatch in most general use today. Water reed from Norfolk is still the most durable material for thatching, but one of the most common materials is straw. There are two types: combed wheat reed which has been harvested by traditional methods with all its impurities combed out, and long straw which also is harvested, but does not have the same neat appearance as combed wheat. Sedge is a marsh grass and although it is a pliable material it is only suitable as a ridge capping.

It is wrong to suppose that all thatching is the same wherever it is found. Almost every region has its own way of thatching and strong regional characteristics have grown up in most areas (see Holden 1998).

Netting is often used to prevent damage by birds and vermin although it is seldom recommended by a thatcher. There are certain disadvantages with netting, such as leaves becoming trapped in the mesh and an impedance in the flow of rainwater which in turn encourages the growth of moss and lichen. Apart from the problem with birds, thatch is particularly prone to fire. There are several fireproof solutions available, but they need periodic renewal. To protect the interior of the building, fire-retarding boards can be fixed to the top of the rafters. This method of fire-proofing can only be used on old roofs where it is necessary to renew the thatch completely.

A surveyor contemplating making extensive alterations to a thatch roof or changing the type of covering should first ascertain whether the property is a listed building, in which case consent to the proposals will have to be obtained from English Heritage or Historic Scotland. The Building Regulations should also be checked with regard to the vulnerability of adjoining buildings to attack by fire from a thatch roof building. The regulations prohibit new thatch on properties less than 12 m from a boundary.

Owners of a thatched building usually approach a local thatcher when a repair or re-thatch is necessary (Brockett and Wright 1986). The surveyor only appears on the scene when a full building survey of the whole building is required or an extension proposed.

It is not always easy to assess the remaining life of an old thatch roof. Thatch decreases in thickness over the years and if the fixings are seen to be close to the surface then replacement is necessary. In most cases complete re-thatching is unnecessary. It is common practice to strip back to a sound base and provide a layer of wheat reed or long straw over the existing thatch.

If the surveyor has little experience in this field he may consider that he needs further expertise to assist him in making a decision and should of course consult a local thatcher or the Master Thatchers Association for names of thatchers in that district (Hall 1988). If extensive repair or re-thatching is required it is always advisable to obtain an approximate estimate for the work. This information can then be incorporated into the client's reports.

Common defects

The following items should be examined when carrying out a survey of a thatched roof:

- An examination of the roof structure is most important. Treatment to eradicate vermin or wood boring insects may be necessary. Ensure that the rafters and battens are correctly spaced to suit the needs of the thatcher. The rafters are usually found to be at 610 to 710 mm centres and the size of the battens 50 × 25 mm at approximately 228 mm centres. Where it is necessary to straighten an old roof, the old rafters can be retained and new rafters laid alongside the old.
- Eaves and verges should be examined. Tilting fillets are usually required at the eaves and verges to force the bottom layer of thatch up and keep rainwater clear of the walls.
- Ensure that trees do not overhang or brush against the roof.
- When carrying out an examination the surveyor should try to avoid climbing on the roof or resting a ladder against the eaves. Damage can occur which will often shorten the life of the material.
- If alterations are contemplated such as new windows, then it is advisable to discuss this with the thatcher on site. Some elements of construction that are essential for the thatcher can then be incorporated into the alteration work.

- While in the roof space the chimney stacks should be examined. Ensure that there are no loose stones or bricks and that the pointing or rendering is sound. Particular attention should be given to the proximity of combustible materials.
- Access to the roof space should be checked to enable easy access to be obtained in case of fire. A hatch should not be less than 914 × 600 mm and have a half hour fire resistance.

All thatched roofs require regular maintenance and provided this is properly carried out the thatch should give many trouble-free years of service (Moir and Letts 1996). A Norfolk thatch, kept in sound condition, could last 80 years. Combed wheat reed could have a life of 40 years and long straw is good for about 20 years.

10.28   Wood shingles

The use of wood shingles dates back to the Romans and Saxons, but due to the shortage of oak the craft fell into disuse. Western red cedar shingles are widely used in America and Canada and have gained popularity in the UK. The timber resists decay and weathers to a pleasant grey colour. The shingles are light in weight and give a certain amount of insulation. In shape, fixing and treatment at junctions etc., they are very similar to slates. The usual size is 400 mm long, of random widths, and they are usually laid to a gauge of 127 mm. Two copper or aluminium nails should be used for each shingle, fixed approximately along the centre line of the shingles. Shingles are usually laid on an underlay of felt or building paper. However, it has been suggested that underlays prevent the circulation of air to the underside of the shingle thus producing conditions conducive to rot. The main objection to wood shingles is that there is a high fire risk from burning fragments from other buildings or chimneys.

When carrying out an examination of oak or cedar wood shingles to a fairly old roof the surveyor should closely examine the timber surfaces for deterioration. If galvanised steel nails have been used they are often attacked by corrosive agents in the timber causing the shingles to slide down the roof slope. If this is the case then large areas may need to be relaid. Shingles should also be checked for moisture movement. Flashings against chimneys and parapets will often be found to be loose and need refixing.

10.29   Roof lights

In many industrial and commercial buildings roof lights provide a source of natural daylight where there is not sufficient light from the side windows of the building. They are usually situated in inaccessible positions and are often forgotten until problems occur such as water penetration.

Roof lights are fitted to both flat and pitched roofs and consist of the following types:

- Fixed roof glazing to pitched roofs is used in conjunction with asbestos sheeting or other types of covering and usually consists of patent glazing bars designed to accommodate single or double glazing secured to the roof purlins (see Fig. 10.12).
- Skylights may be provided in pitched roofs to light a staircase, landing or attic. The skylight may be hinged or fixed (see Fig. 10.13).
- The traditional form of timber lantern light is still efficient and is found on the flat roofs of many old buildings (see Fig. 10.14). During the last forty years they have largely been superseded by dome lights which are manufactured by specialist firms and consist of a combined curb mounted ventilator and roof light unit. The units are constructed throughout in extruded aluminium and are corrosion resistant. The domes are available in glass, acrylic and wire laminated PVC. The aluminium construction reduces roof loading and obviates maintenance. All that is required is an occasional cleaning.

For reasons of safety the glazing to roof lights is usually 6 mm wired glass or cast plate glass. The sheets are from 1 to 3 m in length and 610 mm in width. Where thermal insulation is important double glazing is used with an air space of 12 to 18 mm. In small skylights or lantern lights the

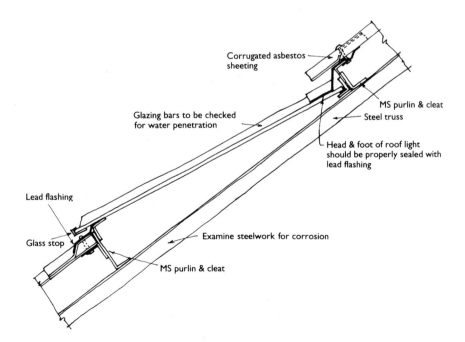

Figure 10.12   Fixed roof glazing.

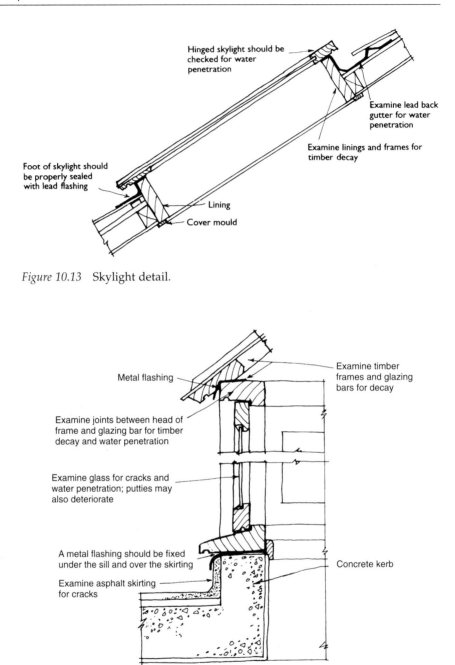

Hinged skylight should be
checked for water
penetration

Examine lead back
gutter for water
penetration

Examine linings and frames for
timber decay

Foot of skylight should
be properly sealed
with lead flashing

Lining

Cover mould

Figure 10.13   Skylight detail.

Metal flashing

Examine timber
frames and glazing
bars for decay

Examine joints between head of
frame and glazing bar for timber
decay and water penetration

Examine glass for cracks and
water penetration; putties may
also deteriorate

A metal flashing should be fixed
under the sill and over the skirting

Concrete kerb

Examine asphalt skirting
for cracks

Figure 10.14   Lantern light.

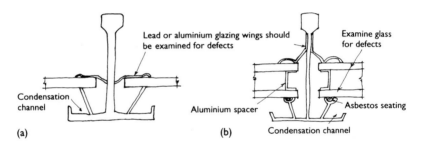

Figure 10.15 (a) Aluminium glazing bar. (b) Aluminium double glazing bar.

glazing bars can be made of softwood or hardwood. They are liable to decay and regular maintenance is necessary especially in very exposed positions. There are several proprietary glazing bars in the form of metal T-sections consisting of steel or aluminium alloy sheathed in lead or PVC. The bars incorporate condensation channels and the glass is secured with aluminium spring cover strips or lead wings. The bars are normally trouble free and produce a durable watertight roof light. Figures 10.15a and 10.15b show examples of single and double glazing bars. The various items affecting the durability of the roof lights and requiring examination may now be considered:

- The head and foot of a fixed roof light or skylight should be examined to ascertain that they are properly sealed with lead flashings and that back gutters have been properly formed. It is often at these points that insufficient lead is provided to make a waterproof joint. The flashings are subject to movement and the fixings often expand and corrode.
- Internal condensation is often a problem and can cause decay in the timber or corrosion of the metal parts including ironmongery. It is advisable to examine the condensation grooves to ensure that they are not blocked.
- Skylights may be hinged or fixed, but if the light is hinged penetration of rainwater is likely.
- All forms of upstands to the lantern lights must be in sound condition and should have an angle fillet between roof and upstand. A metal flashing should be fixed under the sill and over the upstand covering as shown in Fig. 10.14. Poor detailing in this area is a common cause of water penetration.

10.30 Duckboards

Although this is not a roofing material, proper means of access over roofs should be arranged so as to avoid damage to the roofing materials. They are useful on flat roofs and wide parapet or valley gutters, serving the dual purpose of ensuring that snow does not choke the gutter, and avoiding damage to the metal coverings when walked along. Duckboards require careful setting out in order to avoid interference with the flow of water and should be made sectional for easy lifting. When dealing with large roofs or gutters the surveyor may well consider it advisable to recommend the installation of duckboards in his report.

10.31   Metal profile sheeting

Rainwater ingress

A key performance requirement for any building is that it should not leak. Unfortunately, even in recently constructed buildings with profile roof and wall cladding, this problem is not uncommon. When inspecting more modern properties, therefore, the surveyor should be on the lookout for rainwater ingress in this form of building envelope.

Rainwater can enter the fabric of a building via a number of routes and by several mechanisms (Endean 1995). This is especially the case with profile sheeting because of the number of joints and complexity of detailing at junctions, gutters and openings.

Typical entry routes in profile sheeting for water are:

- Defective laps and other joints in sheeting.
- Defective nodes at fixing points.
- Punctures, fractures or other similar damage to the sheeting.
- Inadequate seals in sheeting at vulnerable edges and other locations – e.g. abutments, eaves, ridges, vent pipe and other projections, rooflight flashings, gutters, outlets, etc.
- Blocked or clogged gutters and outlets allowing rainwater to overflow flashings or enter via open edges.
- Inadequate falls in gutters to outlets.

Water can enter the fabric of a building by one or a combination of any of the following mechanisms:

- Gravity.
- Capillary action.
- Kinetic energy.
- Surface tension.
- Air currents.
- Pressure difference (see Endean 1995).

Surface degradation

Since the 1970s, profile metal cladding has been used extensively for roofs and wall panels in many commercial and industrial buildings. The plastic

Figure 10.16 (*a*) Lap Joint Before Treatment; (*b*) Seamsil® Treated Lap Joint; (*c*) Gutter Over-hang Before Treatment; (*d*) Seamsil® Treated Gutter Overhang (courtesy of Delvemade Ltd).

coating of this cladding is usually either PVC (e.g. 'Plastisol') or PVDF (polyvinylidene-fluoride). A typical initial service life for this form of cladding is between 15 and 25 years, before first maintenance is usually needed (Harrison 1996).

The main surface degradation problems are:

- Rust staining – particularly at fixing nodes.
- Pitting of the surface.
- Discoloration or bleaching of surface coating.
- Peeling and blistering of surface coating.
- Delamination of sheeting and insulation.
- Denting or puncturing of sheeting.

Cut edge corrosion

One of the most common problems in metal profile sheeting in modern buildings is cut edge corrosion. It occurs where sheets are cut to length in the factory or on site, thus exposing the unprotected metal substrate. This gives rise to the tell-tale rusting along the joints of the sheeting. Not only is this unsightly, it will also lead to further degradation of the sheeting and substrate leading, eventually causing leakage at the joints.

Increased environmental pollution dissolving the protective zinc oxide layer of the galvanised substrate is the main deterioration mechanism that triggers this problem. This can be exacerbated by the use of less durable organic coatings such as 'Plastisol' or poor workmanship when installing the sheeting (Harrison 1996).

The three main areas of cut edge corrosion attack on a roof detail are:

- Overlaps (including above translucents).
- Gutter edge detail.
- Flashing detail (especially those edges not turned over).

There are now a number of proprietary treatments of edge corrosion available (see Fig. 10.16). Delvemade Ltd (www.delvemade.co.uk) is one such system which uses a silicone (non-acidic) sealer application. It can be used in conjunction with a recoating scheme for the whole external surface of the cladding using a more durable paint system such as PVDF. Consult also technical guides by Weatherproofing Advisers (see website address in bibliography).

11 Fireplaces, Flues and Chimney Stacks

11.1 Introduction

Chimney and fireplace problems usually fall into two categories: those due to physical decay of the structure and those which cause the fireplace to smoke.

Atmospheric pollution caused by gases and fumes from old solid fuel fireplaces is a potential danger to health and the instability of many chimney stacks is a hazard that needs close examination.

A comparison of pre-Second World War housing and post-war housing shows that very few houses from the latter period have chimney stacks. Single flue stacks are common in older domestic properties and are usually found serving a kitchen boiler.

It is difficult to ascertain the number of fireplaces in old dwellings that are no longer required, but a considerable number of dwellings could be improved by having the surrounds and grates removed and the fireplace opening sealed. If the stacks are to be retained but capped off and the fireplace opening sealed it is important to state in the report that the flues are to be ventilated to the external air at head and foot. Alternatively, if the stack is taken down to below roof level and capped there is no need to ventilate the flue. The chimney breasts and flues will normally be retained since to remove them would be a costly operation. In addition to this the owner would probably wish to utilise the flue for a gas heater.

The condition and design of each fireplace and its surround should be carefully considered. Where requirements of floor and wall space permit, good period pieces in sound condition should be retained. This is obviously a matter which should be discussed with the owner who may well require the surround and open fire for decorative purposes. During the past twenty years there has been a revival of interest in the traditional fireplace as a working and decorative feature and a wide range of reproduction fireplace surrounds are now available.

Before making a decision concerning the use of an open fire the surveyor would be wise to consult the local authority with regard to the Clean Air Act 1956 (as amended by the 1968 and 1993 Acts). Under the Act, the local authority with the approval of the Secretary of State may declare the whole or any part of its area to be a smokeless zone. The Act makes it illegal to emit smoke from a chimney within a smokeless zone. There is obviously no point in advising a client to reconstruct a period fireplace for the burning of wood or non-smokeless coal. However, an owner of a dwelling situated in a smokeless zone may apply for a grant towards the cost of converting the fires.

In the following the most common problems associated with domestic fireplaces and chimneys and those connected with industrial plants will be considered.

11.2   Domestic fireplaces and flue entry

The majority of complaints against old fireplaces are usually concerned with their tendency to smoke. The surveyor investigating such a problem will probably find that the cause is due to an inadequate flow of air. The normal action of a fireplace depends on a continuous current of air passing up the flue and carrying with it the gases of combustion. It is, therefore, necessary to have a reasonable supply of air entering the room and passing through the fireplace into the flue and an unobstructed exit for the gases as they leave the chimney. In fact, it is the most important factor in preventing the chimney smoking. If the air supply or chimney is obstructed a check on the up-current of air will occur and down-draught may result.

The leading reformer in fireplace and flue design and the most frequently quoted authority is Benjamin Thompson (1753–1814) better known as Count Rumford (Billington and Roberts 1982). The principles of design that he laid down are still generally applied by the Building Research Establishment. The following are the essential principles regarding fireplace design for open solid fuel burning grates:

- Correct design of the throat, which should be 100 mm from front to back and the width is determined by the splaying of the sides and back of the fireplace.
- To cause a two-way circulation of air in the flue and prevent soot and rain from falling into the fireplace, a smoke shelf should be provided. The shelf should be horizontal and level and should be between 150 and 200 mm above the top of the fireplace opening.
- Immediately above the throat there should be a smoke chamber, the same depth as the flue. The sides should taper inwards at an angle of 60° until the flue is reached.
- The fireplace should have splayed sides to reflect the heat into the room.
- The depth of the fireplace should be between one third and one half of the width of the front opening.
- The flue should have a cross sectional area of one tenth of the area of the fireplace front opening.
- The front of the fireplace opening should have an equal width and height.

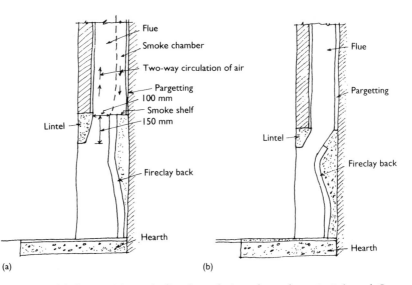

Figure 11.1 (a) Section through fireplace designed on the principles of Count Rumford. (b) Fireplace design commonly adopted in modern work.

- The upper portion of the fireback should be sloped outwards and the sides angled to reflect heat out into the room.

In practice of course, the surveyor will often find that these ideal conditions may not always occur. The fireplace and flue entry shown in Fig. 11.1a has been designed on the Rumford principles (see Billington and Roberts 1982) and Fig. 11.1b shows a section commonly adopted in modern work which is usually successful. The surveyor will often find that dwellings erected in the past forty years have been fitted with precast throat units which ensure that the throat is formed to the correct dimensions.

11.3   Down-draught due to external conditions

A steady blow back of smoke can also be due to unfavourable site conditions. As the wind strikes buildings and trees, areas of air pressure and suction are set up. The area of pressure is on the side which the wind strikes and the area of suction on the opposite side. These areas extend above the roof, whether the roof is flat or pitched. Where a badly placed chimney is in an area of high winds, the pressure may be greater than the upward pressure of the smoke, rising up the flue and therefore a down-draught will result. The main lesson to be learnt from this is that the chimney outlet should be high enough to avoid interference from pressure and suction. Where it is clear that the existing outlet is too low, the almost

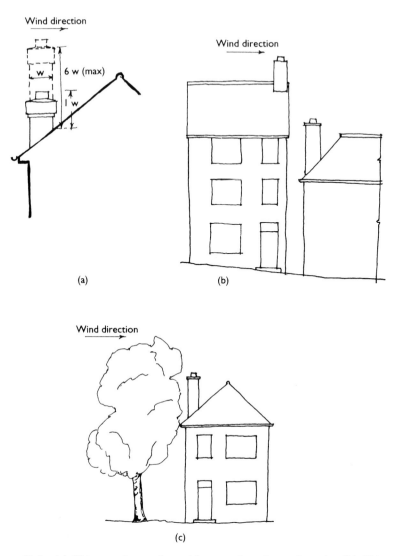

Figure 11.2   (a) Chimney lower than ridge causing down-draught. (b) Chimney too low in relation to higher adjoining building. (c) Tree in too close proximity to chimney causing down-draught.

certain remedy is to raise it to bring the outlet above the disturbed area. If this is done by building up the chimney in the existing materials it is well to remember that no brick chimney should rise more than six times its width (see Fig. 11.2a).

The building regulations require that the chimney outlet should be 1 m above the highest point of the intersection of the chimney stack and roof slope as shown in Fig. 11.3. Adjacent buildings modify the areas of pressure and suction around the building. In some cases it is not sufficient to

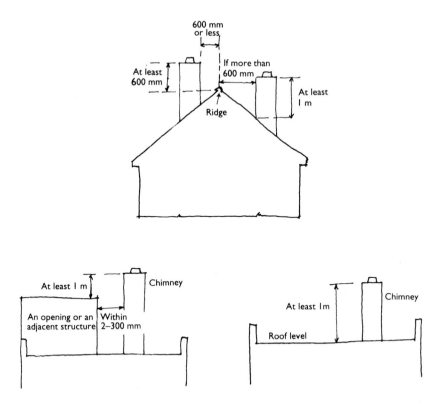

Figure 11.3 Minimum chimney heights laid down in the Building Regulations for solid fuel and oil burning appliances.

carry the chimney up 1 m above the ridge. A single storey building may have the ridge at a much lower level than that of the adjoining buildings. On steep hills one building may be lower than an adjoining building. In Fig. 11.2b the building in the lower position has a chimney which is below the ridge level of the adjoining building although above its own ridge. The lower chimney will probably suffer from down-draught. The remedy is to take the chimney up to a higher level which may make it necessary to install special bracing. A possible alternative is to recommend a change in the form of heating. Tall trees near a building may cause down-draught by creating areas of pressure and suction. A typical case is shown in Fig. 11.2c. If it is impracticable to raise the chimney above the area of pressure, the tree should be lopped.

If wind pressure or suction due to wind blowing down on the chimney outlet is the problem, then a favourite solution is to install one of the many types of cowl that have been designed for this purpose. Surveyors dealing with this problem should remember that cowls are not always the answer to down-draught and smokey fires. The probability is that there will be faults in the fireplace and at the chimney top. Some types of cowl restrict the air flow and often make the problem worse and in other cases the cowl acts as a trap for soot which builds up inside and reduces chimney capacity. The lobster-back cowl is probably best for such situations. It swivels freely in the wind and should create suction at the right point. The tall 'Mancone' cowl works reasonably well, adds about 1 m to the height of the chimney and has no ledges to catch soot. No definite opinion can be expressed as to the effectiveness of any particular type. Everything depends on local circumstances. Cowls should not be recommended until any obvious defect in the fireplace or chimney has been remedied.

In this complex situation it is suggested that the surveyor, with the aid of binoculars, commences his examination of the suspect chimney and roof and works methodically along the roofs of the adjoining buildings or other obstructions compiling notes and sketches as he goes. Examination of the external conditions will tell him a good deal when his inspection is taken in conjunction with his survey notes relating to the interior.

11.4   Flue investigation

Practical conditions usually decide the position of bends. Provided that the change of angle is slight the position appears to be immaterial. The run of the flue can easily be checked with a set of chimney sweep's rods. Great changes in the direction of a flue are certainly not desirable. Under the current Building Regulations, there should not be more than two bends and where a bend is necessary it should make an angle with the vertical of not more than 45°. Flues that are of low inclination will collect soot and mortar droppings causing partial blockage. In older flues blockages may be the result of loose pargetting breaking away from the flue wall and becoming jammed across a bend. When faced with such problems the best course of action is to have the flues swept. Using a scraper at the end of the rods usually clears small loose obstructions. If rodding fails to clear the flue, it will at least enable the surveyor to fix the position of the obstruction by measuring the rodding length. However, in difficult cases where extensive flue damage is suspected then fibre optic devices described in Section 2.5 will enable the surveyor to detect cracks and other concealed failures within the flue. Fibrescopes are only 6 mm thick in diameter and can be passed through a predrilled hole in the brick mortar joint enclosing the flue.

Condensation problems are usually found in flues serving domestic boilers and slow combustion stoves that do not contain a flue liner. Brick chimney stacks situated on the external walls of a dwelling are most vulnerable to this type of attack. As the flue gas flows upward it cools and by the time it reaches the top of the chimney it no longer keeps the side of the flue warm. Condensation will then occur on the cool surface. When the products of combustion condense on the surfaces of the flue they may deposit tarry residues and set up sulphate attack on the pargetting and mortar joints. The sulphate attack causes the mortar to expand and leads to distortion in the chimney stack (see Section 11.10). In cases of tarry compounds forming inside the flue the surveyor will probably find that the decorative finishes on the chimney breasts are likely to be discoloured. If damage is slight the surveyor may consider it possible to recommend the insertion of a flexible metal liner.

Leaking flues are not uncommon and are often difficult to trace. If the lining and joints in the brickwork decay, smoke may leak from one flue to another or from a flue into the roof space. Leakages through the outer walls of the flues are easy to trace, but leakages in the 'withes' causing smoke to pass from one flue to another are often difficult to deal with unless they are near the fireplace opening. One method of detecting leaking joints is by burning special smoke pellets in the fireplace. The type of smoke pellets required for such tests are produced by P H Smoke Products Ltd, Fairfield Works, Glen View Road, Eldwick, Bingley, West Yorkshire.

11.5   Flues serving gas fires

When carrying out surveys of domestic properties the surveyor will often find that gas fires have been fitted to an existing fireplace and chimney. The following points should be checked:

- No opening should be formed in a flue except for cleaning and inspection, when the opening should be fitted with a non-combustible gas-tight cover.
- Problems often occur in old properties when the flues to gas-fired water heaters terminate in unventilated roof spaces. The danger from fumes, condensation and fire makes it essential that such flues are discharged into the open air. If this is not the case then the surveyor should recommend that the flue is carried up to the ridge.
- It has to be remembered that high-rated appliances need flues with stainless steel liners throughout. The masonry flue should be sealed around the liner top and bottom with a clamp sealing plate.
- Active flues should be fitted with terminals and disused flues with a cap. The capping usually consists of a half-round clay ridge tile. It is essential to exclude damp in order to prevent sulphate damage.
- Outlets from flues should be so situated externally that air may pass freely across at all times and should be at least 600 mm from any opening into the building.

Many properties erected during the past fifty years have been fitted with prefabricated block chimney flues which are built into the inner skin of the wall. They are only suitable for gas appliances and are often visible in a brick gable end or party wall in the roof space. (The gas appliances will be dealt with in Sections 14.7 and 14.8.)

11.6   Flues serving oil-fired boilers

Oil-fired boilers are not likely to cause chimney problems unless they are run with too rich a mixture, in which case condensation may be in evidence. The design and condition of the flue terminal should be considered. Gas terminals are often used on oil-burning flues and tend to restrict the exit point. Sticky soot deposits may accumulate around the terminal outlet and block the chimney.

11.7   Hearths

The size and construction of the hearth is important. A constructional hearth must be of incombustible material properly supported and must be at least 125 mm thick; extend at least 150 mm at each end beyond the fireplace opening and project at least 150 mm from the chimney breast. If the hearth is freestanding then the size must be at least 840 × 840 mm. The upper surface must be at or above floor level. However, a point the surveyor should note is that under the current Building Regulations, if the hearth already exists and was built before 1 February 1966 the provisions described above need not be met as regards the projection of the hearth and size of the freestanding hearth.

11.8   Old fireplaces

When carrying out an examination of properties built in the seventeenth and eighteenth centuries the surveyor will find many different types of fireplaces, according to the age of the dwelling. Some early fireplaces of the inglenook type resemble caverns with large straight flues and sometimes the brick divisions or 'withes' between the respective flues have burnt away. The principles stated in Section 11.2 are, therefore, difficult to apply unless the design is to be radically changed by reconstructing the interior of the fireplace. In fireplaces of this type the flues are often too large and therefore require a greater air supply to prevent smoking. The surveyor will often find that the air flow has been reduced by various draught excluders around windows and doors. The air flow can be tested by opening a window or door and seeing if the fireplace still smokes. If this is the case then the remedy is to increase the air supply. This can be done by various methods including the following: .

- With a suspended floor the air supply can be increased by the installation of a metal grille immediately in front of the hearth.
- If the fireplace is situated on an external wall then grilles can be inserted in the walls on either side of the fireplace.
- The problem is more difficult to solve if the floor is solid and the fireplace is situated on an internal wall. In such cases an adequate air supply can often be obtained from an adjoining room or hall by inserting grilles on either side of the fireplace.

Inglenooks usually have excessively wide and high fireplace openings and are often fitted with a small fire basket which means that the fire is no longer in proportion to the chimney. Even with the installation of air vents these large fireplace openings may still emit smoke. This calls for changes in the design of the recess and flue. The opening can be reduced by raising the hearth and providing a metal hood fitted with a flue liner to reduce the chimney size.

When undertaking an examination of an old fireplace opening and flue it is advisable to check that no bressummer, ends of joists or beams are buried in the flue walls and if anything of a dangerous nature is found the details should be described in the surveyor's report.

11.9   Rebuilding

More complex problems concerning fireplaces and flue construction might entail a complete rebuilding. In such cases, it is advisable for the surveyor to consult the approved document J1/2/3 (Heat producing appliances) which is a practical guide to meeting the requirements of the Building Regulations.

11.10   Chimney stacks

Chimney stacks are often a neglected part of a building and being very exposed to the elements are thus susceptible to decay. A good pair of binoculars are sufficient when investigating chimney faults externally. Leaning stacks often have more serious internal defects. The best course of action here is to have the chimney swept and examined by using fibrescopes inserted through a predrilled hole in a mortar joint as described in Section 11.4. In such cases a builder's 'attendance' is usually necessary to provide extension and roof ladders etc.

Common defects

The defects to be considered are those arising from the following causes:

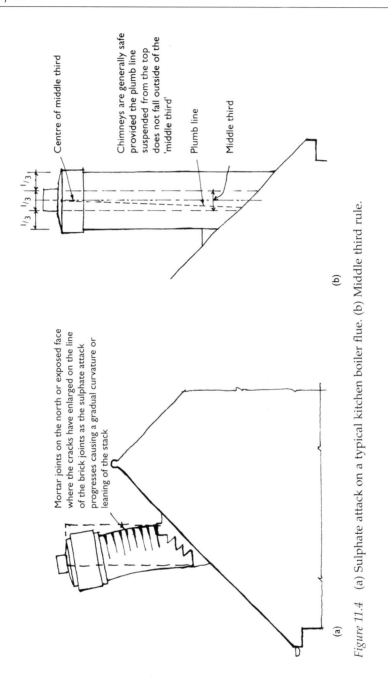

Centre of middle third

Chimneys are generally safe
provided the plumb line
suspended from the top
does not fall outside of the
'middle third'

Plumb line

Middle third

1/3 1/3 1/3

(b)

Mortar joints on the north or exposed face
where the cracks have enlarged on the line
of the brick joints as the sulphate attack
progresses causing a gradual curvature or
leaning of the stack

(a)

Figure 11.4   (a) Sulphate attack on a typical kitchen boiler flue. (b) Middle third rule.

(1)   As already mentioned, sulphate attack causes the mortar to expand and this leads to a gradual curvature or leaning of the stack which can be extensive on rendered stacks. The lining or pargetting can also be attacked by the products of combustion as they rise up the flue. Sulphate attack on a typical kitchen boiler flue is shown in Fig. 11.4a where the shrinkage cracks have enlarged on the horizontal brick joints as the sulphate attack progresses.

   A chimney stack which has begun to lean is generally safe provided a plumb line suspended from the top of the stack does not pass outside the 'middle third' of its area as shown in Fig. 11.4b. If the rate of lean appears excessive the surveyor should recommend that the stack should be demolished to the level at which it appears reasonably perpendicular and rebuilt using a sulphate resisting cement. If the chimney is old and is to be used in conjunction with a gas appliance and is not protected with a flue liner, the rebuilding of the stack will provide a good opportunity to instal a flexible liner.

(2)   Erosion of a brick or stone chimney by the action of wind, rain and frost may lead to crumbling or fractures and a general weakening of the stack. Small defective areas can be repaired, but badly damaged chimneys will often require rebuilding.

(3)   The damp-proof courses, back gutters, soakers and stepped flashings should also be carefully examined. Stacks situated on external walls at eaves level are most likely to give rise to damp patches in rooms just below roof level, due to the short distance between the roof coverings and ceiling.

(4)   Leaning pots indicate a fracture in the flaunching and probably of the brick courses below. These defects should be detectable by using the binoculars. On old buildings tall chimney pots were often used to prevent down-draught. They are frequently unstable and are best removed and other means recommended to prevent down-draught as discussed in Section 11.3. In the seventeenth century and onwards stacks were often finished with decorative chimney pots, which should be retained if they are of aesthetic merit. If badly deteriorated, the pots may have to be dismantled, but minor fractures can sometimes be repaired.

(5)   Where old chimneys are part of a classical design and considered to be of some architectural importance then it is advisable to retain the stack. It may be necessary to carry out a number of repairs, but if the stack is no longer to be used then careful consideration must be given to the matter of ventilation and cleaning of the flues in order to avoid damp, frost and sulphate damage. To ventilate a flue properly it should not be sealed at either top or bottom. When dealing with such cases it is advisable to recommend that some form of capping, such as a half-round ridge tile, or a cast concrete slab is laid over the top of the chimney supported on 300 mm high brick or stone piers at the

corners. A ventilator should also be provided in the blocked up fireplace opening.

(6)   In cases of extensive re-roofing and where the stacks are either redundant or structurally unstable, consideration should be given to demolishing the stack to below roof level and roofing over it and the top of the flue sealed with slate or a concrete slab. In such cases all associated fireplaces should be sealed off without venting the room. However, it is essential to ventilate to the outside redundant stacks situated on the exterior wall (BRE GBG 2).

(7)   When carrying out an examination of a damp stack, the adjacent roof timbers should be checked for wet rot and insect attack.

11.11   Industrial chimney shafts

Tall chimney shafts attached to industrial or large commercial premises may be constructed of brickwork, reinforced concrete or steel (self-supporting or with guys). Brick or concrete shafts are either freestanding or are constructed within a building. Steel chimneys are generally classed as temporary structures which are inspected and licensed for certain periods.

A great number of chimney shafts of all types have been erected in the past, many of which have been demolished. The design of brick and concrete shafts is either calculated by engineers or designed in accordance with the Building Regulations for 'uncalculated' brick shafts. All shafts should be lined with firebricks to protect the main structural materials from the effects of high temperatures.

A preliminary examination with binoculars will indicate the general condition of the external surfaces of the shaft. The points to be checked are as follows:

- The condition of the pointing and jointing.
- Any surface cracks or fractures to the brickwork or concrete.
- The condition of the chimney cap and oversailing courses (if any).

On completion of this preliminary examination the surveyor will be able to obtain a fairly accurate assessment of the situation and the possibility that the services of a specialist in this field may have to be employed to report on the high level aspects and internal condition of the shaft which is outside the professional knowledge of the surveyor. This information should be communicated to the client as soon as possible with some details of the specialist's fee as mentioned in Section 1.3.

There are firms who specialise in the examination of high level structures including chimney shafts. They employ experienced qualified men who examine and test the efficiency of tall chimney shafts using closed circuit cameras, video recording equipment and boro-scopes which enable them to inspect parts of the shaft which are inaccessible by any other means. This examination is followed by a full report and cost of repairs if found necessary.

12 Timber Upper Floors, Floor Coverings, Staircases and Ladders

12.1 Introduction

Suspended ground floors and floors above basements were dealt with in some detail in Sections 8.6 and 8.9. In the following the principal defects found in upper floors and staircases will be dealt with.

The usual type of timber floor consists of a single system of joists to which the floor boarding and other coverings are secured together with the ceiling material secured to the underside. This type of floor is suitable for domestic property and small commercial premises. However, where the span of the joists exceeds about 5 m or the load to be placed on the floors is heavy, the size of joists required would involve an uneconomical use of timber. In such cases the surveyor will no doubt find that the difficulty has been overcome by installing a double floor, that is, dividing the length of the room into bays by means of timber beams or RSJs (Rolled Steel Joists).

As described in Chapter 9 the most serious defect that is likely to occur in a suspended floor is dry rot. However, upper floors are rarely attacked, but outbreaks often occur in pipe ducts and cupboards which, if not detected, will spread into the timber floors. There is little or no danger of dry or wet rot when joists are built into the inner leaf of a well built cavity wall.

Timber upper floors are supported in the following ways:

- Joists bear directly on the wall.
- Joists rest on a timber wall plate built into the wall.
- Joists supported on a mild steel (MS) or wrought iron (WI) bearing bar built into the wall.
- Joists may rest on a wall plate supported on WI or MS corbels built into the wall and are usually spaced at about 700 mm centre to centre.
- In more modern properties the joists are supported on galvanised steel hangers. This method is useful where the joists derive their support from a party wall.

Over the last century considerable attention has been paid to the problem of sound insulation, above or below the floor. The usual type of joist floor supporting floor boarding with a lath and plaster ceiling below, resembles a drain; and sounds of footsteps and other noises may be transmitted to the floor system. It would hardly be right to call this transmission of sound a defect in the floor, but owners of the premises may well consider it a nuisance and may wish to reduce the noise transmission level as much as possible. Noises often reach the room below either as airborne sound, such as conversation, or impact sound such as footsteps.

After removing some sections of the floor boarding it is fairly easy to recognise the various types of insulation to enable a report to be prepared. A simple method is a layer of slag wool about 75 mm thick resting on the ceiling below as shown on the left-hand side of Fig. 12.1a. A better method which is often found in business premises is shown on the right-hand side of Fig. 12.1a and consists of an air space above and below the wool. The wool is supported on boarding which is in turn secured to the joists by timber fillets. A third method is shown in Fig. 12.1b and consists of a heavy plugging of sand on the ceiling with a resilient quilt draped over the joists known as a 'floating floor'. The principle underlying the design of a floating floor is its isolation from any other part of the structure. It is, therefore, important to ensure that the flooring is nailed to the battens to form a raft and not nailed through to the joists. The floor must also be isolated from the surrounding walls. These are points which require checking during the examination.

The surveyor will encounter many different types of staircases constructed of many different materials. Although timber staircases are satisfactory for domestic construction they do not meet the requirements of the Building Regulation regarding fire resistance for many blocks of flats and commercial and industrial buildings. Reinforced concrete stairs which have a higher resistance are commonly used where timber stairs would not satisfy the regulations. Stone stairs are rarely used today, but are often found as an entrance feature to buildings erected between the seventeenth and nineteenth centuries. Staircase defects will be considered in Sections 12.24, 12.25 and 12.26.

12.2 Structural timber floor defects

The surveyor will often find that the floors in unoccupied properties are fairly easy to examine. The removal of one or two boards in the centre of the floor and adjacent to the skirtings will give the surveyor a general idea of the construction. However, there are suspect areas around sanitary fittings and pipe ducts where wet rot defects are possible and should be carefully examined. Floors in occupied properties are often more difficult to examine due to permanent coverings such as fitted carpets or linoleum. Some occupiers will naturally object to large areas of their floor coverings being removed. In such cases it is advisable to look for areas where defects are possible and perhaps look for corners of the room where the lifting of the covering would not cause undue disturbances. Access to the interior of the floor can often be gained from access covers provided in the

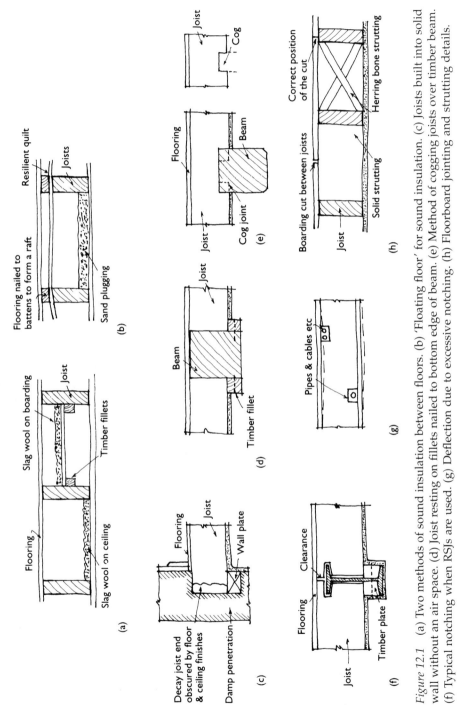

Figure 12.1   (a) Two methods of sound insulation between floors. (b) 'Floating floor' for sound insulation. (c) Joists built into solid wall without an air space. (d) Joist resting on fillets nailed to bottom edge of beam. (e) Method of cogging joists over timber beam. (f) Typical notching when RSJs are used. (g) Deflection due to excessive notching. (h) Floorboard jointing and strutting details.

flooring for gas or electrical services. In cases where there are problems concerning access it is essential to mention the fact in the report.

Common defects

The common defects are as follows:

(1)   One of the principal defects in a timber upper floor is excessive vibration which is easily detected by bouncing under foot. Vibration is usually caused by decay at the ends of the joists and therefore weakening their bearing. Flexing or sagging may be caused by the original joists being too light for the span or by unequal settlement in the external walls or partitions. A floor may need strengthening, not because of any defect in itself, but because it is being made to carry a load greater than that for which it was designed. In such circumstances the floor may sag to a perceptible degree. When dealing with vibration or sagging joists it will be necessary to assess the precise cause of the failure and this can only be done by checking that the joist size and spacing are adequate for the span and load involved. An approximate guide to the depth of timber joists in inches is by using the rule of thumb method of span in metres over 24 plus 50 mm assuming that the thickness of the joists is 50 mm at approximately 400 mm centres. Alternatively, the sizes for domestic loadings can be checked with the spans and spacings given in the Building Regulations. The majority of these floors can be checked by one of these methods, but there are occasions when it is useful to be able to calculate the load that will rest on the members of a flooring system and the size of the joists necessary to carry the load safely.

(2)   If the floor construction has been checked as described in (1) above and is found to be satisfactory then the cause of the defect must be sought elsewhere. A gap between the bottom of a skirting and floor board should be suspect. It is likely that there has been movement in the bearing ends of the joists. The next stage is to open up the floor adjacent to the bearing ends of the joists and ascertain the condition at the point of support. Timber joists or bressummer beams built into solid brick walls of old buildings can rot due to damp penetration. If the joists are built into a wall 225 mm thick without an air space around their ends, they may absorb moisture and, being unventilated, will rot at their seating, although the other parts of the joists may be sound (see Fig. 12.1c). The decay usually affects the portion of the joist built into the wall which can be cut off and the ends of the joist picked up on metal joist hangers.

(3)   As mentioned in Section 12.1 the joists in double floors rest on timber beams or RSJs to reduce the span and the manner in which they do so depends on the type of workmanship that has been put into the construction. A bad form of construction which is sometimes found

in older buildings is a fillet nailed along the bottom edge of the beam with the joist resting on the fillet as shown in Fig. 12.1d. In such cases the nails in the fillets are supporting the joists and if they should fail the joists would collapse. When carrying out an investigation to such a floor, it may be found that the nails have given a little, but not necessarily caused a dangerous condition. If this is so the floor can be strengthened. It is better practice to cog the floor joists over the beam as shown in Fig. 12.1e. Figure 12.1f shows a typical notching where the beam consists of a rolled steel joist. The timber joists are supported on timber plates secured to the web of the RSJ, but are kept clear of the rest of the RSJ to allow for expansion and contraction of the timber. In some cases the top of the joists are flush with the top of the RSJ causing the floorboard situated over the top flange to curl due to the fact that the only fixing is the tongue and groove in the jointing of the floorboard.

(4)   A more serious problem is settlement in a brick or timber stud partition and this is often found in eighteenth and nineteenth century buildings. This defect arises when the partition runs parallel to the front and rear walls and supports the floor joists. The situation may be complicated due to the fact that it may be general throughout a number of floors on either side of the partition. In such cases it may be possible to jack the partition and floors back to their original positions and underpin the lower partition, but this could damage other parts of the building, particularly the roof. The whole problem should be carefully considered before reporting to the client, who would naturally require some detail of costs. Settlement can also occur in a partition that is supported on a single joist which is often too light in the first instance or the timber has deteriorated over the years. The remedial work might well consist of simply doubling the existing joist.

(5)   The centre of the span should be checked for some form of strutting. Although strutting is usually fitted between joists, it may happen that in very old timber floors it has been omitted, in which case undue vibration will occur. This is due to the joist tending to tip sideways under movement, especially if the floorboards have not been nailed to every joist. Also the end wedges between the wall and the last joist may have been omitted or have fallen out. If this is the case, considerable improvement can be effected by taking up the flooring and inserting a row of solid strutting down the centre of the span (see Fig. 12.1h).

(6)   Another problem which could be attributed to deflection in old floor joists is that excessive notching of the joists may have been made to accommodate gas pipes, electrical cables or water pipes (see Fig. 12.1g). If the surveyor feels uncertain on this point it is well to consult the British Standard Code of Practice BS 5268-2:2002 which sets out rules concerning safe notching in joists.

(7)   Trimming around stair wells and fireplace openings should be checked and in many cases, cracks in ceilings will be found following the line of the trimmers. The trimmers are usually 25 mm thicker than other joists and should be connected by a tusk tenon joint. However, the surveyor will often find that the joints are not always constructed in accordance with sound practice and if serious movement has taken place some repair work will be necessary. This operation will no doubt involve renewing portions of the ceiling below.

(8)   In modern properties metal joist hangers have been used extensively. Unfortunately, recent surveys have found instances where the hangers or joists have failed due to one or more of the following causes (see Fig. 12.2a and b).

- Hangers should be tight against the face of the masonry.
- Incorrect grade of hanger used. It is important that hangers are not used with masonry of lower strength than the hanger grade marking. Hangers should comply with BS 6178 Part 1 and are made for minimum masonry crushing strengths.
- Sometimes joists are cut short leaving a large gap between joists and back plate of hanger. This may cause hangers to move and the floor to settle.
- Check to ensure that no damaged or corroded hangers have been used.
- Check that the undersides of the joists are properly notched to accommodate the hanger flanges.

FLOOR COVERINGS

12.3   Introduction

There is a wide variety of floor finishes. Some are only suitable for domestic use; others for commercial situations or where subject to heavy traffic in industrial buildings and therefore only suitable for a concrete sub-floor. In the following several of the most common types are described which the surveyor will meet with during his examination of ground and upper floors.

12.4   Boarded floors

This type of flooring is easily checked by inserting a sharp instrument such as a penknife into the joints to see whether they are tongue and groove or plain edge. The boards are usually 25 or 31 mm thick and 100 to 177 mm wide.

Where traffic over the floor is heaviest and keeps to a more or less definite path, such as at door openings, the top surface of the boarding will

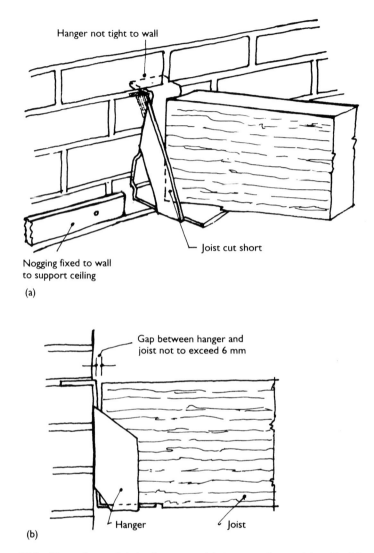

Hanger not tight to wall

Joist cut short

Nogging fixed to wall
to support ceiling

(a)

Gap between hanger and
joist not to exceed 6 mm

Hanger   Joist

(b)

Figure 12.2 Use of metal joist hangers: (a) incorrectly positioned; (b) correct positioning.

be worn more than in other parts of the room. If the wear is considerable the thinness of the boards will cause them to bend under the tread. In softwood flooring, the grain usually runs along the length of the boards, exposing expanses of softer wood divided by lines of hard grain. Foot traffic along the grain wears away these softer portions more quickly than it does the harder, so that the hard grain eventually stands up above the surface of the soft grain.

The surveyor will sometimes find that the ends of the boards meet between joists forming a cantilever as shown on the left-hand side of

Fig. 12.1h and will spring under weight. The cut should be made at the middle of the joists as shown on the right-hand side of Fig. 12.1h.

Another condition that may arise is the curling upwards of the edges of the boards due to the fact that the timber swells when it absorbs moisture and shrinks when the moisture dries out. Thus the board is trying to reduce its width and as the upper surface is exposed to the warm air it dries more quickly than the under surface. Therefore, the board tries to reduce the width of the upper surface to a greater degree than its under-surface and the only way it can do this is by curling. When the boards are covered with linoleum or carpet, curling will cause ridges to appear which will wear the covering more quickly than in other parts. From this point of view the surveyor may consider curling to be a defect which can be remedied by planning or sanding of the raised edges.

12.5 Chipboard flooring

Chipboard flooring provides a satisfactory floor decking if the correct grade is used and the boards are securely fixed. However, several defects have been found during the past few years. The condition of the boards should be checked where practicable. The surveyor may well find it diffi-cult to remove large panels of chipboard without causing damage, and it should be noted where this is not possible. As described above for boarded floors, the ends of the boards will sometimes be found to meet between joists forming a cantilever and will often 'spring' under weight. Particu-lar attention should be given to chipboard floors in wet areas such as kitchens and bathrooms. Boards in these areas should be protected and be moisture-resistant and marked BS 5669 type 11/111. If the boards are per-manently wet then expansion and loss of strength will occur.

12.6 Hardwood strip flooring

The boards are usually cut from one of the more decorative hardwoods and are in widths of 75 to 100 m. When laid on a suspended timber floor, a strip flooring is not likely to suffer from the effects of damp, but if defects are noticed, they will probably be due to ordinary wear. However, if the flooring is laid on a solid floor at ground level defects may be caused by damp rising from the ground and much will depend on the manner in which the fillets have been laid. Flooring in contact with a concrete base needs to be protected by a damp-proof membrane and many failures have occurred in older buildings due to the omission of such damp-proofing. It will often be found that ventilation has not been provided, as indeed it cannot be, if the fillets are embedded in the concrete or the spaces filled in with fine concrete. If on opening up the floor the fillets are found to be resting on top of the sub-floor without any filling and no form of ventilation, then a recommendation should be made in the report that small openings are cut out of the fillets at frequent intervals to allow air to pass under the floor. Gratings can be fitted in the skirtings or at the extreme edge of the floor to allow the air from the room to pass under the flooring. Should such flooring need to be taken up and relaid because of an outbreak of dry rot then this provision of ventilation should be made.

12.7　Wood block

Wood blocks may be laid in a variety of patterns, a very common one being the herringbone pattern, and they are available in several varieties of hardwood such as oak and teak.

Defects in wood block flooring are usually the result of long and hard wear, although a common defect is moisture penetration causing expansion of the blocks. The blocks are dry when laid and are, therefore, ready to absorb any moisture in the atmosphere or through failure to provide an adequate DPM under the blocks. This causes the blocks to expand and if there is no room for expansion the flooring will lift and arch up. It is, therefore, important to investigate the base. The blocks should also be examined for any sign of fungal attack. It is also necessary to check that expansion joints have been provided, for example, a cork strip around the perimeter of the floor. The skirting or kicking fillet will cover the expansion joint.

12.8　Floor screeds

In order to apply the final floor finishes to a concrete floor it is necessary to have a smooth level surface and this is done by the application of a cement and sand screed. The screeds can be monolithic, that is, laid on the concrete base whilst it is still green, or superimposed after the concrete has dried. Problems can arise with the 'superimposed' screeds. As the screeds dry out they tend to shrink and if not restrained by a good key to the concrete base they will crack. Cracking is the most common fault experienced with all screeds accompanied with hollowness. The problem can usually be detected by tapping after removing the loose tiles and timber flooring.

12.9　Granolithic paving

Where hardwearing qualities are essential granolithic paving is widely used, particularly in industrial premises. The usual mix is two parts Portland cement to five parts granite chippings. The surface is often treated with an application of sodium silicate to prevent dusting and carborundum is sprinkled over the surface to obtain a non-slip finish. As with the cement screed, granolithic requires to be well bonded to the concrete base. If the bond is inadequate, then cracking and lifting of the paving will occur. In extreme cases it may be necessary to recommend taking up all the old granolithic and replacing it with new.

12.10   Terrazzo

This is a special form of finish using coloured cement and crushed marble chippings. Terrazzo is easy to clean and has a high resistance to abrasion. However, the surveyor may find a few problems and the most common of these is crazing or cracking due to differential shrinkage in the screed base and the terrazzo. This defect is prevalent in terrazzo laid *in situ* and to avoid such problems there has been a greater use of terrazzo tiles.

12.11   Cork tiles

Cork tiles are manufactured by moulding cork granules compressed hydraulically under heat into tiles. They are bedded in an appropriate adhesive and treated with a sealer or wax polished. They have good wearing qualities but can be affected by grease. Lifting may occur on softwood flooring where an underlay of hardboard has been omitted.

12.12   Linoleum

In the past linoleum was by far the most commonly used floor covering particularly in domestic work, but during the past fifty years it has been largely superseded by carpets, PVC and vinyl asbestos tiles. Sheet linoleum should comply with BS 810 or BS 1863 for felt backed linoleum, the latter type having a greater moisture stability. Defects found in linoleum are principally cracking and lifting, apart, of course, from wear due to long use. If the sub-floor is of timber boarding which tends to curl at the edges as described in Section 12.4, then the linoleum is forced up into ridges which soon crack and wear through under traffic. If the linoleum is laid on a cement and sand screed then it must be bonded to it; a damp-proof membrane is, therefore, essential. If damp reaches the surface the adhesive is affected, adhesion broken, and the linoleum will lift at the edges. In such cases a permanent cure cannot be expected unless the cause of dampness has been removed, or its effects prevented.

12.13   Rubber flooring

Rubber flooring should comply with BS 1171 and may be in sheet form or tiles. It is made in varying thicknesses and is fixed with an adhesive. Rubber flooring is vulnerable to grease, oil, fat and petrol. However, there are a number of synthetic rubber floorings which are resistant to these substances. Rubber has good wearing and sound absorption qualities. Troubles that may be found are the usual ones of lifting at the edges or general loosening. It is not easy to obtain good adhesion between rubber and a cement screed which is in the least damp. Any problem with rising damp through a concrete ground floor slab will cause moisture to collect between the top of the concrete and the underside of the rubber. This moisture will break down the adhesion and it is pointless to stick down the lifting edges of the rubber if the cause of the trouble has not been removed.

12.14   Thermoplastic, PVC and vinyl asbestos tiles

Thermoplastic floor tiles should comply with BS 2592 and PVC (vinyl) tiles with BS 3261. Thermoplastic tiles should be examined for any signs of deterioration due to being in contact with oil or grease. The PVC floor tiles are more resistant to oils and grease and to abrasion, but can be damaged by cigarette burns. Vinyl asbestos tiles should comply with BS 3260. They have reasonable resistance to oil and grease but can be marked by cigarette burns and any hot object placed on the tiling. However, given the health risks associated with asbestos, the surveyor should recommend that if these tiles are present, they should be overlaid or removed (by an approved specialist asbestos removal contractor).

All these finishes are usually applied to cement and sand screeds and the commonest cause of failure is due to moisture penetration passing through the screed. The moisture contains alkalis derived from the concrete which attack the adhesives used to secure the tile flooring to the screed. Tiles are often more marked by rubber footwear or castors on the legs of furniture.

12.15   Clay floor tiles

Clay tiles are described in BS 1286 and should be laid in accordance with British Standard Code of Practice BS 5385-3:1989. Two types of tiles are specified in BS 1286: Type 'A' floor quarries and Type 'B' semi-vitreous and vitreous. Many failures are manifested by arching of the tiles due to the differential movement between the screed and the tile. Clay and concrete have different coefficients of thermal expansion which causes different movements with temperature changes. The worst conditions occur when the tiles are bonded direct to new screeds and insufficient time has been allowed for the screed to dry out. When dealing with such cases the surveyor may consider it advisable to recommend that the tiles should be replaced over a separating layer consisting of building paper or polythene sheeting. The purpose of the separating layer is to isolate the tiles and bedding from the screed and thus prevent stresses in the latter from affecting the floor tiles. An important point to ascertain is whether or not expansion joints have been provided around the perimeter of the tiled floors either between the tiles and wall or if a coved skirting has been installed between the tiles and cove.

12.16 Concrete tiles

Concrete tiles are made under pressure in moulds and are cured and dried under controlled conditions. The tiles usually have a wearing layer approximately 6 mm thick with a backing of fine concrete. For non-industrial use, fine pigmented mixtures are used to give a colourful appearance. For industrial use where a tough wearing surface is required, hard natural aggregates or metallic aggregates are often used. The tiles are laid on a bedding of mortar. In some cases, the surveyor will find that the tiles have lifted due to the fact that they have been bonded direct to the base without the provision of a separating layer such as building paper or polythene sheeting. The quality of the jointing should also be examined to ensure it is firm and completely filled.

12.17 Magnesite flooring

This type of flooring is not commonly used, but is not yet extinct and is often found in commercial and industrial buildings (see Pye and Harrison 1997). The material is a mixture of magnesium oxychloride and fillers such as sawdust, wood flour, ground silica, talc or powdered asbestos and should comply with BS 776. The flooring is available in various colours and in mottled or grained effects. The thickness varies from 10 mm for single coat work to 50 mm for two or three coat work. Magnesium oxychloride is particularly susceptible to moisture and will soon deteriorate if damp penetrates the finish from below. This defect will cause cracks and the breaking up of the surface, usually due to the omission of a damp-proof membrane. Magnesite floors also take up moisture from the air and thus have a tendency to sweat. Metal is liable to corrode when in contact with this type of flooring. The surveyor should carefully check service pipes for corrosion if in contact with magnesite flooring. In serious cases of disintegration it will be necessary to recommend complete replacement and for the service pipes to be protected with bitumen or coal-tar composition.

12.18   Mastic asphalt and pitch mastic paving

Mastic asphalt for flooring should comply with BS 1410 (natural rock asphalt), BS 1076 (limestone aggregate) or BS 1451 coloured mastic asphalt (limestone aggregate). Other grades are available for special industrial purposes such as acid resisting construction.

Mastic asphalt provides a jointless floor which is impervious to moisture. However, concentrated loads which are often found in commercial or industrial premises may cause indentations in the surface. It is, therefore, advisable to recommend supports for heavy point loads. This can be done by the use of hardwood blocks laid on the sub-floor and set in the paving. Mastic asphalt is also liable to soften with prolonged contact with oil, petrol or grease. Polishes containing oils are injurious and may cause softening and affect the colour. Pitch mastic paving should comply with BS 1450 for black pitch mastic paving or BS 3672 for coloured pitch mastic paving. The material is suitable for many conditions from domestic premises to heavy duty industrial paving. Treatment of the base and conditions generally are similar to those for mastic asphalt.

If the mastic asphalt or pitch mastic paving is badly cracked or has lifted from the sub-floor and requires renewing, then it is possible that there are problems with the concrete slab. In such cases it is advisable to seek the advice of an asphalt specialist, with regard to the preparation of the sub-floor and the right grade of asphalt to be used. Where the floors are liable to become hot e.g. over boiler chambers, they tend to soften due to excessive heat. In such cases it may be necessary to recommend some form of insulation below the floor. All asphalt pavings should have an insulating membrane of black sheathing felt to overcome the effect of movement especially on timber or porous bases.

12.19   Rubber latex cement flooring

This type of flooring is composed of hydraulic cement, aggregate or fillers of fine chippings, granules of cork or wood chips. Aggregates for harder grades consist of crushed marble or granite. The mixture is gauged on site with a stabilised aqueous emulsion of rubber latex and usually laid to give a thickness of about 6 mm. If mixed with high alumina cement the flooring has good resistance to attack by sulphates, weak acids and sugar solutions. The flooring may be laid on a timber base only if it is completely rigid. Any movement in the timber structure will result in cracking. This type of flooring requires regular cleaning and polishing.

12.20   Metal tiles

Metal tiles are usually about 300 mm square consisting of 10 gauge steel plates and are mainly used for heavy duty industrial floors. They are pressed into newly laid concrete. The tiles are hard wearing and resist all forms of abrasion and impact. Oil and grease may leave stains, but regular cleaning is all that is necessary. If repairs are required they are easily replaced, but in such cases the concrete sub-floor may need some attention.

12.21   Slate

Natural slate for flooring can be cut to various sizes and thicknesses. Slates up to 380 mm square are usually 19 to 25 mm thick and are laid upon a screed bedded and jointed in cement mortar. Slate flooring is extremely hard wearing and requires the minimum upkeep. Dirt cannot penetrate the surface, but oil and grease leave stains upon the surface if not regularly cleaned. If some slates are loose it is usually due to shrinkage of the concrete base.

12.22   Marble in tile or slab form

Natural marble is usually 19 to 25 mm thick and tiles or slabs are, are up to 900 mm square and are laid as described above for slate. They are impervious to water and have good resistance to oil, fat and alkalis. They are obtainable in a wide range of colours and form a very durable surface. As for slates, if the slabs are loose it is usually due to shrinkage of the base.

12.23   Conclusion

The condition of all floor coverings should be clearly stated in the surveyor's report together with advice on any remedial measures which he considers necessary. Particular attention must be given to defects caused by damp penetration where floor coverings are attached to solid floors. The most important factor to be considered is to ensure that there is a link with the DPC (see Fig. 8.2c). This is often omitted and has been the cause of damp problems in many post-war buildings. Underfloor heating systems often cause a high shrinkage rate which seriously increases the risk of cracking. This defect is prevalent where concrete tiles have been used. The surveyor will often find that the surfaces are covered with fine cracks. Granolithic can also be affected by underfloor heating if not laid monolithically.

STAIRCASES AND LADDERS

12.24   Timber staircases

Generally the best constructed piece of joinery in any building is the stair. They seldom develop any serious structural defect, repairs being confined to renewing or making good worn members. The majority of domestic staircases are constructed of timber. Generally, one string abuts against a wall, while the other, called the outer string, is tenoned to the newel posts which support the handrail. If the width of the stair is over a metre an intermediate support should be inserted in the form of a carriage piece, to which are nailed brackets to support the centre of each tread. The treads and risers should be tongued and grooved together with their edges glued and screwed, although in cheaper work they are merely butted together and nailed. The ends of the treads and risers should be housed and securely wedged into the strings. The nosings are often supported by a scotia moulding housed into the tread or butted against its surface. To give additional rigidity to the treads all internal angles on the underside of a stair should be fitted with glued angle blocks. If the soffit of the staircase is open the surveyor will find that the construction is fairly easy to examine. Plaster soffits should be checked to ensure that there is proper adhesion to the underside of the staircase.

Common defects

The following points should be checked:

- Creaking in stairs is due to the rubbing of loose members and can generally be traced to loose or missing angle blocks between treads and risers. The wedges securing risers and treads let into the strings should be tight and glued and the glued angle blocks between the treads and risers should also be checked.
- The tongued joint between tread and riser should be screwed from below. If this joint is not secure there is a tendency for the stair to squeak.
- Damaged treads and nosings should be carefully examined and reported. Damaged nosings are particularly dangerous.
- Strings should be at least 40 mm thick and properly housed to the newel posts.
- Balustrades should be checked for stability. The main defects found are cracked or loose balusters and handrails. Newels are often found to be loose due to poor fixings to strings and aprons.
- Handrails fixed to wall surfaces are usually supported by metal handrail brackets plugged and screwed to the wall. The fixings should be checked for rigidity.
- The surveyor will often find that the bottom riser of a staircase in an old basement will be affected by dry rot or rising damp from the wall

or floor. The bottom riser should be carefully examined with a bradawl or knife.

12.25   Metal staircases and ladders

These are usually found in industrial premises and are used as a means of access to roofs and other high places and as a means of escape. In most instances the design and erection is entrusted to special firms. As with structural frames the chief point of interest to the surveyor is the rigidity of the joints and the protection provided against corrosion. Care should be taken that all bolts are secure and that the method of protection is in accordance with the recommendations of the British Standard Code of Practice (BS 5395-1:1977).

12.26   Reinforced concrete stairs

Reinforced concrete stairs which have a higher fire resistance are commonly used where timber stairs would not satisfy the regulations. The size and spacing of the reinforcement is calculated according to the required conditions of load and span. Defects in reinforced concrete have been dealt with in Chapter 7. Sometimes a small surface crack in a concrete stair need not give rise for concern, but if a step has cracked through it can be a sign of structural failure and will require investigation. It must first be examined to ascertain if it is spanning between walls or other supports as a beam. The stair should be relieved of its load by shoring up the underside. Damaged concrete can then be hacked away and the bars exposed for examination. The area should be carefully examined to see that the adhesion between steel and concrete has not been destroyed.

The back of the tread is usually covered with one of the materials mentioned under 'floor coverings' and finished at the front edge with an extruded aluminium nosing with non-slip inserts. The treads should be in good condition without excessive wear. Balustrades and handrails are usually of metal or plastic coated metal, and should be checked for stability and decorative condition.

13 Finishes and Joinery Externally and Internally

13.1 Introduction

The subject of finishes is so complex and changes so rapidly with the invention of new materials and methods of application, that it is not possible within the limits of this book to treat it comprehensively. It seems better, therefore, to identify the nature and cause of the defects occurring in individual situations and materials in common use at the present time.

When dealing with finishes it is the physical condition of the building which must be the surveyor's main interest and not the aesthetic effect of colour etc., except where the decorations form an important feature. For example, decorative wall panelling or an ornamental plaster ceiling may well be considered important to the purchaser, and should, therefore, be examined and its condition reported. Finishes sometimes cover defects below the surface, and should be examined critically to see if there is an underlying defect. In such cases the condition of the surface decoration must be ignored.

PLASTER

13.2 Types of plaster

Plaster defects are met with in all types of plasterwork, mainly because the leisurely technique of former times has given way to a demand for speed in all building operations. The more common defects which the surveyor will encounter are discussed in Section 13.5, but first a few brief notes on the materials used are given which may help to clarify the explanation of the defects.

Plastering provides a smooth, hard and hygienic surface to walls and ceilings. It can also be used to increase fire resistance and to improve the thermal insulation and acoustic properties of a building.

When carrying out a survey of an old building the surveyor will no doubt find that the walls and ceilings are lined with lime plaster consisting of lime and sand applied in two or three coats. Animal hair was included in the mix to restrain the shrinkage. The mix, however, was weak, easily damaged and would shrink on setting. Properties built from the 1930s onwards will probably be plastered with gypsum plaster and a wide range of plasters of this type are now available to suit different backgrounds, and to meet special requirements concerning sound absorption, condensation and protection against X-rays.

13.3 Plasterboards and wallboards

Both boards are well established as a substitute for the old type of timber lath and plaster. They can be finished with one or two coats of plaster or decorated direct. Plasterboard consists of gypsum plaster encased in a double paper liner and is used as an internal liner for walls and ceilings, and for encasing columns and beams. There are seven basic types of board available:

- Wallboard and plank is available with two types of edge: tapered for smooth seamless jointing and square for cover-strip jointing or plastering.
- Gypsum lath and baseboard are gypsum plasterboards designed as a base for gypsum plaster. Both types are supplied in relatively small sheets and are used for lining timber joists, studs and battens.
- Gyproc vapour check wallboard is a normal plasterboard with a metallised polyester film backing which provides water vapour resistance and helps prevent interstitial condensation.
- Gyproc moisture resistant board is for external use in protected situations or where exposure to external conditions exists during the construction process.
- Gyproc thermal board is a laminate composed of gyproc wallboard bonded to expanded polystyrene. It is used as an insulating lining to walls and ceilings.
- Gyproc 'Fireline' board is a gypsum plasterboard with an addition of glass fibre to improve the fire protection properties of the board. It can be used as a suspended ceiling or floor construction to increase their fire resistance.
- Industrial grade board consists of gyproc plasterboard faced with white PVC film to improve the thermal insulation.

13.4 Metal lathing

Metal lathing often referred to as 'expanded metal lathing' has now superseded wood lathing. There are several proprietary designs, most being pressed out of sheet steel to form a small mesh. The lathing should be protected by bituminous paint or by galvanising. For suspended ceilings spanning up to 1.5 m a self-supporting type of metal lath is used having V-shaped ribs formed in the sheets. Special plasters are used for metal lathing incorporating a rust inhibitor.

13.5   Plaster wall and ceiling defects

Some common defects in new and old plastering due to a breakdown in the plaster itself are given below:

Popping or blowing. Small blisters can appear on the surface and subsequently fall away leaving pits in the finished surface. It will be found that at the base of every pit there is a small fragment of lime or sand which has expanded after the plaster has set.

Plaster soft and powdery. Slow setting plasters dry before the setting process is complete. This may happen generally or in patches. This trouble is usually associated with strong finishing coats. In several cases the whole of the finishing coat may lose adhesion.

Fine hair cracks. Small hair cracks are often caused by the use of loamy sand if the work is in gypsum plaster or excess lime in the finishing coat. Plastering mixes containing cement or lime shrink as they dry and calcium sulphate mixes expand as they set causing differential shrinkage between backing and setting coats. If cracks follow a definite line particularly on building boards it is usually due to poor treatment at the joints or shrinkage of timber joists or battens. Similar effects are sometimes due to thermal movement.

Loss of adhesion. This defect is often caused by a strong final coat over a weak undercoat or an inadequate 'key'. Occasionally, the wrong type of board is used, such as a plasterboard intended for direct decoration.

Presence of soluble salts. The salts are usually seen on the plaster face having been brought forward from the background to which the plaster has been applied.

Mould growth. This type of growth usually occurs on new plaster. The spores will develop if dampness is present, but when the plaster has dried out the growth will stop and any mould can be removed.

Rust staining. Sometimes due to application of unsuitable plaster to metal lathing or plaster in contact with corrodible ferrous metal in damp conditions.

Flaking or lifting of top coat. Excessive moisture penetration through the background.

Chalkiness. This defect can result from several different causes. The final coat is subjected to either excessive heat, draughts during setting, excessive suction of the undercoat or undue thinness of the final coat.

The surveyor will often find that plastering defects may be due to causes other than the use of faulty materials or workmanship. Damp penetration through a solid external wall may cause flaking, blistering or a breakdown of the plaster 'key'. An examination externally with the aid of the moisture meter will usually provide an answer to the problem.

Plasterboard linings to the inner face of the external walls are often used as a substitute for plaster. They can be finished with one or two coats of plaster or decorated direct. In all cases the quality of the work depends upon the degree of care exercised in fixing the boards and making the joints. Given dry conditions, the method of battening out is satisfactory if the spacing is carried out in accordance with the manufacturer's instructions. In addition the bearers must be sufficiently rigid to prevent movement at the joints caused by impact or pressure. This method of lining a wall is perfectly sound, but in thin solid walls in an exposed position, it can prove to be ruinous when damp penetrates the wall. If no ventilation has been provided behind the board dry rot will no doubt develop. Failures of this type are most likely in older properties and pressure and tapping of the plasterboard surface will indicate whether the battening has deteriorated; if suspect it is advisable to remove one of the panels for further examination.

A defect which sometimes occurs within a year of plastering is often due to shrinkage. Blockwork tends to shrink as it dries out, especially if it is wet when laid. Plaster undercoats consisting of cement, lime and sand tend to shrink with the blockwork causing hollowness. Drying shrinkage also occurs in the finishing coat of gypsum plaster causing it to break away from the undercoat. Joints between structural columns and beams and internal partitions also tend to crack due to differential movement between the different backing materials. When carrying out an examination of some older commercial or industrial premises the surveyor will often find that the external angles of plastered walls or columns have been damaged by impact. In such cases it is advisable to recommend the provision of a metal or plastic angle piece to protect the plaster.

Much decorative plasterwork in large public buildings is often carried out in fibrous plaster sometimes known as 'stick-and-rag' which is prepared in a precast form on a suitable backing and fixed in units on the job. The moulds are filled with plaster of Paris reinforced with fabric to the back of which timber laths have been attached to facilitate fixing. Apart from a few hairline cracks the external face of the plaster usually appears sound, but if the fabric or timber laths become damp due to water penetration from leaking roofs or defective water pipes then the units may have become detached from their backing. If the jointing material between the units is loose or there are signs of damp staining on the plaster face, then in all probability there are faults in the backing material which should be investigated by removing one of the units.

13.6   Old plaster ceilings

To express an opinion on an old plaster ceiling can present difficulties for the surveyor because many show no visible sign of any defect when examined from below, and the plaster face is often obscured by lining paper. Plaster ceilings tend to bulge or crack more readily than wall plaster, mainly due to structural movement in the supporting timbers or to fungal decay. The construction of the ceiling will have been established during the inspection of the floor or roof timbers. In older properties built before 1930, timber lath and plaster will predominate, replaced after that date by plasterboard or gypsum lath. A plastered ceiling may crack because of excessive deflection in the floor or ceiling joists. The plaster not being elastic cannot give with the movement, and is, therefore, inclined to crack. Any indication of bulging or fractures should lead to examination of the ceiling from above. The backs of old ceilings are usually covered with dust; the first operation is therefore a thorough cleaning. Hand cleaning can be done with a soft-haired brush, but a small vacuum cleaner is much better if one is available. After the cleaning operation the cause of the defect can then be fully recorded. Failure in old lath and plaster ceilings is usually due to one or more of the following causes:

- Breaking of the 'key' in an old lath and plaster ceiling is rare. The majority are sound because the hair was generously applied. Also, the first coat of plaster wraps itself securely over the laths making the ceiling 'one piece' with them. Occasionally, laths are found to be set at too close a spacing which does not allow sufficient plaster to be forced through the laths to form a hook shaped key.
- Lathing nails if not galvanised will sometimes cause the laths to become detached from the joists.
- Failure is often due to decay in the laths either by damp penetration or woodworm.

One of the problems the surveyor will encounter is the failure of an ornamental ceiling or cornice in a 'listed building'. In such cases the owners are under a statutory duty to preserve it. An examination from below and above is essential and the steps to be taken concerning any necessary repairs must be carefully considered.

As previously mentioned in Section 3.2 plaster walls and ceilings which have lost their 'key' should be carefully examined by tapping the surface for 'hollow' areas and the defective parts marked out. If large areas are loose it is often found more economical to replaster a complete ceiling or wall.

SHEET LININGS

13.7   Types of board

The types of board now available vary considerably in their properties and behaviour. Different types of sheet have certain characteristic advantages, such as durability, fire resistance, sound insulation and resistance to the effects of moisture. Some of the principal types the surveyor will encounter are briefly reviewed below:

(1)   Hardwood and pine panelling are traditional internal finishes and are often used in domestic work for their aesthetic and acoustic qualities.

It is one of the major crafts in building and many fine examples will be found in old buildings.

(2)   Fibre building boards have considerable advantages. They have good insulating value and when fixed with an air space to ordinary walling the heat insulation is greatly improved. The lighter boards are known as insulation boards and the heavier as hardboards. Hardboards are denser and thinner than insulation boards and have a smooth glossy surface on one side and are suitable for dadoes and panelling. Super grades have greater strength and additional moisture resistance which makes them particularly suitable for use under the most exacting conditions.

(3)   Resin bonded chipboard is much used in joinery work and can be used as a wall or partition lining. The boards are made of woodchips bonded with synthetic resin. Chipboard is a rigid board and can be used in large panels for partitioning. It is obtainable with a melamine face made as an integral part of it, and is widely used for kitchen units, sliding doors and panellings.

(4)   Plywoods are of three types:
- normal commercial grades and those faced with selected veneers
- metal faced plywood
- resin bonded variety which does not disintegrate under damp conditions and is much stronger than the normal grade (WBP – water and boil proof)

One of the advantages of plywood is that it can be faced with veneers of the more expensive timbers such as oak or mahogany. Panelling in this material can be indistinguishable from the solid timber panels described in (1) above.

(5)   There are several lightweight non-combustible building boards for use in a wide range of internal and semi-exposed applications. The boards are composed of calcium silicate reinforced with selected fibres and fillers. The boards provide a half hour fire resistance, but some high performance boards can provide up to four hours' fire resistance. This type of board is also available as a bevelled edge panel for suspended ceilings using a concealed grid.

(6)   Phenolic foam boards complying with BS 3927 have an exceptional fire performance and provide an excellent lining for upgrading the insulation values of brick, timber or concrete structures. The boards are applied to most walls by a range of mechanical and adhesive fixings to suit various conditions.

(7)   Woodwool slabs which consist of tangled wood shavings bound together with a cement slurry give a high degree of thermal insulation. Normal plastering can be applied readily, but it is essential to fix the slab rigidly and cover the joints with a hessian scrim. The thin slabs are mostly used for insulation; nailed to studding, held in metal

channels or laid as roof decking. They are sometimes used unplas-
tered to the underside of roofs for acoustic reasons.

(8)   Asbestos cement sheets or asbestos insulation boards are no longer
employed as a lining material. Present factory regulations recognise
asbestosis as an industrial disease which could arise from the inhala-
tion of asbestos dust. However, the surveyor engaged in the survey
of an older building will occasionally find asbestos panels with a
glazed finish.

13.8   Common defects

Modern sheet linings are carefully manufactured and checked so that it is
unlikely that defects will occur in the material. The most usual troubles
associated with sheet linings are concerned with jointing, fixings or damp
penetration through external walls and are described below:

- In old buildings the surveyor may well find that the back of the
panelling, being unventilated and close to the wall surface, may be
affected by damp finally leading to an outbreak of dry rot. Many properties built during the past sixty years are likely to be panelled with
plywood finished with a hardwood veneer. It is advisable to make a
close examination of the plywood which may have become corrugated
by damp or dry rot.
- The condition of any cover strips and skirtings should be noted and
any damage included in the report.
- When dealing with modern sheet linings the surveyor must ensure
that the panels are properly fixed to the battens or studs in accordance
with the manufacturer's instructions. In some cases the panels are
secured with adhesive applied in dabs which may have become
detached. Gentle pressure on the panels will indicate whether or not
the linings are securely fixed.
- Mechanical damage to linings due to wilful or accidental fracture or
abrasion by excessive wear does not present a difficult problem. In
general, the procedure is to extract the damaged panel and replace it
with new. If the damage is extensive, reference should be made to the
manufacturer's fixing instructions, and a note to this effect should be
included in the report.

WALL TILING

13.9   Types of tiling and fixings

Ceramic wall tiles are a very popular finish and are available in a wide
range of colours and several different facings. Internally they provide a

durable hygienic finish to bathrooms, kitchens, cloakrooms and public toilets. The use of lightweight building blocks in modern buildings forms a suitable background provided the rendering to receive the tiles complies with Code of Practice BS 5385-2: 1991. There are special mastic adhesives for different types of background which should be used in accordance with the manufacturer's instructions. An opaque coloured glass known as Vitrolite is often used as a decorative wall lining in hotels, commercial buildings and shop fitting. The panels are secured with mastic applied to the back of the panels, fairly evenly spaced and jointed similarly to glazed wall tiling. Frost resistant tiles and mosaic are often specified for exterior work and are attached by mortars or adhesives rather than by mechanical fixings. The general principle is to select an adhesive appropriate to the background and external conditions.

13.10   Common defects in finishes

This work is usually entrusted to specialist firms, but nevertheless problems do occur, and the principal defects arising are given below. They should be carefully noted and recommendations for their treatment should be made in the report:

- In the past, tiles have been bedded in cement mortar. Lack of adhesion of mortar to tiles is usually caused by a failure to properly adjust the suction of the tile before applying the mortar. Wall tiles have a very high suction which is reduced by wetting, but if the tile is saturated, this will again result in lack of adhesion. The use of unsuitable sands in the rendering of the inner wall surface tends to increase drying shrinkage. Failure of adhesion in old wall tiling or opaque glass is also caused by the high temperature from heating systems causing shrinkage between bedding mortar and the backing material. Failure can also occur when tiling is applied to brick walls subject to moisture penetration.
  In such cases the trouble is caused by the penetrating water carrying sulphate salts in solution to the cement mortar backing. In all these cases large areas of tiles will be found to be loose and will easily fall away. The new mastic adhesives mentioned above provide a more resilient bedding and obviate all these shrinkage problems.
- The cracking of wall tiles may have no more serious consequences than disfigurement of the wall surfaces. Cracking of the glazed surface occurs as a result of shrinkage movements in the backing.
- Crazing on the tile face is again caused by the irregular application of the cement mortar. This results in a slight distortion of the tiles due to non-uniformity in drying shrinkage. Fine cracks then develop in the glaze which in ordinary circumstances are not very noticeable.
- Many defects have occurred in recent years where ceramic tiles or mosaic sheets have been used externally. Problems are often caused

by an inadequate 'key' on the background material. The differential movements of the structure and the tiling can also cause the tiles to fall away due to thermal or moisture effects. Damp penetration reaching bedding mortars and freezing has also contributed to failures of the bond. When carrying out an examination it is advisable to check that any movement joints incorporated in the structure are not tiled over and are carried through to the face of the tiling.

LIGHTWEIGHT AND DEMOUNTABLE PARTITIONS

13.11   Types of partition and finish

Since 1939 various types of patent non-load-bearing demountable partitions have been introduced, and have mainly replaced the traditional timber-framed partition lined with lath and plaster. In most cases the surveyor will not find it too difficult to discover whether or not a partition is non-load-bearing. However, when dealing with public or commercial buildings this is often a more difficult task than it might appear, and in such cases it may be necessary to check the original plans of the building as discussed in Section 6.12. If plans are not available then measured drawings of the various floors will have to be prepared showing each partition and how they play their part in the framework of the building. This is particularly important where alterations are contemplated. The construction of this type of partition is usually based on a timber batten on a metal channel frame with various infill panels consisting of wallboard, insulation board, or chipboard. A typical type of lightweight dry partition is known as 'Paramount'. It is constructed of prefabricated panels which consists of two Gyproc wallboards separated by, and bonded to, a fibrous honeycomb core. The partitions are readily adaptable to meet individual requirements and are easily erected. Provided the partitions are installed in accordance with the manufacturer's instructions there should be no faults. Any damage caused is usually due to misuse as described in Section 13.8.

JOINERY

13.12   Doors and windows

When examining joinery, a sound knowledge of architectural form and material as well as familiarity with the orthodox methods of joinery construction is essential. Many joinery items for domestic work are selected from manufacturers' catalogues and are purchased ready made. The surveyor must become acquainted with the various types, standard sizes and construction laid down in the British Standard Specification (e.g. BS 6375-

1: 2004). Doors, door frames and windows in particular can be obtained in stock pattern sizes and in many designs.

Old woodwork, especially if covered with several coats of paint, is very deceptive in appearance. Some parts of the timber may appear quite sound on the surface, but as soon as a portion is cut away, many defects may be exposed. In all types of joinery which require considerable repair it may be more economical to replace the door or window completely than to recommend repairs, but the surveyor should state clearly and concisely his reason for doing so. On the other hand, the joinery to be repaired may be of great value, in which case the timber should be examined closely and every endeavour made to retain as much of the existing timber as possible, and also preserve the true character of the original work.

Sometimes the surveyor has to deal with abnormal conditions such as those which arise during the installation of a new heating system or in the restoration of an old building which has stood empty for some time. The vast changes in humidity which take place are extremely detrimental to the joinery, and often cause warping and opening of joints. The most serious defect in joinery is decay or rot in the material. This matter has been dealt with in some detail in Chapter 9. It may suffice here to state that under certain conditions wood will decay and unless the decay is eradicated at an early stage it may spread over the entire woodwork of the building. It is not wise to assume that because a building has been erected fairly recently the doors and windows are free of decay. Decay in external joinery has been widely reported in many comparatively new buildings. Rainwater penetration through the joints and internal condensation have been the main cause of problems that do occur.

Casement windows can be constructed of timber, plastic, metal or metal in a timber frame. They may be hinged at the top, side or bottom; be centre-hung on pivots, arranged to slide horizontally, or to fold. Modern casement windows are made with several wind checks and throatings on the edges of the sashes, and on the faces of the rebates to prevent driving rain penetrating inside the frame. These weather resisting devices do not rely on the close fitting of the sashes; instead, spaces are left between the sash and its frame. The sashes therefore, do not stick and only require easing occasionally.

Metal windows are obtainable in many standard sizes and designs complete with all fittings and fastenings and in some cases include locking devices. From the 1920s onwards, hot-dip-galvanised metal windows were extensively used on all types of property. The galvanised coating did not always prevent corrosion and condensation. Opening lights were frequently draughty due to poor fitting between sash and frame.

In recent years this material has been superseded by anodised aluminium, often as a replacement for older defective timber frames. These windows have several advantages such as better draught-proofing, double glazing to reduce heat loss, and low maintenance.

Metal windows within hardwood or softwood frames provide a more satisfactory construction and give greater rigidity. Hardwoods have been selected because of their resistance to decay and attractive appearance. Wood frames have excellent insulation properties which is useful where condensation is present.

Windows constructed of uPVC are now widely used where replacement windows are required. They have good insulating properties and maintenance costs are low, although installation and repair costs are higher. In many commercial and industrial buildings the surveyor will find that the metal frames are fixed direct to brickwork. In such cases sealants between the frame and brickwork often form the only defence against rain penetration and should be carefully checked.

Boxed frames and double-hung sashes are invariably found in older buildings. The surveyor should be familiar with the various parts of the construction. Both sashes move vertically in grooves formed in the box frame and are suspended by either sash cords or chains which pass over pulleys housed in the members called pulley styles, against which the sashes slide. They are balanced by cast iron weights contained within the boxes formed by the frame or by spiral sash balances.

External and internal doors whether panelled or flush are made up in a variety of arrangements and designs, but the principle of construction is common to all arrangements. This consists of providing a rigid frame constructed of stiles, rails and muntins of uniform thickness, and filling the spaces between them with thin timber panels or glass. External doors will often be found to be protected from the weather by porches or balconies. On the other hand panelled doors that are unprotected receive more wetting and with their numerous joints provide possible places of entry for rainwater. Large doors to garages and industrial premises are seldom protected and their large size makes them more liable to mechanical damage. These problems will be discussed below under 'Examination of defects'.

Doors are hinged to solid wood frames of various types. For external doors the solid type is invariably used with a 13 mm rebate on the inner face to form a stop. When frames are close to the outer face of the wall they should be protected by a weatherboard secured to the head. A contributory cause of rain penetration through doors has been due to the absence of a weatherboard at the bottom of the door and a galvanised weather bar inserted in the threshold.

For internal doors, jamb linings are generally used to form the door surround. They are usually made from 38 mm thick material equal in width to the thickness of the partition plus the thickness of the plaster on both sides. The stops are often a single piece planted on the lining.

Sliding, or sliding and folding, doors will be found in many public buildings, schools and garages. There are many variations to satisfy almost all conditions. Some of the early types had problems with the mechanism which was often caused by neglect. During recent years the mechanism for this type of door has greatly improved and the problems with maintenance are better understood. The weight of the door is usually taken on wheels or rollers at the bottom or suspended from hangers on a track at the top. The heavier the door, the larger the track, but in general the top track support is the most efficient and free from maintenance problems.

'Up and over' doors are often fitted to garages and are most useful in a confined space. The mechanism consists of springs or balance weights. The door has to be of rigid timber construction or aluminium sheeting on a mild steel frame.

13.13 Skirtings, architraves and picture rails

Skirting boards are fixed round the walls of a room to protect the wall surface at a point subject to hard wear. In many cases they will be found to be of the same material as the floor covering. The present trend in design is towards simple shapes which can easily be cleaned, but the surveyor dealing with an old building will often find more ornate types of skirting which have been damaged by impact or by damp penetration from external sources. The latter defect has been dealt with in Chapter 8. In commercial and industrial premises the skirting will often consist of a hollow metal casing to accommodate electrical or telephone cables. They are usually detachable to enable maintenance or alteration work to be carried out easily.

To cover the joint between timber linings and plaster an architrave or cover mould is fitted round the opening. The vertical and horizontal sections should be secured at the head of the frame by means of a mitred joint. The general appearance and quality of the work is dependent on the degree of care taken when fixing these finishes.

The present trend is to remove old picture rails, but many will still be found in older domestic properties. They are generally fixed at the same height as the architrave on the door head to form a frieze. The rails are nailed either to plugs or to fixing bricks built into the wall.

13.14 Cupboard fitments

A wide variety of standard fitments are now available for kitchens, bathrooms, bedrooms and public toilets. In some cases they have been made for a specific purpose to fit a particular position and include storage cupboards and shelf units of all kinds. The principal construction material for interior fittings is chipboard, blockboard or plywood with a decorative melamine plastic surface, or the interior may be simply left with a timber facing prepared for painting. However, there has recently been a return to the traditional form of cupboard fitting consisting of raised and fielded panel doors with turned knobs. The units are made under stringent performance specifications in a variety of timbers with many variations of grain and colour. This type of fitting may still be found in the older domestic and commercial properties consisting of stiles, rails and muntins, and apart from some splitting in the decorative treatment due to successive layers of paint, the timber is usually of sturdy construction.

13.15   Ironmongery

Door and window fittings include all the ironmongery necessary to secure the door or window when closed. The strongest fittings are made of steel although some wearing parts are made of bronze. However, much iron-mongery is made of brass with a plated finish. During the past few years, anodised aluminium has become very popular due to its pleasing appearance and freedom from corrosion. Stainless steel is often used for high class work.

It is probably no surprise to learn that crime is still on the increase. Nine out of ten break-ins are through insecure doors and windows. By thoroughly checking the locks and fittings the surveyor will soon discover any defects or weaknesses. A common fault with door fastenings is the displacement of the striking plate causing the latch to spring open. This defect is generally caused by the door drooping. Hinge binding is another common trouble with a door, and is often responsible for loose hinges and worn bolts. In new joinery the trouble can often be traced to shrinkage in the stiles, which leaves the ends of the tenons on the rails protruding. Particular attention should be given to the examination of ironmongery attached to fire-resisting doors. Precautions to limit the spread of fire normally include the provision of fire-resisting doors in corridors, entrances to staircases and lobbies.

13.16   Examination of defects

For efficient treatment of defective joinery the cause of the damage should be carefully considered so that a repetition can be avoided. Joinery is liable to mechanical damage, shrinkage, warping and decay of the material. The soundness of wood can be tested by stabbing the surface with the point of a penknife; decayed wood can be pierced without difficulty while the fibres of sound wood grip the point. Decayed timber is often dished and paint surfaces are generally peeling and cracked. The principal defects that are found during a survey are given below. They should be carefully noted and recommendations for their treatment should be described in the surveyor's report.

Common defects

The common defects are as follows:

- Decay in external joinery does not often produce a growth of wet rot and is usually confined to those portions of joinery which are continually damp. The omission of throatings and drips is largely responsible for this condition. All sills should have an adequate 'run-off' and the sealants between frames and adjoining masonry should be checked to ensure that a watertight joint has been formed.

Doors

- Entry of moisture at the joints causes weakening due to failure of the glue leading to loosening of the joints and distortion of the doors.
- Entry of moisture may also occur when glazed door panels are secured with putty or glazing beads. Moisture will penetrate through loose or cracked putty, or the beads may come away from the glass.
- The absence of a weatherboard on the external face of a door usually means that rainwater is not thrown clear of the gap under the door. The absence of a weather bar in the sill will aggravate the problem.
- It is sometimes found that internal grade plywood has been used on external doors. In such cases the plywood becomes wrinkled and splits at the edges.
- If the temperature conditions are markedly different on the two sides of an external door, and it is poorly protected due to inadequate painting then distortion is likely to occur.

Windows

- The problem concerning joints described above can also occur with timber windows. Some sashes are not sufficiently robust and have distorted through moisture movement, permitting the entry of rainwater.
- Bottom rails of sashes are particularly vulnerable to decay if the putty fillet becomes loose or if the paint blisters.
- A problem which seems to be increasing is condensation on the inner face of the glass. (Condensation has been dealt with in Chapter 8.) It may suffice here to state that condensation running down the glass collects on the surface of the back putty slowly penetrating the timber rail. Many windows lack condensation channels and tubes to pass the water to the outside. Windows in habitable rooms should each have 8000 mm^2 of trickle ventilation (Oliver 1997).
- The most frequent type of defect when dealing with double-hung sashes concerns broken sash cords or chains. If breakages are reported the sashes should be carefully examined. The problem may be due to badly fitting pulleys or to the weights jamming in the casing.
- Metal windows, especially those installed between the two World Wars were not always galvanised, leading to corrosion of the metal. Failure of the metal may well result in rainwater entering the building. Metal frames have a high thermal conductivity thus providing a cold bridge between the exterior and interior of the building. This can often lead to severe condensation problems.

- Jammed sashes may be due to the sash drooping, the wood swelling or the hinges becoming disturbed.

Skirtings and architraves

- Good practice requires skirting boards to be selected from rift sawn material to prevent bowing outwards, but too often this has not been done, and the skirting tends to curl away leaving an extremely unsightly gap at the top edge. A wide skirting board is liable to shrink and expose an open joint between the bottom edge and the floor. Occasionally, this gap is likely to be aggravated by the shrinkage of the floor joists and settlement in the wall.

  The condition of the architraves, cover mouldings and window boards should be examined to ensure that they are free from infestation or organic defects. This type of finish is often found to have been damaged by impact or abrasion which should be noted.

Cupboard fitments

- An important point when examining kitchen fitments, shelving units or fitted cupboards is to take particular care that all horizontal members are level and all vertical members are plumb. Doors and drawers should be examined for ease of movement and a note made of any defective fittings and furniture.

EXTERNAL RENDERING AND POINTING

13.17   Introduction

There are several materials used for external walling that require further treatment such as lightweight concrete blocks, solid brickwork and no-fines concrete. The rendering gives a waterproof coating and a more finished appearance to unsightly material. The following are the more common types of finish:

- Pebble dash sometimes known as dry dashing consists of throwing on a final coat of small pebbles or stone chippings immediately after the application of the previous coat.
- Rough cast is usually more resistant to penetration than one of a smoother texture. The final coat contains fine stone chippings or gravel to replace the sand which is 'thrown-on' instead of being applied by trowel.
- Textured and scraped finishes should be chosen in relation to atmospheric conditions. In a contaminated atmosphere, a less rough finish is preferable.

- Smooth floated finishes produced by the use of a wood or felt faced float.
- Machine applied finishing coats vary in texture with the type of machine and materials used. Proprietary materials are supplied ready for mixing and are suitable for all conditions.

The main object of external rendering is to prevent moisture penetration. In view of this fact, there is often a temptation to apply a dense impervious mix which frequently defeats its purpose. Dense cement rendering has a high drying shrinkage and the greatest movement occurs after the rendering has set and hardened. This results in the formation of cracks, the spacing of which is dependent upon the strength of the backing, the adhesion obtained by the rendering, and the rate of drying. Cracks in dense rendering act as capillary paths to the inner part of the wall, and if the wall is constructed of relatively porous material it may absorb sufficient moisture through the cracks to cause dampness on the internal face. Moisture penetration through cracks in a dense rendering usually occurs where the rendering has a smooth surface. Rough cast and pebble dash finishes shed much of the water that falls on them, and are less likely to cause dampness. Relatively pervious renderings are usually made with cement, lime and sand in the proportions of 1:1:6 or 1:1:5, the strength of which can be varied to suit the nature of the backing and conditions of exposure. The quality of the rendering will often depend upon a good sound base of brick or blockwork. It does not matter how rough the surface of the wall is, provided it is firm. There must be no progressive movement, or decay of the wall material which would cause disruption of the rendering. If the backing is suspect, the first step is to cut out a portion of the rendering and examine the wall material.

To carry out a satisfactory investigation it may be necessary for the surveyor to check the area of rendering from ladders in order to establish the exact cause of the defects, paying particular attention to any 'hollow areas'. He can then decide whether or not the rendering should be totally renewed or simply repaired in parts.

13.18   Rendering defects

The principal defects associated with rendering are described below:

- Sulphate attack is a problem on rendered walls and can occur through salts derived from wet brickwork. It is manifested by horizontal cracking corresponding to the mortar joints in the brickwork below.
- Rendering consisting of cement and sand without the inclusion of lime and applied with a floated finish will often produce 'map' patterns. In exposed conditions this type of rendering will lead to loss of adhesion due to frost action and provide damp conditions necessary for sul-

Figure 13.1   Cracked rendering as a result of sulphate action.

phate attack. Figure 13.1 shows cracking of rendering as a result of sulphate action.

- Small shrinkage cracks may be seen on the surface and are usually due to differential movements.
- One of the problems with rendering that is difficult to diagnose is when moisture collects behind the render coats. This is often the result of capillary action moving upwards from the ground or downwards from the eaves. The trapped moisture is unable to evaporate from the

surface and will eventually soften the undercoat causing a loss of bond between undercoat and finishing coat.

- A defect often found in older buildings is when the rendering is taken down to ground level, thus bridging the damp course and providing a path for rising damp. If the damp course has been extended through the full thickness of the rendering it can easily be seen. This problem is relatively easy to diagnose.
- Where rendered brick walls support flat concrete roof slabs, the area immediately under the eaves will often contain horizontal cracks. The cracks are caused by differential movement between brickwork and slab mainly due to variations of temperature.
- Cement or masonry paint finishes are often applied to rendering to reduce the risk of moisture penetration and for appearance. If properly applied they give reasonable protection for periods of five to six years. Conditions of exposure differ, but the basic cause of deterioration is the presence of moisture penetrating the paint film. All decorative work of this nature should be examined to ensure that the material is firmly adhering to the rendering. This examination is easily done by rubbing a paint stripping knife over the surface to see if the material will 'lift'.
- Too strong a finishing coat can cause flaking and may require its complete removal, and its replacement by a mix no stronger than the undercoat.
- Much moisture penetration into rendering can be caused by faulty protection to edges, omission of protective flashings and damp-proof courses below copings and window sills. The areas of rendering affected by these defects may have become detached. In such cases it is often necessary to remove large areas of rendering to enable the background to dry out before the making good is carried out.
- During his examination the surveyor should give particular attention to architectural features i.e. sills, copings and string courses where streaking may have occurred due to poor detailing.

Any of the defects described above should be carefully noted and any recommendations for their repair described in the report. Defects in rendering are not always easy to diagnose particularly between sulphate and frost attack. The surveyor must, therefore, diagnose the problem carefully before prescribing any treatment. In cases of doubt it is advisable to recommend chemical tests to establish the presence of sulphate attack.

13.19 Pointing

It is of the utmost importance that the pointing of the brick or stonework should be in good condition. If the pointing has perished or has broken away due to frost and rain penetration it must be renewed as soon as possible. If the pointing mortar is much stronger than the bricks, the flow of rainwater into the brickwork and the subsequent drying out will take place through the brick and not through the mortar joint. It sometimes happens that where re-pointing has been carried out using a strong cement mortar the edges of the bricks will break off and the brick surface will be eroded leaving the pointing projecting. This is due to the crystallisation of salts contained in the mortar or brickwork. Care should be taken to examine the pointing below the damp course level. If a weak mortar has been used in a damp situation it will often be reduced to a soft powder. If the bulk of the pointing is in poor condition it is advisable to recommend that the whole of the brickwork should be repointed using a lime-based mortar for older buildings.

PAINTING AND DECORATING

13.20   Defects due to poor application or unsuitable backgrounds

There is now a wide variety of decorative finishes available to suit various backgrounds and different forms of usage. Modern decorating materials are carefully manufactured and failures due to the paint itself are very rare. However, all paints are susceptible to deterioration if not properly applied on a well prepared background.

Cracking usually results when the undercoats remain softer than the final coat. This may be due to an excess of oil, the presence of dirt or grease, or insufficient time allowance for proper hardening between coats.

Peeling and chipping are frequently caused by lack of 'key' between surface and paint. Any surface which is not stable eventually breaks down under the weight of superimposed coats of paint.

Blistering is usually caused by the action of heat upon trapped moisture or solvents below the paint film. In this connection the surveyor must realise that although the surfaces of wood or plaster may appear to be dry it is possible that moisture may be present below the surface. This moisture may be drawn to the surface after the application of the paint, and cause blistering.

Ginning is caused when the undercoat shows through the finishing coat; the coat not having sufficient obliterating power or an undercoat of unsuitable colour.

Loss of gloss is frequently caused when the finishing coat is applied before the undercoat is sufficiently dry or the paint not properly stirred. Atmospheric conditions such as dampness, mist or condensation can also cause loss of gloss.

Wrinkling usually results when the paint is applied too thickly to a horizontal surface.

Mould growth which develops on painted surfaces is distinguished by the appearance of black or brown spots consisting of minute growths of fungi. Mould growths grow under damp conditions (>70% relative humidity) and are often caused by condensation problems or even by penetrating damp, which causes localised staining.

Crazing consists of an irregular cracking of the finishing coat due to old age or a hard drying paint being applied to a soft or oily undercoat.

Chalking consists of a powering on the paint surface and is usually due to porous surfaces being insufficiently sealed.

Bleeding is a discoloration of the finishing coat caused by the medium dissolving material in the background, such as bitumen, tar, asphalt or creosote, which has not been properly sealed.

Poor opacity is usually caused by overthinning the paint or failure to stir the paint thoroughly.

Many minor painting defects are easily solved by an experienced surveyor. On the other hand, due to the complicated nature of some defects he may consider the problem to be outside his professional knowledge and may require the assistance of a consultant or chemist to carry out an examination and laboratory analysis.

13.21 Interior finishes

Internal decorations are often a personal matter, particular in domestic work, and tastes vary considerably in the type of decoration required. It is, therefore, no use marking a detailed schedule describing the decorative condition if the client proposes to carry out a complete redecoration. Also it is unwise to include the cost of any redecoration when preparing the report as most clients have a fairly good idea of the cost of redecorating a room or office, and in these days will often do the work themselves. In cases concerning large industrial buildings or office blocks advice may be unnecessary, particularly if the owners employ their own maintenance staff. In such cases only a simple paragraph stating that the decorative condition of the room is 'fairly good' or in 'poor condition' is all that is needed. However, there are instances where the internal finishes are very poor and the client has given explicit instructions to carefully examine the decorative condition of the property. It is, therefore, necessary for the surveyor to recognise the various defects most commonly found in decorative finishes and have some knowledge of their cause (see Section 13.20).

When carrying out a survey of commercial or industrial premises it must be remembered that a considerable amount of cleaning and painting is compulsory because of legal requirements. The Factories Act 1961 and the Offices, Shops and Railway Premises Act 1963 both require cleaning and painting to be carried out at prescribed intervals to protect staff from risk. In such cases particular care must be taken to ascertain the condition of the paintwork in order to advise the client when the next repainting operation is likely to be required. Hospitals, clinics and food producing industries usually require a smooth painted gloss surface which is easily maintained.

Wall and ceiling papers are usually textured or decorative; and the defects associated with this type of decoration are usually concerned with workmanship. The lengths of paper should be closely butt jointed and the pattern matching through, and not lapped. The papers should be neatly trimmed around the door architraves and windows etc. If the papers are dirty, loose or badly marked then they need to be stripped off. Heavy papers may conceal defective plaster which will fall away with the paper. In such cases it is wise to examine this type of paper in conjunction with the plaster defects.

There are specialised finishes which have a heavy bodied stiff composition, similar to plaster, which can be worked on by forming patterns with combs and stipplers. The coatings will mask small irregularities, are durable, easily maintained, and normally need no further decoration. The writer has found that this type of finish is popular on domestic ceilings and rarely shows evidence of any defects. The only problem with this type of finish is when it is applied to an old plaster ceiling which may have lost its 'key'. Defects may be difficult to detect, but by carefully tapping the surface the 'hollow areas' can be detected.

Destructive influences to internal paintings are usually limited to the action of condensation which has a definite solvent action upon paintwork of window frames. In serious cases of condensation mould growths can occur causing disfiguration of the decorated surface. Disfiguration of the internal finishes of a building may occur due to the differential deposition of dust causing darkening of the plaster surface which virtually forms a complete replica in light and shade of the floor joists etc. It may also occur over the heads of nails used for fixing ceiling or wall boards, or reinforced concrete, in hollow tile roof slabs and hollow block partitions. The cause of the trouble is that dust collects more easily on cold surfaces than those which are warm. Therefore, where any form of thermal insulation occurs which is not continuous, dust is likely to collect on those parts of the structure which are less well insulated. The surveyor should carefully inspect the fixing materials used for the purposes of insulation to ensure continuity. Improvements can be made by recommending a form of heating which will keep the surfaces at a uniform temperature, e.g. a radiant heating system, or by adding insulation where the heat flow is high (Douglas 2002).

13.22   Exterior paintwork

The effects of weathering upon external paintwork are somewhat complex (BRE IP 16/87). The sun is often a destructive force, and the effects of acid rain plus accumulations of dirt and chemical impurities are also damaging. Extremes of temperature cause a considerable amount of contraction and expansion especially in woodwork. The surveyor will often find that modern hardwood windows and doors have been treated with clear preservative finishes in order that the natural timber shows and will simply require recoating.

In the past masonry was often painted as a protection, and if the adhesion is sound then no attempt should be made to remove the paint. Removing paint from brick or stone will sometimes damage the base material. Old paint is often difficult to remove from porous brick or stone. Alkaline strippers and air abrasive treatments should be avoided. The writer considers that the best course would be to allow the paintwork to disintegrate naturally. However, if moisture is trapped behind the paint film and is unable to evaporate there is a risk of cracking or blistering, each causing disfigurement. The effect of frost and salts can also cause further decay. Decay of walling behind the paint film can also result in flaking. There is also the possibility that flashings and damp-courses are defective which will cause deterioration of the paint surfaces. Care should be exercised where blocked gutters and hopper heads have saturated the wall behind an impervious layer of paint.

13.23   Metal surfaces

Iron and steel will corrode rapidly in damp conditions both internally and externally and the severity of the attack varies according to atmospheric conditions. Iron and steel pipes are subject to corrosion under almost any conditions of exposure and should be protected by galvanising or by painting. Condensation on pipework, though occasional, may occur for sufficiently long periods to cause the paint film to crack. In coastal areas the concentration of common salt in the atmosphere absorbs moisture forming strong solutions which can cause corrosion if these collect in metal crevices. All defects in metalwork should be carefully noted and described in the report. Close inspection is particularly desirable in cases of corrosion where costly preparation is required.

13.24   Historic buildings

When carrying out an inspection of an old building the surveyor will often come across paintwork, marbled work and graining dating back to the eighteenth and nineteenth centuries. The decoration may have been redone several times, but if found to be in fairly good condition it deserves to be protected. The Society for the Protection of Ancient Buildings has always advocated that where old paintwork on the interior of a building is sound it is best left undisturbed. The adhesion between the coats of paint and between paint and background is thought to decrease once paint strippers have been applied prior to redecorating. Old oak panelling was usually left in its natural colour, silver grey, and should remain so although an application of softened beeswax is often used to preserve the timber. Staining and varnishing old oak, a custom often followed in the past has now almost disappeared.

GLAZING AND LEADED LIGHTS

13.25   Introduction

The range of glasses often changes and if the surveyor is in doubt as to the type and quality of a particular glass, reference should be made to manufacturer's information, but for general information and examination some particulars are given below.

13.26   Symptoms and defects

Glass is a durable material and is seldom affected by the various agencies of deterioration mentioned in previous chapters. Cracked panes are readily detected and are usually due to accidents or vandalism or the rusting of metal windows. However, if proper allowance has not been made to accommodate differential movement then thermal stresses can cause cracking. Alkali from paint removers can cause surface etching.

The condition of the putty or glazing beads should be noted and any defects described in the report. Old putty often cracks and falls away from the glass causing moisture to enter and soak into the framing. Timber and metal glazing beads should be puttied between glass and bead. Glass used for lightweight cladding or in metal windows can crack due to a higher coefficient of thermal expansion in the metal. This defect can only occur if there is insufficient clearance between glass and frame.

Plate glass is used for large glazing areas such as shop-fronts and is fixed in rebates with beads. Owing to the weight of the glass and the possibility of fracture through excessive vibration the glass should be supported on felt or lead. If metal beads are used they should be fixed with proper angle clips.

Old leaded glazing should be carefully removed. The small panes of glass are held between lead cames which often become weak through age, but can be repaired by turning back the lead and re-cementing the flanges.

Before the nineteenth century most clear glass was 'crown' glass. It was made from blown glass spun at high speed until it was spread by centrifugal force to a large thin disc. Crown glass has a slightly bellied surface and when viewed from the outside plays with the light giving an attractive colourful effect. This type of glass is seldom made today, but a very close imitation is obtainable from the manufactures when a sufficient quantity is ordered.

Wired glass is normally used in roof lights or where fire-resistance is important and has been dealt with in Section 10.29.

A development that has gained momentum over the past twenty years is that of the purpose-made window and door consisting of double glazed units with hermetically sealed air space. The frames are usually made from hardwood with anodised aluminium or white polyester sashes manufactured and installed by specialist firms. The demand for such windows mainly arises from the desire to reduce heat loss through glass areas and to eliminate draughts. However, the surveyor should note that while double glazing reduces the risk of condensation on the glass it will not necessarily prevent condensation on the metal sashes and frames. This will depend on the internal humidity and temperature. In the writer's experience this type of window is produced under stringent conditions and it is unlikely that defects will be found. However, the following points should be checked:

- Sills should be properly weathered and throated.
- Friction stays and hinges should be in sound working order.
- Inner and outer aluminium sections should be totally isolated from one another throughout by a continuous polyurethane resin barrier.

14 Services

14.1 Introduction

Building services have become a major item in modern buildings, and are costly to maintain and operate i.e. large blocks of flats or commercial properties. Hence particular care should be taken in the examination to ensure that there are no defects, and that maintenance can be carried out easily and economically.

In Section 2.2 it was suggested that in view of the complicated nature of many of the services it is advisable that specialists should be employed to test various installations and plant. This point must again be stressed because no clear opinion can be given unless these services have been properly examined. This is particularly important where gas or oil-fired boilers are installed and where electrically operated pumps and thermostatic controls are fitted. It is essential to make sure that any specialist employed is well aware of his responsibilities and that he must submit a true and unbiased report. It is also essential that the specialist must include an approximte estimate for any remedial work required to the installation. Some specialists' reports tend to be couched in technical language which often confuses the layman and leads to misunderstanding. Specialists who are accustomed to writing this type of report will usually understand the situation, but if not, they should be persuaded to rewrite in a simple fashion, and avoid long technical words and expressions. Although a full specialist service is preferable for most schemes it is realised that many surveyors dealing with small simple schemes i.e. small domestic properties and shops etc. may well have sufficient knowledge of building services and would feel confident in carrying out an examination themselves. Therefore, the descriptions and list of defects in the following paragraphs refer to small properties.

COLD WATER SUPPLY

14.2 Types of pipework

Cold water storage tanks in roof spaces and tank rooms have been dealt with in Section 10.8. It is unusual to expose underground pipes for examination except where leaks or damage have been reported. The main service from the water authority's main, and in fact any pipe below ground should be at least 900 mm underground in order to be clear of frost action. A stop valve is situated in a brick or concrete pit just outside the boundary of the premises and is the property of the water authority. The valve is used to cut off the main supply when required. The depth of the pipe can be checked by lifting the cast-iron cover at ground level, and any signs of leakage at the joints should be noted.

Lead was usually specified for underground pipes in old buildings, but deteriorates with age and is now prohibited. Copper is still used today and is more durable, but can be readily attacked when the trench is being back filled with rubble and will often need to be protected with bitumen impregnated tapes and wrappings. Polybutylene and unplasticised PVC are now used for economy and do not corrode although they can be damaged by careless handling. If problems are encountered in old installations it is often due to the corrosive action of the soil.

The corrosiveness of soil is a complex function. General experience is that certain soils such as moist acid clay, wet acid peat, red marl or clinkers are corrosive to most metals. If chlorides or nitrates are present in the soil, they usually lead to corrosion. If it is necessary to replace pipes in soils known to be corrosive then new pipes should be wrapped to provide a damp-proof covering as described above.

The distribution of cold water through a building so that draw-off points can be adequately served depends on the requirements of the local water authority and the available pressure on the mains.

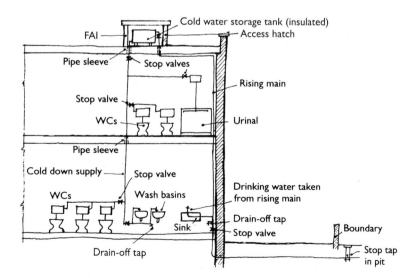

Figure 14.1 Layout of cold water supply to small office block.

14.3   Guide to checking cold water installations

The following is intended to be a guide to the surveyor when checking the cold water installations.

(1)   At the point of entry into the building the rising main should be provided with a drain-off tap and stop valve. The drain-off tap is for the sole purpose of emptying the main pipe beyond where the water is shut-off by the stop valve, although in many old properties the drain-off tap is absent, and this should be noted when preparing the report. In older properties the stop valve will be found in a pit externally adjacent to the point of entry.

(2)   The internal cold supplies should be fixed on internal partitions or walls, rather than on the inside of the colder external wall. The temperature is not likely to be so low on internal partitions, particularly in occupied buildings. If the pipes must be fixed in positions subject to frost they should be suitably protected.

(3)   No water for consumption should be taken from a storage tank. At least one branch should be taken from the rising main to serve a convenient water draw-off point which invariably supplies a sink unit.

(4)   Pipe runs feeding the various sanitary fittings should be so designed that the whole system can be drained of water if necessary to enable repairs to be executed conveniently. The supply pipes that feed downwards from a cistern should be devoid of dips, and should fall regularly towards the draw-off points.

(5)   Trace the pipe through the building and where possible check joints, taps, valves and drain cocks for leaks. Particular care should be taken to examine all exposed pipes for pinhole leaks. The surveyor may consider draining the system himself if there is a delay before a plumber can call.

(6)   The installation should be planned to provide adequate protection to safeguard the pipes from damage by impact, expansion, contraction, or corrosive action. Attention should be given to the degree of accessibility for inspection and repairs. In a well designed layout, it is customary to arrange pipe runs so that they are contained mainly in recesses or fitted in ducts with access panels. Pipes should not be buried within the fabric of a wall or solid partition. However, in certain instances pipes have to be bedded within the thickness of solid concrete floors. In such cases, the pipes should be inserted in a metal sleeve to protect them against damage so that no joint is included in the encased section.

(7)   Horizontal pipe runs in timber floors, internal angles between wall and ceiling or contained in horizontal ducts should be secured with suitable fixing devices. For copper piping the recommended maximum spacing for horizontal fixings for bores of 12, 19 and 25 mm is 1.220, 1.850 and 1.850 m, respectively. The maximum hori-

zontal spacing for galvanised mild steel pipes of similar bore is required to be 1.850, 2.450 and 2.450 m.

(8) Water hammer in cold supply systems is produced by a concussive effect caused by the quick shutting of a tap or ball valve transmitting a shock through the wall of the pipe. The noise is usually caused by a tap with a loose jumper, or defective washer in the ball valve of the cold water storage tank. The pushing out of the jumper causes the ball of the valve to bounce on the water and to start the valve reverberating which is the noise referred to as 'water hammer'. If taps and washers are in good condition the surveyor should recommend the installation of an air vessel of a larger bore fitted at the inlet to the storage tank. The concussion is then taken up compressing the air in the vessel. It may also be remedied by fitting an equilibrium ball valve, in which the shock would be exerted equally on each end of the piston and obviate the movement of the ball.

(9) In older buildings, the surveyor will often find that the piping consists of dissimilar materials. The pipework should be carefully examined, particularly when a copper pipe is connected to a galvanised steel tank. Electrolytic action between the copper and zinc coating can cause the zinc to deteriorate and ultimately perforate the steel tank.

14.4   Old lead pipes

Lead pipes are still found in old buildings and the majority will have physically deteriorated and the surveyor should recommend replacement. The following points should be noted when carrying out an examination of lead pipework:

- The sagging of horizontal lead pipes between supports or the pull of a long vertical pipe on its top fixing may lead to fracture at the point of support. The maximum spacing for horizontal fixings for lead pipes of all sizes should be 600 mm.
- Pinhole leaks in old pipes are often due to the effect of lime on the lead, but are often difficult to trace particularly where pipes are concealed. Lead becomes brittle with age and is also vulnerable to damage by frost action.

HOT WATER AND HEATING INSTALLATIONS

14.5   Direct and indirect systems

Hot water supplies to small properties may be supplied by one of the following systems:

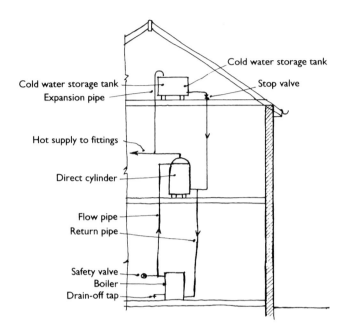

Figure 14.2   Layout of direct heating system.

Direct heating systems

In direct heating systems the principal components are boiler; flow and return pipes; hot water storage cylinder; cold water supply pipe; expansion pipe and draw-offs (see Fig. 14.2). Furring of pipes in this type of system is common in 'hard water' areas.

Indirect heating systems

An indirect system of hot water supply is one in which the water used is heated indirectly by hot water in a calorifier placed in the hot water vessel. In this system the hot water to the fittings does not pass through the boiler. Two separate cisterns are necessary. One is the feed and expansion cistern and the other is the cold water storage cistern for the building. Both act as expansion cisterns. The advantages of the system are that only one boiler is necessary for both installations and after the initial deposit has taken place in the boiler the water constantly recirculates between boiler and calorifier. The water in the calorifier is hardly ever replaced, thus there is very little furring of pipes (see Fig. 14.3).

Both direct and indirect systems can be heated by a free standing slow burning solid fuel boiler or solid fuel back boiler connected to a suitable

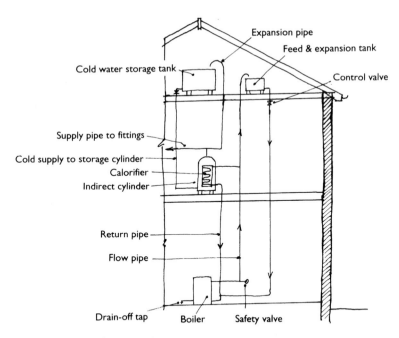

Figure 14.3 Layout of indirect heating system.

flue with an adjustable damper as well as the more common oil or gas-fired boilers.

14.6 Oil-fired boilers

Oil-fired boilers always require a suitable flue. The building regulations require the oil to be stored in a suitable fuel tank on the premises. The building regulations are stringent regarding safety, and the position of the oil storage tank in relation to the remainder of the premises is important. The surveyor may have a sound knowledge of the regulations, but if in doubt it is advisable to instruct the heating engineer to examine the entire heating installation. Oil-fired installations can be a problem when it comes to testing the safety measures, such as valves controlled by fusible links to check the flow of oil to the boiler.

14.7 Gas-fired boilers

A gas-fired boiler can be free standing or wall mounted connected to a conventional or balanced flue (see Section 14.13). Some have a permanent pilot light, others have automatic electric ignition to light the pilot which

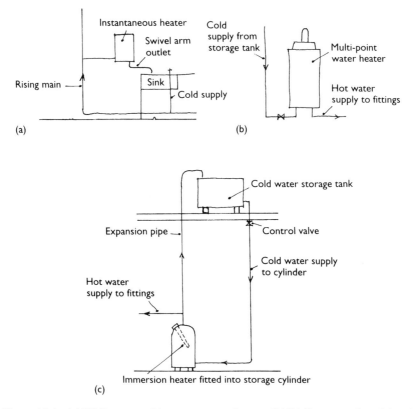

Figure 14.4   (a) Wall mounted instantaneous heater. (b) Wall mounted multi-point heater. (c) Heating water by electric immersion heater.

in turn lights the main burners when the boiler starts operating. Gas room-sealed boilers in domestic property are usually installed in a kitchen area. If the layout of the dwelling will not permit this and the heating system layout dictates that the boiler be installed within an existing cupboard, then adequate ventilation must be provided to this compartment. There should be two permanent air vents, one at low level and one at high level, both communicating either directly with outside air or with a room which is ventilated.

14.8   Wall mounted water heaters

Wall mounted water heaters can be heated by gas or electricity. The instantaneous type is usually found above a sink unit in domestic properties and does not take up valuable floor space. The water is heated at a single point and is supplied through a swivel arm outlet. The multi-point heaters can be heated by gas or electricity. They receive the cold supply from a cold water storage tank which then supplies hot water to several fittings (see Fig. 14.4b).

14.9   Gas fires

Gas fires have been used extensively for space heating and are frequently fixed to an open fireplace flue as described in Section 11.5. Now that permanent ventilation is not generally essential it is advisable to check that some permanent ventilation has been provided for introducing air to all rooms where gas fires have been installed. The Building Regulations 2000 (Heat producing appliances) require that a room containing an open flued appliance shall contain some form of permanent ventilation to the external air as follows:

- For a decorative appliance an area of $1800\,mm^2$ for each kW of rated input over 2 kW.
- Any other open-flued appliance an area of $450\,mm^2$ for each kW of rated input over 7 kW.

Provided the products of combustion are conveyed efficiently to the flue and the connections to the appliance are properly made, there should be little cause for concern.

14.10   Items to check

Old or badly fitted gas water heaters can be dangerous and the following items should be checked and noted:

- Older type properties may have one of the old type open-flued water heaters in the bathroom or kitchen. Like all heaters, if it has not been regularly serviced there is a possibility of danger. It is now illegal to install this type of heater in a bathroom. An existing heater of this type should be checked by the supply authority.
- The danger signs on water heaters are cracked and blocked flues, and orange or yellow flame. Gas fires also need regular servicing and cleaning and the following points should be noted.
    (i) The appliance should have a void behind it to accommodate any fall of soot or pargetting. This void is referred to as the catchment space.
   (ii) A closure plate should be provided to seal the gas fire to the catchment space, and must have a slot at the base to allow air to enter and to clean out the catchment space.
  (iii) Check that the appliance operates at the various settings.
  (iv) Ensure that the burners are clean and in sound working order.

14.11   Immersion heaters

Another method of water heating is by electric immersion heater fitted into a copper cylinder. To prevent overheating, the immersion heater is thermostatically controlled. In hard water districts there is the possibility of scale forming on the inner surfaces of the cylinder. Cylinder capacities vary; the standard size is about 125 l, but larger sizes are available if there is sufficient space and heating capacity. The cylinder should be well insulated, otherwise the heat loss will be considerable.

In modern buildings the surveyor will often find that the cylinders are coated with sprayed on foam insulation in lieu of jackets. This type of insulation is very efficient and has a neat appearance. All pipes connected to the cylinder should also be insulated (see Fig. 14.4c).

The majority of immersion heaters are 3 kW rating and are fitted with an adjustable thermostat. The method of testing is as follows:

- After turning the heater off check the rate at which the meter disc revolves.
- Remove the immersion heater cover plate and raise the setting to the high position. In this position the thermostat cut-out will not switch off the heating element.
- Switch the heater on and note the rate of disc revolutions which should considerably increase if the heater is in sound condition. If the speed does not increase, then the immersion heater is defective.

14.12   Storage heaters

Off-peak storage heaters consist of a metal cabinet containing blocks heated by electrical elements from the power supply. The blocks absorb the heat during the hours of darkness, gradually emitting the heat throughout the following day. To examine a storage heater the casing should be removed and the blocks checked for any signs of cracking which would impair the efficiency of the heater. A report on a storage heater should also cover the condition of the heating element (see electrical installations Sections 14.28 to 14.31).

14.13   Boiler flues

Flues and chimney stacks serving gas- and oil-fired boilers have been dealt with in Sections 11.5 and 11.6. In the event of an existing flue being used by a gas-fired boiler an adequate supply of air is needed in the room for the purposes of combustion, and the flue should be lined as previously described.

When dealing with a fairly modern building the surveyor will often find that a gas-fired boiler operates in conjunction with a balanced flue. Balanced flues are used where there is no chimney at all or where the existing chimney is in the wrong position or perhaps is defective. The flue consists of a short horizontal metal duct which links the boiler through one part for combustion and the other lets out the products of combustion. Whatever external wind pressure is on the flue terminal, the inlet and outlet have the same balanced pressure. Obstructions in the air inlet and outlet of the balanced flue would seriously affect the correct functioning of the boiler. It is therefore important to check the inlet and outlet to ensure there is no obstruction. It is also important that the flue terminals are correctly positioned, and the following points should be noted:

- Flue terminals should not be tight under eaves soffits.
- Not placed under opening windows, and at least 600 mm from any opening or projection in the building.
- Protected with a terminal guard if close to a footpath where people are passing.

A very popular method in present day installations is the combination of solid fuel or gas boiler with an electric immersion heater fitted into the domestic hot water cylinder for summer or emergency use.

CENTRAL HEATING

14.14 Introduction

Many domestic properties now combine the hot water supply with a central heating system, although in some instances the surveyor will find that the central heating system has been added at a later date and is, therefore, completely separate. A typical basic system can have many variations. Apart from radiators the heat-emitters can be convectors of different sorts and there are a number of ways to control the amount of heat for each room and the time the heat is switched on.

Here again, the surveyor may have a working knowledge of the system, but as mentioned in Section 14.1 it is advisable for a specialist to deal with the entire heating system. Only a complete test of the system will give a clear indication as to the efficiency of the boiler; whether the system is adequate for the heating of the rooms and the supply of hot water to the various draw-off points; and finally the system's safety in operation. However, it is the surveyor's responsibility to give the specialist an adequate briefing as to what is required and also to satisfy himself that the installation is aesthetically acceptable.

14.15   Common defects

In the following the most common defects that are found in hot water and heating systems will be considered.

Poor flow

This problem is usually due to air being drawn into the system through a vent pipe or an insufficient head of water. The defect can be remedied by raising the storage tank or inserting a larger supply pipe.

Check the system for leaks

Trace the pipe runs through the building and where possible check joints, taps, valves and drain cocks for leaks. Particular care should be taken to examine all exposed pipes. If severe leaks are found it is advisable to switch off the electricity supply at the main. Outbuildings are particularly vulnerable, and any pipes should be well lagged. A note should be made of all leaking taps, particularly in baths, where they cause stains. Taps leak when the washer becomes worn or when the metal seatings become eroded.

Check the controls

For any type of hot water or central heating system, check that the controls are adequate. The system should have at least a two-period timer for intermittent heating and a thermostat. Where hot water and heating are combined the hot water should run independently during the warm season.

Lack of hot water

This problem may be caused by an excessive length of primary flow and return pipes, air locks, poor quality fuel or by an inadequate sized boiler.

Pipe noises

Knocking may occur in the primary flow and return pipes resulting from furring, corrosion or freezing of the water. In such cases the pipes may require descaling or renewing.

Oil-fired boilers

When dealing with an oil-fired boiler which is not operating correctly the surveyor can make the following simple checks:

- Check that the oil tank is not empty and that the vent is clear.
- Check that the oil filter is not blocked, and that there is no air lock in the oil supply pipe from tank to boiler by opening the vent.
- Occasionally, the burner goes out, but there may be a reset button which has to be pressed.
- Ensure that the electrical switches and fuses are in order.

Gas boiler

With a gas boiler make sure that all gas cocks and valves are in the open position, and that the pilot light is working. The burner cannot come on if the pilot light is not properly lit. If the pilot light is unstable the cause could be a faulty thermocouple or too low gas pressure. The system should be checked and maintained by a CORGI registered gas engineer.

Solid fuel boilers

This type of boiler may go out because there is not enough draught.

- The chimney should be checked to ensure that it is not blocked.
- Conversely, a boiler may go out due to too much draught through the boiler when the dampers are closed. In such cases all the seals on the fuel and ashpit doors should be checked and if necessary renewed.
- Dampers should also be checked and the chimney examined for excessive draught.
- Boilers will not burn well if incorrect fuel is used or the fuel is damp. This is mainly a matter of experience and trial and error with different amounts of fuel.

Water supply

Check that the pump is switched on and running. When a pump is running correctly it produces a slight humming sound, which can be detected by holding a screwdriver or similar object against the pump with the handle close to the ear. Poor circulation can be checked by feeling the connections on the flow and return to each radiator. If the flow connection is hot and the return only slightly warm, this indicates a faulty pipe, wrong pipe sizes or a blocked pipe. If after releasing the air from all the radiators and switching the pump back on, there is still no circulation of hot water, then a specialist must be consulted. Check that there is sufficient water in the feed and expansion cistern, and that the ball valve is working properly.

If a radiator is warm at the bottom and cool at the top, air is present and can be released by slightly opening the air vent with an air vent key.

SANITARY FITTINGS

14.16   Materials

When carrying out an examination of sanitary fittings it is advisable to consider the type and age of the fitting and not just its condition. Obsolete types will still be found in older buildings, but are now considered to be inefficient and insanitary. Sanitary fittings must be non-porous, durable and easily cleaned. Vitreous china is probably the highest quality in sanitary ware, but fireclay, stainless steel and enamelled steel are also used. Ceramic materials are sometimes difficult to control during manufacture. The material tends to warp during drying and firing.

14.17   Wash basins and shower trays

These fittings are available in fireclay and Perspex. Perspex trays are supported on a timber or steel angle cradle. The surfaces of these fittings should be closely examined and if cracked, pitted, crazed or irretrievably stained, then the defects should be noted and replacement recommended. Wash basins can either be supported on a pedestal, cantilever towel rail brackets, concealed hangers or legs and brackets. All brackets and pedestals should be checked to see that the fixings are secure. Wash basins of stainless steel are usually found in commercial and industrial buildings. They are strongly made to resist damage by accident or misuse.

14.18   Baths

Baths are usually of cast-iron with a porcelain enamelled finish, but in some more expensive properties ceramic ware is sometimes found. Reinforced Perspex baths are often used in more modern properties and are supported on a tubular steel frame with adjustable fixings.

Baths suffer the same faults as wash basins and any defects should be carefully noted. In older properties the surveyor will often find the roll-top edge bath some of which have been timber panelled. Many of these baths have become insanitary and the panels affected by wet rot should be replaced. A lot of damage to baths is undoubtedly occasioned by impact damage which has often been 'touched-up' with various types of paint. This type of repair is simply temporary, and the fault is discovered with use.

Bath panels may be of enamelled hardboard, asbestos, or Vitrolite, or in some cases the sides have been tiled to match the wall tiling. Access can be difficult with panelled baths, and this is an area where defects are likely, such as leaking traps or pipes which may have remained undetected for many years causing defects in the timber floor. If access cannot be obtained without causing damage to the property then this should be clearly stated in the surveyor's report.

14.19 Bidets

Bidets are usually cast in vitreous china in a fairly wide range of colours and are available with single or double taphole punching. They are fitted with flushing rim, overflow, ascending spray, pop-up waste or chain waste. The hot and cold supplies are entered by the flushing rim. Most of the faults described in Section 14.17 for wash basins are applicable to bidets, and should be carefully noted including an examination of the joints to the pipes and fittings for any tell-tale leaks.

14.20 Sinks

Sinks are made of fireclay or stainless steel and are supported on cantilever brackets or on the framework of a kitchen floor unit. Fireclay sinks are susceptible to damage as described above for wash basins. Stainless steel will be found in more modern buildings where it is moulded to form a combined sink and drainer unit, or twin washing-up bowls. It provides a hygienic surface and only requires general cleaning. White enamelled steel sinks are also formed as a combined sink and drainer. If the enamelled surface becomes chipped or cracked the metal can deteriorate fairly rapidly.

14.21 Taps

There is often some confusion as to the meaning of the various types of tap. There are three basic types in general use:

- Bib taps have a horizontal inlet and free outlet and usually project above a sink. A hose tap is similar, but has a union for attachment of a hose.
- A pillar tap as used on baths has a vertical inlet and horizontal free outlet. Some have an inclined high waisted pillar.
- In modern buildings the baths and sinks are often fitted with a pillar mixer and swivel nozzle in the centre.

There are many kinds of special taps for different purposes. Spring loaded taps operated by a push button of the non-concussive type, which prevents 'water hammer' noise, are much used in industrial premises in an endeavour to save water. There are also pillar mixer fittings with swivel nozzle and quarter turn levers suitable for hospital fittings. There are many kinds of non-splash taps designed to enable re-washering without turning off the water supply.

A large majority of taps are of the screw-down variety. The principle of the screw-down tap is that a spindle, on the end of which is a washer, is screwed down until the washer engages on to a brass seating. All types of washers wear with constant use and the sign of a worn washer is when the tap continues to run after being turned off. It should also be remembered that seatings wear and become pitted and are often the cause of constant re-washering. If water discharges around the top of the tap spindle, this is due to faulty packing or a worn spindle.

14.22   Water closets and cisterns

Both these fittings require close attention. Cased in WC pans of the old valve operated type or the obsolete hopper type will still be found in old properties, but are now considered insanitary and should be noted for replacement. The surveyor should carefully note any pans that are chipped or cracked. Defects of this type harbour germs and are therefore insanitary. The majority of pans found in domestic properties are single trapped. The two-trap syphonic pans are usually found in larger buildings and are becoming increasingly common. They are far more efficient and silent in action.

The surveyor should carefully examine the joints between the fitting and waste pipe including the floorboards for signs of a 'slow leak'. Damp floorboards are frequently the cause of wet rot around the base of a WC.

High level flushing cisterns in older properties are frequently made of cast-iron with a bell-type action. The iron casing and valve fittings will often be found to have deteriorated especially in older commercial and industrial buildings. They are also a source of noise and annoyance. In such cases it is advisable to recommend replacement. Ceramic or plastic cisterns incorporating a piston actuated flush are usually fitted to modern low level suites, and if properly installed they function with far less noise. The cistern is controlled by a high pressure inlet valve connected to a ball float arm. The capacity is normally 9 litres to give a reasonable flush and this is the maximum permitted by some water authorities. The cistern should be fitted with an overflow pipe to give warning of a fault and must discharge to the outside of the building. The water level in the cistern should be approximately 25 mm below the overflow pipe. The overflow pipe should be larger than the inlet pipe, usually 19 mm in diameter if a 13 mm inlet pipe is fitted. Flush pipes can be of copper, plastic or steel with a telescopic joint.

The following items should be checked:

- The cistern brackets should be securely fixed to the wall.
- The ball valves should be watertight and not punctured.
- Inlet valves for faulty washers.

- Flush pipe connections to the WC pan. The joints should be made with a rubber or plastic connector.
- The lever action for loose or worn fittings.

14.23   Urinals

Urinals are the most difficult sanitary fittings to keep in sound condition. There are three basic forms in general use; stall urinals where the slab is integral with the channel; the flat back slab type set upon a separate channel and urinal bowls fixed to a wall. The weak point in the slab and stall type is the joint. Many of the older type urinals were installed with a cement joint, and if the jointing is porous or loose, absorption may be high with a risk of seepage into the floor below. This is a point which should be carefully noted when carrying out an examination especially where defects in the floor have been reported.

In more modern installations the joints have usually been treated with a polysulphide rubber-based flexible sealer. This material has a high resistance to attack by acids and alkalis. Many urinals are now made of stainless steel of one-piece construction eliminating risk of leaking joints. All types of urinals should be fitted with a plastic or fireclay automatic cistern supplying 4.5 litres of water per stall at maximum intervals of 25 min. The cistern distributes water through a stainless steel or copper sparge pipe or CP button spreader with the pipework in a duct. Apart from the jointing problems mentioned above the following points should be checked:

- The floor adjacent to the channel should be impervious and fall towards the channel.
- Channel outlets should be properly trapped and access provided by removable inlet gratings.
- Urinals made of fireclay suffer the same faults as WCs and basins when chips or cracks occur.
- Urinal bowls need to be well secured to the wall with cast-iron hanger and bracket supports. All bowls must be fitted with trapped outlets and sealed joints. Ensure that the outlets are not blocked.

WASTE AND SOIL PIPE INSTALLATIONS

14.24   Introduction

The basic function of any soil and waste system is to immediately accept the flow from any sanitary fitting and discharge it efficiently into the drainage system and in such a manner that no nuisance is caused. The surveyor engaged in the examination of waste and soil systems will find that most modern installations use the single stack system. The detailed design of this system is governed by the general principles set out in BRE Digests 248 and 249 and the DOE advisory leaflet No. 73 'Single stack plumbing'. Before 1965 the Building Byelaws divided waste and soil pipes into two parts, external soil and ventilating pipes and internal waste plumbing. Under the 1965 Building Regulations for a building of four or more storeys the soil and waste pipes must be fitted internally. For a building up to three storeys the soil and waste pipes may be inside or outside of the external wall. However, the surveyor will no doubt be concerned with all types of property so all the various methods of soil and waste disposal will be described below.

14.25   Types of soil and waste disposal systems

Two pipe system

In this system the soil and waste water are piped separately with or without ventilating pipes according to the size of the installation. The waste is discharged to the drain through a trapped back inlet gully, and the discharge from the water closets is direct to the drain. This system has a disadvantage in regard to the present day use of detergents and is now seldom used (see Fig. 14.5a).

One pipe system

In this system the soil and waste are conveyed together in one pipe with ventilating pipes connected to each fitting and sized according to the loading of the installation. The main stack is connected directly to the drain (see Fig. 14.5b).

Single stack system

This is a much simpler system, does not employ separate ventilating pipes and is commonly used although it is based on strict rules which must be observed e.g. deep seal traps must be used on all fittings (see Fig. 14.6).

Obsolete systems

In many domestic properties built before the Second World War the surveyor will find that the waste pipes from fittings above ground level, such as baths, basins and showers, will discharge into hopper heads and the sinks at ground level discharging over a trapped gully. The WC is usually connected to a ventilated soil pipe. The waste pipe and hopper are not a satisfactory method and will often be found to be partially blocked with slimy soap and hair etc., and in very cold weather the waste water will freeze and possibly fracture the pipes (see Fig. 14.7).

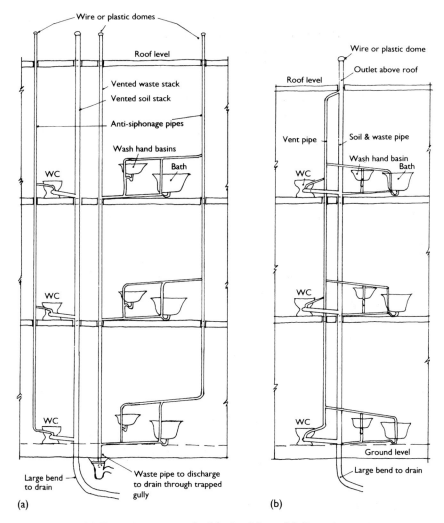

Figure 14.5 (a) Two pipe system for block of flats. (b) One pipe system.

14.26 Pipes

Generally, both for external work and internal work in ducts, heavy duty cast-iron was used exclusively in the past and is still used for stacks in high buildings. Branches will also be in cast-iron, but for intricate work in older buildings they are often in lead. Plastic soil and vent pipes and fittings jointed with preformed butyl seal suction type joints are now commonly used in modern buildings. Copper soil and vent pipes will be found in older buildings. They are suitable for internal sanitary plumbing systems, and are strong, rigid and light. Pitch fibre pipes have been used

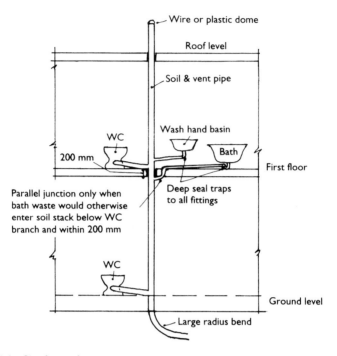

Figure 14.6   Single stack system.

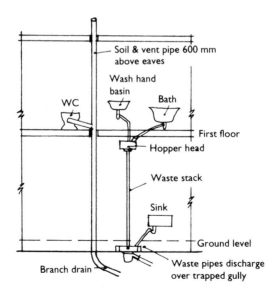

Figure 14.7   Disposal of soil and waste water in older domestic properties.

in vertical stacks. The material is tough and flexible but experience of their use is limited.

14.27   Common defects

Defects in waste and soil pipes may stem from one or more of the following factors:

- Traps should be self-cleaning, readily accessible and if possible they should be easily removed by the use of two piece traps or be fitted with a cleaning eye. WC traps should have a minimum water seal of 50 mm. Traps of other appliances should have a 75 mm water seal.
- Where pipes have been enclosed in ducts the surveyor may find difficulty in obtaining access, but if the access covers are fitted with brass covers and screws it should be a fairly easy matter to remove the cover. The ducts should provide ready access for maintenance, testing and cleaning, and should be constructed appropriately for fire resistance and sound insulation. Access panels should be carefully positioned to allow sufficient elbow room for cleaning rods.
- Pipe joints and traps etc. should be checked for leakage. Defective joints which allow waste water to penetrate into a floor may possibly lead to an outbreak of dry rot.
- When dealing with one pipe or two pipe systems the position of the ventilating pipes should be checked. The branch discharge pipes should not be less than 75 mm from the crown of the trap and should be carried upwards and connected to the main vertical anti-siphonage pipe. Both the soil and vent pipes should be carried up above the eaves and fitted with wire or plastic domes.
- Stacks should always be straight; there should be no offsets on stacks below the topmost appliances unless venting is provided to relieve any back pressure.
- Access covers should be provided to soil branch connections except on the ground floor. The covers should be readily accessible with adequate clearance for the entry of clearing rods.
- Where the vertical stack is of somewhat complicated nature e.g. a block of flats with shops at ground level and a basement below, soil stacks have to be taken horizontally and are often secured to the basement ceiling. If a single stack system is used throughout the flats then it is necessary to ensure that the advantages of the single stack are used with modifications necessary to maintain proper ventilation of the seal of all the traps in the system. Stoppages occur where the pipework is complicated by sharp offsets or knuckle bends. In old buildings access to horizontal pipes is often inadequate and badly sited.

- Vertical plastic soil stacks in high buildings have been known to settle and crack at the socket collar due to thermal movements and not being properly supported. The surveyor should check that every length of SVP is securely supported and that the spigots are free to expand into the lower socket.
- Plastic waste pipes fixed on external wall surfaces should be properly supported at intervals no greater than the following:
  Pipes up to 40 mm dia: 0.5 m maximum.
  Pipes up to 50 mm dia: 0.6 m.
  Check that 'push-fit' joints have been used with clearance for expansion.
- Exposed polypropylene pipes (to BS 5254) should be carefully checked. Unless they are fully protected from sunlight they become brittle and eventually fail. If found to be unprotected a recommendation should be made for the pipes to be coated with a suitable gloss paint.

ELECTRICAL INSTALLATIONS

14.28 Introduction

Electrical systems in buildings must be designed and installed properly for health and safety reasons as well as to satisfy performance requirements. For example, faulty wiring or loose connections can trigger a fire or give an electric shock to anyone who comes into contact with certain parts of the installation. If incorrectly installed, the wiring or fittings may be unsightly or may be inadequate for the needs of the user (e.g. power points too few or positioned too low for disabled users) as well as being unsafe.

Electricity began to supersede gas for lighting purposes in the early 1910s. A considerable number of buildings equipped with the original electrical system have already been re-wired once and are no doubt due for a second test. Any property over 25 years old is probably in need of rewiring especially if TRS cables have been used. During the past 30 years there has been an increase in the number of electrical appliances used, especially in offices and kitchens. Besides power and lighting systems, there are internal telephones, security systems, cable television, fire protection and alarm systems. It is, therefore, important to the prospective occupier that the existing installation is adequate for their requirements and this can only be ascertained by testing. A further fact which the surveyor must recognise is that electricity when abused can be lethal. It rarely gives warning, the first indication something is wrong being when a person receives an electric shock, which in some cases can prove fatal or inflict serious injury.

14.29   Wiring systems

There are several methods of wiring in common use some of which are now obsolete and often dangerous. It is, therefore, of the greatest importance that the examination is carried out in an atmosphere of 'safety first' and this recommendation cannot be too strongly stressed. The details of the various wiring systems are described below:

- Cable installed in conduit is regarded as a high quality job, and can be laid in floor screeds or chased into walls which protects the wiring from nail and screw damage. The system permits rewiring to be carried out more easily. However, it is susceptible to corrosion and condensation from external sources. PVC insulated cables are now used in many modern installations. There is no risk of corrosion, but cables can be damaged by nail and screw fixings. Earthing is provided by an earth wire. If VIR (vulcanised india rubber) cables are found in metal conduit then renewal is due.
- PVC (polyvinyl chloride) sheathed and insulated cable has been used in most modern buildings since about 1950 and is the most commonly used system at the present time. The advantage of PVC sheathed cable is that it can be adapted easily for many types of installations.
- Polythene sheathed cable is similar to PVC and is slightly cheaper. It sometimes becomes softer at lower temperatures.
- TRS (tough rubber sheathing) has been used since about 1925, but from about 1960 has been practically superseded by PVC. This system has vulcanised rubber insulted conductors surrounded with tough rubber sheathing. The cables are usually concealed in floors and walls, but if placed on the surface are generally drawn through the conduit. If this type of wiring is found it is no doubt due for renewal.
- MIMS (mineral insulated metal sheathed) is a very reliable system and consists of solid copper or aluminum conductors surrounded by a compressed insulating mineral, but often has a copper or PVC outer sheath. The cable can be used in difficult situations. It is non-inflammable, and impervious to oil, water and condensation. The cables can be buried in concrete or plaster fixed with copper clips and saddles.
- Many mid-Victorian and Edwardian buildings had electrical installations installed between 1910 and 1920. The wiring is lead sheathed and fixed on the surface enclosed in a wooden trough. This type of wiring has been obsolete for many years and is seldom found today.

14.30   The ring circuits

The surveyor will find that the ring circuit is now the recognised method of supplying socket outlets. The ring circuit consists of a pair of conductors and an earth wire which commences its journey at a 30 amp fuseway in the consumer unit and returns to the single terminal on the same fuseway. Miniature circuit breakers (MCB) are now used as well as fuses. The latest IEE wiring regulations state that ring circuits can feed an unlimited number of socket outlets, but are limited to an area serving 100 m^2. A simple check on how many different circuits there are is to examine the fuseboard to see how many fuses are being used. In a three bedroom house with gas central heating there are usually three or four fuses; one for the ring main, one for the lighting circuit, one for the immersion heater and one for the cooker point.

14.31   Testing and inspecting installations

Wiring regulations

Every installation and every major alteration to an existing installation shall, on completion, be inspected and tested in accordance with the 16th Edition of the Wiring Regulations or BS 7621: 2001 and subsequent amendments at least every five years (Scaddan 2001). Some tests rely upon the soundness of other parts of the installation for their effectiveness. Testing simple installations is not difficult for the surveyor experienced in this type of work provided he has the right equipment. Circuits are tested by means of a 'megohmeter' which is used to ensure that the insulation resistance is sound enough to prevent leakage. The instrument is connected to the circuit terminals, but before commencing the test the fuses must be removed and the main switch at the consumer unit turned off. The megohmeter will also indicate defects due to age or damp conditions. Each circuit should be tested from phase to earth, neutral to earth and between phase and neutral. Megohmeter readings should not fall below 1 megohm.

An earth loop tester is used for testing the earthing of the installation. The instrument is plugged into the socket outlets and the main switch in the 'on' position. If the reading is above 1 ohm the earthing is defective.

The supply authority have now adopted a method of earthing known as 'Protective Multiple Earthing' (PME) in lieu of the old connections to the gas and water mains. Now that so much plastic is being used for gas and water services these pipes can no longer be relied on as an earthing conductor. The surveyor's report should cover not only the insulation resistance and earthing of the installation, but also an opinion on the following items:

- Condition of the switch gear.
- Circuit wiring to cooker controls, water heaters and immersion heaters.

- Condition of the socket outlets which should be three pin incorporating an earth.
- Condition of the pendant drops, including ceiling roses and lamp holders.
- Off-peak storage heaters, electric fires and portable appliances (including associated wiring) in public places should have safety tests.
- An indication of the condition of the wiring may be obtained by unscrewing a switch plate from the wall and examining the wiring attached thereto.
- In timber joist construction, the cables should run in the direction of the joists and be fixed to the sides and not the top of the timbers. Where cables run in the opposite direction they should be passed through the 'top third' of the joists and not laid on top where nailing down of floor coverings could penetrate the cable.
- Bathroom lighting and high level electric fires should be controlled by pull cord switches.
- In order to establish that all fuses are on the right circuit, all the fuses should be removed from the consumer unit and the MCBs switched to 'off' position except the circuit being checked. Then check that everything on that circuit is working satisfactorily. Some fittings have fuses fitted in them such as immersion heaters, shaver points, cooker sockets and storage heaters.

The above notes are for general guidance only and, 'safety' should always be the first priority when carrying out an examination of the electrical installation. If in doubt it is advisable to consult a competent electrical engineer or the supply authority.

Tests must be carried out in a particular order because the test itself may cause danger when applied if the system is faulty and the protective device does not function properly. A recommended sequence of tests is as follows:

(1)   Ring circuit continuity.
(2)   Protective conductor continuity.
(3)   Measurement of earth electrode resistance.
(4)   Measurement of insulation resistance.
(5)   Assurance of protective efficiency of barriers and enclosures.
(6)   Measurement of the resistance of non-conducting floors and walls.
(7)   Verification of polarity.
(8)   Measurement of the earth fault loop impedence.
(9)   Testing of earth leakage circuit breakers.

For example, the possible causes of problems such as too high an impedance are: bad connections in the wires used, imperfect bonding of metal cable sheaths and loose joints between steel conduits. Locating such sources of high impedance may be time-consuming and laborious, but unless rectified the condition would prevent satisfactory operation of the protective device under fault.

Surveyors can use simple electrical testing tools such as a 13 amp socket tester described in Chapter 2. This plug-in device can highlight faults such as no fuse, correct fuse, live/neutral reverse, neutral fault, live/fuse fault – but it will not detect earth/neutral reversal of wiring. If any of these are identified, the surveyor can then indicate that a qualified electrician or electrical engineer should check the installation.

The full and proper testing of an electrical installation is a specialised job, which should only be undertaken by a qualified person (e.g. certified electrician or chartered electrical engineer). However, a visual examination of an electrical installation might form part of a building surveyor's inspection of a property. The report could then advise on the overall condition of the installation. As the inspection is of a visual nature the wiring is not tested, although some parts of the installation should be checked – such as the power points using the plug test device referred to earlier.

Other inspection checks

Apart from the above items the CEM (2003) suggest that the following checks should be made:

- The face of each socket should be felt for any signs of undue warmth. Feel the outside of any plug that is in a socket, again for any signs of excessive heat. As a result of any undue warmth, switch off the electrical installation and advise that it is checked by a qualified electrician.
- Make an examination of each socket outlet; look for any signs of burns or scorch marks around the pin positions on the face of the socket.
- Locate a plug socket in the warmest location of the house; this would probably be found nearest the fire or near to the boiler or other heat source. Remove the front cover and examine the cable insulation near the terminal. Being in the warmest location it will probably be in the poorest condition. Report any cracking in the sleeve or insulation, or if it is breaking away from the wires.
- Access the roof space and examine the location of the wires. Identify whether the wires are located beneath or within the insulation. This can lead to a build up in heat within the wiring and the rapid deterioration of the insulation. If the insulation is polystyrene granules it reacts with the plastic coating of the wiring and results in migration. This can lead to the deterioration in the quality of the insulation. If any junction boxes have been used, check the cover of the insulation runs into the junction box. Check for any bare wires around the junction box or connections within the roof space.

- Check the wires that run from the mains supply to the electricity meter. The original wires are usually rubber sheathed – if they are dull pink they will need to be replaced. Replacement wires will be red and black sheathed in PVC and should be of an appropriate rating.
- The mains supply fuse should be checked for its capacity. The capacity should be written on the side of the container. If it is less than 60 amps it is not sufficient for a modern domestic installation.
- The fuse and wire ratings of any additions such as for a shower unit should be checked for capacity and tightness of connections. There is a risk of overheating in wiring of shower units over 9kW with a fuse rating of less than 40 amps, especially if the connections are loose or the cable size is inadequate.
- Check the wires in any pendant drop. If they are covered in fabric or cotton braid it will indicate old wiring. If the property has been rewired then it may suggest parts of the rewiring have been skimped.
- Check the condition of all fixed appliances, e.g. extractor fans and night storage heaters. Check that the items work correctly and don't make any undue noise; look for signs of overheating.
- Check the size of the mains fuse at the main entry; it may not be adequate for the number of outlets in the property.
- Check the condition of the wiring on the tails of the meter board. This wiring may not have been changed when the rest of the property was rewired; report if the wiring is of poor quality.
- Check the length of the cables within the property supplying electrical fittings. Where these cables are in excess of 2.5m it can be an indication of inadequate sockets or inconvenient location.
- Check that no power socket is provided in the bathroom, other than a shaver socket, and if the light switch is within the bathroom it should be the pull type.

Earth bonding

The IEE Wiring Regulations state the following should be earth bonded:

- Hot and cold water supply on kitchen sink.
- Gas installation on the household side of the supply.
- Hot and cold supplies to baths, basins and bidets and cold supply to toilet.
- The connections running to and out of the boiler.
- Water heaters and towel rails.

Completion

The Wiring Regulations require that after completion of electrical installation work the contractor must give the person ordering the work a completion certificate. They also require that any defects and omissions revealed by an inspection or test must be made good before a completion certificate is issued.

The completion certificate should be signed by one of the following:

- A chartered electrical engineer.
- A member of the Electrical Contractors' Association.
- A member of the Electrical Contractors' Association of Scotland.
- An approved contractor of the National Inspection Council for Electrical Installation Contracting (NICEIC).
- A suitably qualified person acting on behalf of one of the above, in which case the person for whom action is taken should be stated.

14.32   Regulation of electrical installation work in dwellings

Background

As from 1 January 2005, tighter controls have been imposed by government legislation on the standard of electrical installation work in residential properties across England and Wales. These controls are analogous to those required for gas installations by the Council for Registered Gas Installers (CORGI).

The controls covering the standard of electrical installation work in dwellings are incorporated in Part P of the Building Regulations 2000. The Building Standards (Scotland) Regulations 2004, which are in force as from 1 May 2005, are likely to follow this development (in Section 4: Safety). The primary effect is to make electrical safety in housing a legal requirement.

Scope

Approved Document P (Electrical Safety) applies to all fixed installations after the distributor's meter in buildings or parts of buildings comprising dwellings and related property. It is therefore a legal requirement for anyone undertaking electrical work in residential buildings to comply with Part P of the Regulations by adhering to the fundamental principles for achieving safety delineated in BS 7671: 2001. Unlike this British Standard, however, Part P does not cover the inspection and testing of existing electrical installations (NICEIC 2004).

All electrical installation work in dwellings must be notified or self-certified by a prescribed competent person, even if only 'minor works'. This includes special installations and locations in and around dwellings, such as bathrooms, kitchens, saunas, solar power systems, etc.

Survey implications

Surveyors should note any alterations or additions to the electrical installation when undertaking a domestic building inspection. They should check that any such work has been certified as complying with Part P and other relevant parts of the Building Regulations 2000 that have been affected (as mentioned in NICEIC 2004), such as Part B (fire safety), Part L1 (energy conservation), and Part M (accessibility).

GAS INSTALLATIONS

14.33 Introduction

The preparation of an accurate report on the gas installation is sometimes difficult to obtain. This is often due to the fact that the pipework is concealed in the structure and cannot be examined visually. In the past surveyors have simply made a superficial examination of the gas taps and meter, but have not carried out tests of the gas piping as it is sometimes known. Perhaps the reason for this is that in the past the installation in the average domestic property consisted of one run of piping between meter and cooker. This simple form of usage seldom caused problems. Serious faults, if they occurred, were usually due to defects in the service pipe from the main to the meter or perhaps a faulty meter both of which are the responsibility of the supply authority. However, in recent years the position has changed considerably due to improvements in heating appliances and the demand for gas central heating and hot water installation coupled with the increasing cost of oil supplies.

Older type properties, say 70 to 100 years old are obviously suspect, and no doubt the gas installation has never been tested. Over the years the service or installation pipes may have become corroded or clogged. These defects, together with the increase in the number of gas appliances, make it advisable to point out to the client that the supply authority should carry out an inspection and test the service pipes and the appliances. (Hot water and central heating boilers have been dealt with in Sections 14.5 to 14.15.)

14.34 Checking defects

During the initial survey, the surveyor should check the following points:

- Gas leaks should be investigated by a CORGI registered gas engineer at once but may not necessarily mean that a fault has developed. There could be a gas tap not quite turned off on fires or cookers, or a pilot jet blown out. Leaks in pipework are best located by smell, the odour getting stronger as you approach the leak. Use a torch in dark places, never a naked flame! When the leak has been located, the lever at the side of the meter must be turned off, but make sure that all appliances, including pilot lights are also turned off. The supply authority should then be informed. The supply authority should also be told if gas can be smelt inside the building, but cannot be traced, as it may be percolating through the earth.

- Care must be taken that control valves are not situated in easily accessible positions where they may be vandalised. They should also be checked for ease of action.
- Check that all installation pipes where exposed are properly supported and protected against accidental damage.
- Sharp bends on the pipework should be as few as possible. They produce a loss of pressure. Pipes through walls should be in sleeves.
- Gas pipes should be kept well away from electric wiring and should not touch other service pipes.

14.35   Gas meter location

When dealing with domestic properties erected during the past ten years, the surveyor will often find that the gas supply authority has installed the meter externally in a reinforced plastic box. The box is fitted on the outside wall of the dwelling facing the gas main or if practicable on an adjacent wall not more than 2 m from the front wall. In new work the box is usually installed by the main contractor in accordance with the supply authority's requirements and is built in as the work proceeds. Generally, the bottom of the meter box must be at least 500 mm and not more than 1 m above the finished ground level. Gas and electricity meters should not be fitted within 150 mm of one another unless adequate fire-resisting and electrically insulated material is placed between them.

LIFTS AND HOISTS

14.36   Introduction

The design of lifts and hoists is a highly specialised form of engineering. Most manufacturers have developed their own systems which embody a number of patents, and many of their parts are not interchangeable. During his first preliminary visit the surveyor will no doubt find that the lift will appear to be working satisfactorily and that the owner or occupier of the building has a maintenance contract with the installer of the equipment. The contract usually states that the lift will be periodically examined and be kept in sound mechanical order.

Lift installations almost without exception consist of two quite separate but essential elements. Firstly, the mechanical equipment for operating the lift, and secondly the building structure supporting the equipment and lift car. Normally, when carrying out a periodical examination the engineer is not concerned with the structure. However, it must be remembered that these two elements must be considered collectively in order that the surveyor can produce a full report on the installation. It is, therefore, of paramount importance that the surveyor examines the building structure supporting or enclosing the lift, and this operation can only be carried out with a lift engineer in attendance.

There are three main areas requiring the surveyor's attention: the lift pit, shaft and machine room. The first operation on site will involve the lift engineer who will shut off all push button controls at all levels so that the surveyor and engineer can use the lift without interference from the landing call points (see Fig. 14.8).

14.37 Lift pit

The procedure is as follows. The engineer will raise the car to the first floor level and open the ground floor entrance doors. With the aid of a short ladder the surveyor will be able to examine the lift pit. The following points should be observed. The base of the pit should be equipped with buffers and guide bases secured to the pit floor. The sides and base should be constructed of waterproof concrete or, if constructed of brickwork, the inner face must be lined with waterproof rendering. It is important that the pit is impervious to damp penetration.

14.38 Lift shaft

At ground floor level the surveyor should examine the shaft between ground and first floor. The inside of the lift shaft should be smooth and apart from the projection of the landing nosing should be free from ledges and recesses etc. Check that there are no signs of structural movement in the brickwork particularly at the internal angles. Examine the shaft between first floor level and the underside of the machine room floor. This operation can only be satisfactorily carried out by the surveyor and lift engineer standing on top of the car. The engineer will bring the top of the car to first floor level, the surveyor and engineer will then enter through the outer lift doors. At first sight this manoeuvre may appear to be dangerous but the writer has done this many times, and if the necessary precautions are taken there is no danger. Before entering the lift shaft it is advisable that the lift engineer demonstrates that adequate safety

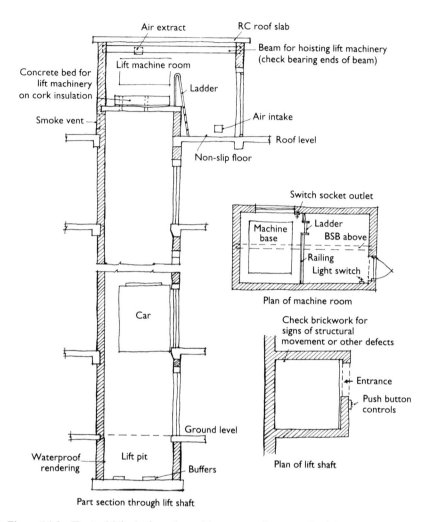

Figure 14.8 Typical lift shaft and machine room showing 'Builders' work'.

measures have been taken and that the surveyor must pay due regard to the importance of any instructions the lift engineer may give him so as to ensure the safe operation of the lift.

The surveyor should also ensure that his safety helmet is worn at all times during a lift examination. Failure in certain cases to wear a safety helmet is likely to lead to enforcement action being taken by the Health and Safety Executive under sections 2 and 7 of the Health and Safety at Work Act 1974, against either the lift engineer or the surveyor. The engineer can easily control the lift mechanism from this position leaving the surveyor to examine the walls of the shaft keeping a look-out for the various points mentioned above. A lamp is usually provided as part of the equipment on top of the lift car, but it is also advisable to take a torch for close-up work.

Apart from making notes of any structural defects the surveyor must also check the ventilation. Lift shafts must be permanently ventilated at the top to allow smoke to dispense into the atmosphere. The minimum unobstructed area shall not be less than $0.1\,m^2$ for each lift in the shaft, and must open directly to the external air or via the machine room through trunking. The vent should be louvred or protected to prevent rain, snow or birds entering the lift shaft. Check the opening to ensure that it is not blocked.

14.39   Machine room

The lift machine room is normally placed above the lift. The following 'building items' should be checked by the surveyor.

- Ventilation by air bricks at high and low level.
- A BSB (British Steel Beam) for hoisting tackle is usually situated just below the ceiling and directly above the lift machinery. Check bearing ends for any signs of movement. Steel embedded in brickwork may corrode and cause the brickwork to crack.
- The wall and ceiling surfaces should be smooth, clean and preferably plastered although this is not essential. Floor surfaces should be of non-slip material.
- Doors and windows should be examined for any signs of decay or corrosion as described in Chapter 13. If trapdoors have been installed in the floor they are usually half-hour fire resisting and it is therefore important that the timber and ironmongery are in accordance with the current Codes of Practice.
- Check the concrete base supporting the lift machinery bed plate. The base usually sits on a 76 mm thick bed of cork insulation covered with bitumen felt. Even so, noise from the lift machinery is difficult to eliminate.
- Metal work such as ladders and railings should be examined for stability and any signs of corrosion.
- Care must be taken to ensure that all statutory requirements in respect of means of escape and the building regulations governing the construction of the machine room are complied with. It is important that the machine room is separated from the lift shaft with non-combustible materials except for minimum openings necessary for the passage of wires and cables.
- Permanent electric lighting above the machinery is essential. The light switches should be positioned adjacent to the personnel access doors. A switched socket outlet should be provided for a wandering lead or for the lift engineer's power tools. These circuits must be independent

of the lift supplies and should be tested by the electrical engineer as described in Section 14.31.

14.40   Prevention of damage

One of the problems facing those responsible for the care and maintenance of lifts is the current trend towards vandalism particularly on housing estaes. The writer does not wish to give an opinion on the sociological aspects of the problem, but to suggest methods of reducing the incidence of vandalism by selecting materials which offer greater resistance to damage. Lift engineers are well aware of the problem and will usually cooperate with the surveyor's suggestions. The following fittings have been adopted on many housing estate lifts, some of which have been recommended by the BRE (see Digest 238).

- Various finishes can be adopted for decorative purposes and to give protection against wilful damage to the lift car interior. One such lining is stucco embossed stainless steel sheet. The sheet is roller embossed to produce an irregular raised pattern finish. The advantage of embossing is that it is not so easily damaged by writing or scratching.
- Damage caused by tradesmen wedging lift doors open when delivering goods is best resolved by the provision of stop buttons.
- Plastic control buttons can be burnt or prised out and should be replaced with flush metal buttons.
- Fouling of the lift floor can be a problem and to reduce this risk it has been suggested that the design of the flooring shall allow it to be renewed without disturbing the car body work. The flooring shall be made of 25 mm thick WPB bonded plywood covered with epoxy resin mortar screed 12 mm thick, and the edges butting up to the car skirting panel shall be coved to a height of 35 mm above floor level.
- The car operating panel should not be opened from the inside of the lift car. The panel is to be made from stainless steel and fitted with a locking arrangement to ensure that the panel shall spring open when released by an opening handle fitted to the car roof.
- Armour plate glass covers to lighting fittings are advised as plastic covers are easily broken.

14.41   Small service lifts

These are mainly used for food lifts in restaurants or hotels and as light goods lifts from basements serving one or two floors. The entrances consist of a rise and fall shutter fitted with fully automatic push button controls for call and despatch. In modern installations the lift compartment is usually constructed of enamelled mild steel or stainless steel, and operates inside a self-supporting metal frame. This type of lift does not require a 'load bearing' shaft or separate motor room. The winding unit consists of an electric motor and reduction gear-box and is mounted in a small compartment at the top of the lift shaft. The external cladding to the supporting frame and access door to the winding unit should be protected to a minimum standard of half-hour fire resistance, but in larger public buildings the local authority may require a one-hour fire resistance. It is important for the surveyor to familiarise himself with the fire regulations in respect of the building, and check that the lift enclosure complies with the Building Regulations. As with all lifts the equipment should be regularly serviced by a competent person whose report must be examined by the surveyor.

14.42 Hand power hoists

This type of hoist is similar to those described above and is usually found in older buildings for the conveyance of food in restaurants, hotels or large houses. The operation of the hoist is by a hauling rope which passes outside the enclosure. The self-sustaining gear is automatic in action, locking in any position as soon as the hauling rope is not pulled. Many of these hoists including the enclosure are made of timber and are not always protected by fire-resisting materials. Although this type of hoist is now obsolete they are extremely simple installations and usually work satisfactorily. However, if after a detailed examination the surveyor considers that there is a fire risk, the occupier or purchaser must be warned of the problem.

14.43 Stair lifts

It is now recognised that lifts need not be just for public and commercial buildings, but can be utilised to help the elderly or handicapped people who find stairs difficult to negotiate. The surveyor will find that electrically controlled stair lifts have now been installed in many private residences, nursing homes and residential care homes.

The stairlifts are compact, easy to install and can be fitted to an ordinary stair whether straight or curved. The seat arm is fitted with constant pressure push-buttons and automatic stop and release. The rail is simply attached to the stair by the installers and no 'builders' work' is involved. However, the surveyor should check the stair construction as described in Section 12.24. Stairlifts are usually examined and tested periodically. Copies of their certificates should be in the owner's possession, and the surveyor should make a note of the details for his report.

VENTILATION AND AIR CONDITIONING

14.44   Natural ventilation

It is a requirement of the Building Regulations that any habitable room shall (unless it is ventilated by mechanical means) have one or more ventilation openings. The total area of the openable part of a window, hinged panel or adjustable louvre must be 1/20 of the floor area of the room served and open directly to the external air. The opening portion must not be less than 1.75 m above the floor. A door opening directly to the external air may be utilised if it contains a ventilation opening having a total area of not less than 0.01 m^2 which can be opened when the door is shut.

When dealing with older buildings the surveyor will often find that habitable rooms have permanent ventilators in the form of terracotta air bricks or patent horizontal window ventilators fitted within the glazing rebates at the top of the metal or timber window. Both these methods are effective provided they are properly positioned and do not result in unpleasant draughts. Having carried out many building surveys of domestic and commercial properties the writer has found that the occupiers have often blocked the vent apertures with rolled up pieces of paper, thus rendering this type of ventilation ineffective. Whichever system is found, only accidental damage can occur.

14.45   Mechanical ventilation

Mechanical ventilation is an essential requirement for internal bathrooms, WCs and kitchens where there is no natural ventilation at all. The equipment consists of metal duct extractor fans allowing three complete changes of air per hour and must discharge directly into the external air. Some extractor fans are controlled by the lighting switch operated by the person entering the compartment. The Building Regulations require that lavatories must be approached through a ventilated lobby, but corridors and staircase landings are permissible approaches. A duct must be provided from the outside air into an internal lobby, whether mechanical extraction is used or not.

These simple mechanical systems rarely give trouble if the fan motor is functioning satisfactorily. The installers of this equipment will usually supply the occupiers with maintenance instructions that are simple to follow and easy to carry out. If there are doubts concerning the motor's performance then it is advisable to add this to the list of items to be checked by the electrical engineer.

14.46   Air conditioning

Air conditioning is becoming increasingly important both for large manufacturing processes and for office buildings where it has become more of a necessity and less of a luxury. The design of an air conditioning system for a large building is a very complex study, particularly when selecting positions for inlets and outlets. In the following only a brief outline is necessary in order to give some of the factors involved.

In a well designed system there are five important elements:

- Temperature.
- Humidity.
- Cleanliness.
- Distribution of air.
- Noise control.

Of these five elements it is temperature control which is predominantly important and provides comfortable conditions for the people working in the building.

There are several different types of air conditioning systems in current use each having its value for a particular circumstance. For the surveyor's purpose it is not necessary to consider the comparative merits of the various types. The important thing to note is that the system has been properly maintained. Whatever type of installation has been installed a good maintenance programme is vital and this includes regular inspections by the engineer responsible for the operation and maintenance. In the case of large installations the heating and ventilation system will be the responsibility of the plant engineer. A good engineer will always keep a comprehensive operating manual and planned maintenance scheme including details of the operating programme which the system is to perform. Taking into consideration all these various factors the surveyor should be able to formulate a basis upon which to prepare his report.

15 External Works

SOIL AND SURFACE WATER DRAINAGE

15.1 Introduction

The drainage systems of a building are of two kinds: foul water and surface water. Foul water points include WCs, wash hand basins, urinals, kitchen sinks, baths and slop sinks. Surface water is rainwater drained from roofs and pavings. The foul and surface water drains are connected to the public sewage system or in rural areas to a cesspool or septic tank. The various systems are described below in Sections 15.2 and 15.3. Before undertaking an examination of an existing drainage system, the various installations and conditions likely to be found should be studied.

15.2 Property erected before 1900

If inspection chambers are not visible on site, the possibility that they may have been covered up by raised flower beds, pavings or rubbish should not be ruled out. If, however, inspection chambers cannot be located, there is a strong possibility that the building is still equipped with the original system which was probably installed some time between 1850 and 1880. Such a system was usually laid in four- or six-inch salt glazed pipes with clay joints and without any form of concrete bed. In the case of terraced houses, the most common layout was a straight run under the house from the outside WC to the sewer. This meant that the soil branch drain connected to the WC outlet and vented to the open air and went directly to the main drain while a second pipe took the waste water from the kitchen sink to an outside trapped gulley. Sometimes the WC connection entered the main run with a 'Y' junction, the main run then being extended via an easy-bend up to yard level where it was capped off with a rodding eye. However, it was fairly common practice to form a junction by cementing the end of a 4″ branch pipe into a hole cut into a 6″ main drain (see Fig. 15.1).

Alternatively, where several houses were in one ownership, the drains were collected into a main run at the rear of the terrace before being discharged under the last house in the terrace and thence to the sewer (see Fig. 5.2). Towards the end of the century the local authorities usually

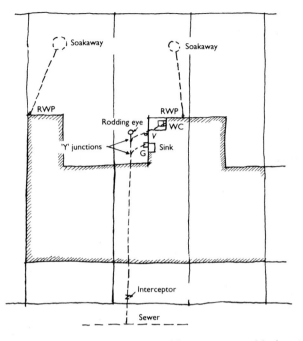

Figure 15.1 Soil drainage system to terraced housing erected before 1900.

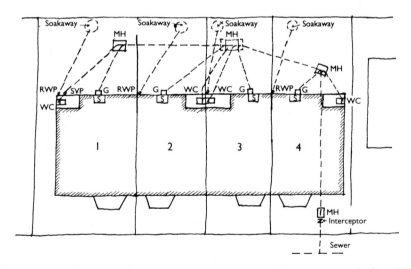

Figure 15.2 Soil and surface water system to terraced housing erected after 1900.

insisted on inspection chambers at every change of direction, but this was not always adopted and pipes were often laid in curves. The Public Health Act 1875 was the first Act to regulate the construction and ventilation of sewers. It was, therefore, considered necessary to disconnect the house drain from the sewer by means of an intercepting trap. Some of the early types of trapped gulleys and interceptors introduced were inefficient and simply consisted of a shallow bend with an unreliable water seal. These fittings often became blocked and were difficult to rod. Joints between pipes were often badly formed and in surface water drains they were almost nonexistent.

Drainage systems of the kind shown in Fig. 15.1 are seldom found today. They do not satisfy any of the requirements of the current Building Regulations in that they are unlikely to be watertight and not laid to a self-cleansing fall. The surveyor can safely assume that, if no inspection chambers can be found, a new drainage system must be provided as the system will certainly be condemned by the Environmental Health Officer. Where inspection chambers are found on nineteenth century properties similar to that shown in Figure 15.2 it is usually an indication that the drains have been relaid at a more recent date. An examination of the type and condition of the brickwork and channels can give an indication of the age of the system.

15.3   Property erected after 1900

No difficulties should be found in locating drain runs, inspection chambers and interceptors etc. During this period more reliance can be placed on the deposited plans and the date of construction can be verified. However, it will often be found that unauthorised additions have taken place since the plans were deposited. Since about 1950 drainage installations have been of the 'one-pipe' system as described in Section 14.25. An important feature of this system is that waste pipes must discharge below the level of the gulley grille thus avoiding blocked gulleys as shown in Fig. 15.9b.

It may well be found that the surface water drainage is on a 'separate system' either to a surface water sewer or to a soakaway within the curtilage of the site as shown in Figs 15.1, 15.2 and 15.3. Care must be taken not to get the systems inadvertently interconnected. 'Separate systems' were adopted by many local authorities in large towns in order to avoid overloading the sewage treatment plant. They will not allow rainwater pipes or yard drainage to be connected to the foul sewer. However, some local authorities allow the soil and surface water to discharge into the same sewer as shown in Fig. 15.4.

In some cases it will be found that, although there is no inspection chamber on the property, the drains are in fact connected into an inspection chamber in an adjoining property, as shown in houses 2 and 3 in

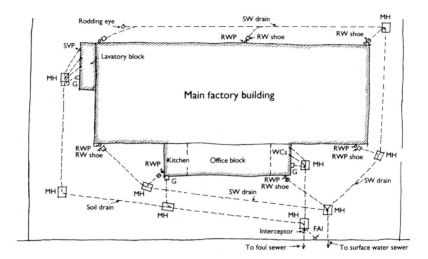

Figure 15.3   Typical separate system layout for a small factory.

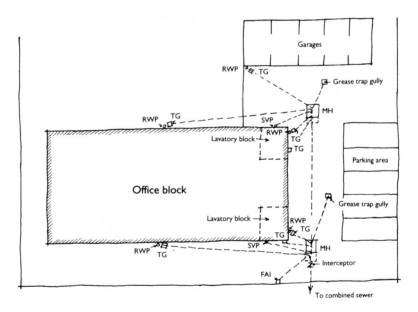

Figure 15.4   Typical combined system for small commercial building.

Fig. 15.2. No particular problems should arise from this unless the properties belong to different owners. Should this be the case a 'drainage agreement' will have to be entered into by the various owners, whereby joint responsibility is accepted for the repair of common sections of the drain. It is advisable to check this point with the owner when carrying out a drainage survey.

ASSESSMENT OF THE CONDITION OF EXISTING SOIL DRAINS, GULLEYS AND INSPECTION CHAMBERS

15.4   Sketch plan of the system

At this point a simple sketch plan of the drainage layout should be made showing the following information.

- The position of all surface water and soil gulleys giving a reference number or letter and note any waste pipes discharging into them and the fittings they serve.
- Each inspection chamber should be shown and numbered and a note made of its size.
- Soil and vent pipes should be noted and provided with a reference number.
- The drain pipe runs should be indicated. The sizes and direction can be ascertained from an examination of the inspection chambers.
- The position of fresh air inlets and interceptors should be noted.
- Septic tanks or cesspools together with their inspection chamber covers should be indicated.
- Positions of surface water drains and soakaways are often difficult to locate and can only be traced by rodding from a gulley or rainwater shoe.

When the inspection chamber covers are removed and the surveyor finds that the layout is not immediately obvious from the arrangement of channels, it is sometimes possible to identify branches by having an assistant to turn on taps or flush a WC. In more complicated installations where the runs are close together, pieces of coloured paper can be flushed down gulleys or WCs.

15.5   Pipes and fittings

It is not intended here to describe the procedure to be adopted in laying drains. This matter is covered by the various Codes of Practice. However, before dealing with the most common defects and tests required, a brief description of the various types of pipework in common use follows.

Clayware drain pipes and fittings to BS 65 are the conventional and most familiar pipes which the surveyor will find when carrying out a drainage survey. The pipes may be given a grooved spigot and socket joint for jointing with tarred gasket and cement mortar or they may be equipped with a patent jointing system such as 'Hepseal' and depend upon a polypropylene collar or natural rubber sealing ring. Pipes with patent joints are often used on drainage systems which can be laid straight into a trench bottom or in 'granular filling', since the patent joints allow a degree of flexibility which can be of use where ground movements may be encountered.

Cast-iron drains and fittings to BS 1211 class 'B' are available with either rigid joints for jointing with tarred gaskin and molten lead or with patent flexible joints. Cast iron pipes are normally used under buildings or in unstable ground or where drains are laid in shallow trenches below roads and pavings. A note of the changes in material as between cast iron and clayware pipes should be made when carrying out an examination of the inspection chambers. Cast-iron with its longer lengths tends to simplify the work and having fewer joints there is less likelihood of defects in the pipes and joints. Pitch-fibre pipes, couplings and fittings to BS 2760 have a great degree of inherent flexibility and do not need concrete beds. The saving on concrete beds together with a simple coupling joint make for an effective economy in their use. Although the pipes will withstand corrosive attack they are not very amenable to industrial drainage systems where continuous hot water is present or wastes containing pitch solvents such as petrol, oil and fats are used.

Apart from definitely identifying the function of the various branches it is essential to carry out a test on the system where these materials are being discharged. Concrete pipes reinforced or unreinforced to BS 556 and 4101 are manufactured in several strength classes according to size. They are mainly used for surface water drains, but can be used for soil drains if acceptable. They should not be used in industrial properties where acid effluent is discharged into the drainage system nor laid in soils in which concrete is liable to be attacked unless suitable precautions have been taken. The pipes can be supplied with flexible joints consisting of a rubber joint ring. This type of joint is particularly suitable to combat traffic vibration and subsidence.

Unreinforced concrete pipes to BS 4101 are often manufactured with ogee joints and should only be used to carry surface water.

Unplasticised polyvinyl chloride pipes are not widely used for underground drainage, but technically there is no reason why PVC cannot be used for this purpose. For drainage work it is recommended that pipes not lighter than those specified in BS 4660 and BS 5481 be employed using a simple spigot and socket joint. PVC's remarkable resistance to corrosion and chemical attack make it ideally suitable for chemical and laboratory drainage systems. However, the material softens at high temperatures and should not, therefore, be used where the temperature of the effluent is likely to exceed 60°C (149°F). At very low temperatures the material can become brittle.

Asbestos cement pipes and fittings for drainage systems are manufactured to BS 3656. The pipes are available coated with bitumen and are suitable for domestic sewage and most trade effluents. The joints allow flexibility and consist of rubber rings and a sleeve. The exceptional length of asbestos pipes reduces jointing and facilitates laying. The pipes are highly resistant to aggressive soils and can be safely laid in sulphate clays, alluvial soils, peat and made up ground.

VISUAL INSPECTION AND TESTING

15.6   Drainage defects

The principal defects the surveyor will encounter when carrying out a drainage survey are as follows:

- Leaky pipe joints which may require rejointing. Old drains are almost certain to have leaky joints. Cement joints often shrink and crack.
- Movement of the ground through settlement or shrinkage may disturb the pipes and often the joints. This type of defect is illustrated in Fig. 15.5.
- Cement and debris left in pipes causing obstruction.
- Obstruction due to solid matter being washed down the drains.
- Lack of proper ventilation causing bad smells in the system.
- Bad smells through leaks, faulty fittings, or blocked traps.

Other indications on the site can well give warning of possible drainage problems, for example:

- Trees growing over or close to a drain run.
- Shallow drains in clay soil which would be subject to moisture movement. Branches would be vulnerable and more particularly the bed at the top of the branch where it connects to a soil pipe or other fitting.
- Indications of settlement in the building, boundary walls or pavings, all of which point to ground movement.

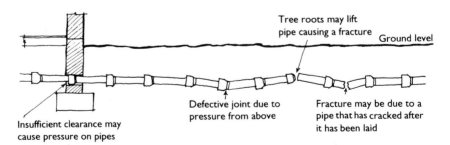

Figure 15.5   Defects due to movement of ground through settlement, shrinkage or tree roots.

15.7 Testing

Today there is a tendency for a comprehensive drain test to be carried out as part of a building survey, particularly when dealing with large commercial or industrial properties. Many of these drainage systems were installed in late Victorian or Edwardian times and are now under far greater strain than they were originally designed for.

If the surveyor has reason to suspect a length of drain is blocked or defective the following simple tests should be used. By flushing the WC pans or turning on wash basin taps it is possible to identify the function of the various branches from the open inspection chamber. It is also possible to see how well the water runs away down the main channel and whether there is any sign of a backward surge, indicating a partial blockage or settlement. Sluggishness in clearing, which may again be due to a blockage or to a poor fall. The water should run away cleanly leaving a virtually dry channel.

A simple test to ensure that no obstructions have occurred can be done by rolling a ball, slightly smaller in diameter than the pipes, through the various lengths. In the absence of any obstruction the ball should run freely down the invert of the pipe. This method will only indicate obstructions and will not necessarily pinpoint the exact position of the blockage. If the exact position of the blockage is required then the builder should be instructed to rod from both sides of the blockage. The rods can then be measured and so establish location of the blockage.

More serious defects such as fractures or settlement problems can be detected by a reflection test between two inspection chambers. The mirrors are placed at the ends of a straight length of drain at a suitable angle. The interior of the drain will be clearly seen and any defect located.

In the case of extensive systems when a considerable amount of money could be involved, a closed circuit television camera (CCTV) survey can be carried out by specialists. The CCTV system of inspection allows the surveyor to record each section of the drainage system and any branches or 'Y' junctions which are often found in old systems. This can be an expensive operation and is not normally necessary. The surveyor should not instruct a specialist firm to carry out a TV survey unless he is convinced that it is essential and has received a definite quotation from the firm selected. Normally, the following tests are satisfactory and can be carried out by most builders or drainage repair firms.

15.8 Water test

The only certain method of demonstrating that a drain will hold water is to fill it with water and note if the level drops. This can usually be done by gaining access to the drain at the lowest inspection chamber and stop-

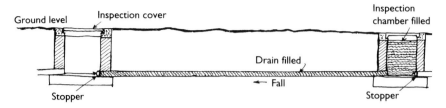

Figure 15.6 Water test applied to section of drain between two inspection chambers.

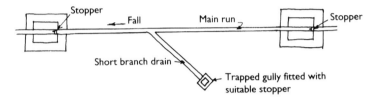

Figure 15.7   Water test applied to main and branch drain.

ping the inlet with an expanding stopper. Another stopper should be fitted to the inlet (or to each inlet if there are more than one) at the next inspection chamber higher up. These positions are shown in Fig. 15.6 which is a section through a drain being tested. The higher inspection chamber is then filled with water and the distance from the top of the cover frame to the top of the water carefully measured and then left for about an hour and a half to two hours. If the water level in the chamber goes down to any appreciable extent a leakage is indicated. There will be a slight lowering of the water level even if the drains are watertight due to absorption and possibly to some evaporation. If there is any doubt as to the leakage the drain must be emptied by removing the lower stopper and the test repeated. At the second test there should be practically no loss of water caused by absorption.

Short branches connected to a main drain between inspection chambers as shown in Fig. 15.7 should be tested with the main drain. The gulley should be fitted with a suitable stopper and any confined air in the gulley should be drawn off as the pipe fills with water, using a rubber tube passed through the seal of the gulley. The shorter the lengths tested at one time the better, as the job of opening the ground and searching for defects can be confined to those sections in which defects are indicated by test.

This test is particularly suitable for new drains, but if applied to an existing drain it would very likely damage the joints and give rise to a leaking drain when previously a sound drain existed. Nevertheless, the function of a drain is to convey liquids and in the event of a blockage some water pressure would be exerted. It would, therefore, seem reasonable to apply a not over-rigorous water test which, while it will simulate blockage conditions, will not damage the jointing material.

It is, therefore, suggested that an adequate test for an existing drain can be carried out by filling the plugged drain with no more water than is necessary to fill the pipe under test. The drain should not be allowed to remain under pressure for longer than the ten minutes or so necessary to ensure that no water is being lost. Once satisfied that the drain is holding water the surveyor should watch the inspection chamber while the plug is removed. The opportunity can then be taken to observe whether or not the drain empties clearly.

15.9 Air and smoke test

Other testing methods generally depend on some means of applying air-pressure to the drain and monitoring the drain's ability to hold the pressure. A loss of pressure indicates leakage in the drain. When this type of test is applied to underground drains it is often inconclusive and will not indicate the actual position of the leak. Another method is to use a smoke generating machine, which consists of a container in which a smoke producing material such as cotton waste or roofing felt is burnt and hand-worked bellows used to drive the smoke into the drains. The smoke pipe is connected to the drain through an expanding stopper fitted with a suitable inlet. An attempt to use smoke to indicate the position of a leak is not as a rule successful, especially in the case of deeply covered drains, as the smoke may not rise through the ground. On urban sites the presence of pavings, wet or clay soil and trenches for other services generally cause any emission of smoke from the drain to be either lost or to emerge from an irrelevant position.

Other tests sometimes recommend the use of smoke rockets or flushing strongly scented essences, such as oil of peppermint or cloves down the drain. These tests do not seem to have much relevance to the problem, and are useless for long lengths of covered drains, as the pressure is very slight.

Clearly a blocked drain is always a matter of urgency and the owner of the property should be informed as soon as possible. Obstructions can usually be cleared by a builder or drain clearing specialist using flexible rods. If the blockage is immovable then the pipe run is probably disrupted by settlement or weak joints. In the case of large industrial drainage systems where heavy deposits have blocked the pipe run, there are a number of modern techniques available to clear them. A high pressure water jet is one of the most common methods used today. Another method used for breaking up hard deposits on pipe walls is electro-mechanical cutting equipment which is suitable for most types of pipework in industrial and commercial buildings.

15.10   Adjoining owners' drains

The surveyor should be careful to note that under no circumstances should an adjoining owner's drain be subjected to any form of water test without the consent, and preferably with the co-operation of his professional adviser. The surveyor should also check with the local authority or the deeds of the property as to whose responsibility the drainage system is.

INSPECTION CHAMBERS, INTERCEPTORS, COVERS AND FRAMES

15.11   Inspection chambers

Access to the drains is usually provided by inspection chambers. The guiding principle in the location of the inspection chambers is that they should be situated so as to allow every length of drain to be accessible for maintenance. Brick inspection chambers should not be rendered on the inside face. Watertightness should be obtained by the use of bricks complying with BS 3921 class 'B' Engineering quality laid in English bond 229 mm thick with the joints finished flush. However, shallow inspection chambers will often be found to be rendered internally. In such cases it is safe to assume that the wall thickness is 112 mm thick brickwork. The chambers can also be formed in preformed prelocking concrete sections which can be circular or rectangular and are often backed up with 150 mm thick concrete. The bases are constructed with a precast invert and benching. In recent years all-plastic chambers have become common.

It is important when carrying out an examination of an inspection chamber not to forget its purpose and to remember that in the event of a blockage a man will have to get into the chamber (unless it is very shallow), assemble drain rods and subsequently withdraw them.

The interior of each inspection chamber should be carefully examined and the construction and condition of the following parts noted (see Fig. 15.8).

- The connections to the chamber are made with half-round channel bends and junctions. These should curve into the chamber in the direction of the flow and should be designed so that they do not greatly retard the flow.
- The type and condition of the channels and whether they are damaged or not. The composition of the drain pipes should be noted.
- Benchings around channels and bends are usually formed in fine concrete and should be examined for cracks.
- Rendering to walls should be examined for 'hollow' areas and cracks. If the interiors are not rendered, the brickwork must be watertight and

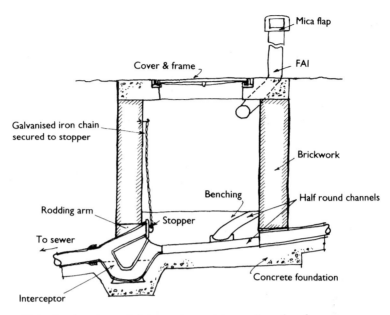

Figure 15.8 Section through interceptor and inspection chamber.

the joints finished flush. It will sometimes be found that a bond is used where the 'through joints' are broken in order to prevent leakage. It is of little importance what bond is used provided that all joints are completely filled with mortar.

- During the investigation it is advisable to measure the invert depth in each chamber, so that on measuring the length of drain between chambers the fall can be checked in relation to the diameter of the drain.
- Inspection chambers over 900 mm deep should have step-irons built into the wall at every fourth course or at 300 mm intervals. Unless they are of the straight bar corner type, the step-irons should be set staggered in two vertical runs at 300 mm centres horizontally. The top step-iron should be 450 mm below the top of the chamber cover and the lowest not more than 300 mm above the benching.
- If water is found to be lying in the main channel, this is usually an indication that the drain is either partially blocked or that a settlement has taken place.
- Where pipes pass through walls or sides of an inspection chamber there should be a flexible joint in order to prevent damage by differential movement.

15.12 Deep inspection chambers

Deep chambers will be found where the surface level to invert is greater than 3.3 m. The roofing usually consists of 150 mm thick reinforced concrete slab to support the ground above. Pipes over 150 mm diameter should have a brick-on-edge arch formed over them. It is recommended that ladders be used in chambers over 4.5 m deep.

15.13 Back-drop chambers

An inspection chamber incorporating a vertical drop for the purpose of connecting to a sewer or drain at high level to one at a lower level is known as a back-drop inspection chamber, or tumbling bay. The vertical drop is usually constructed outside the chamber, but may be constructed inside provided there is sufficient space in the chamber as shown in Fig. 15.9a. If the vertical drop is constructed outside the chamber an access branch should be provided through the inspection chamber wall. If the vertical drop is constructed inside the chamber then a cast-iron access bend should be provided at the top of the vertical pipe. When dealing with both deep

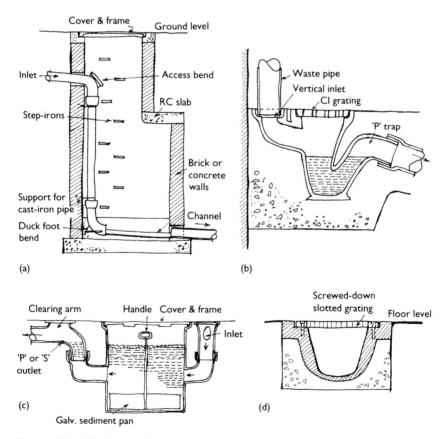

Figure 15.9 (a) Section through back-drop inspection chamber. (b) Back inlet gulley. (c) Grease trap. (d) Drainage channel.

and back-drop chambers the surveyor should note the details of construction and any defects that may have occurred as described in Section 15.11.

15.14   Interceptors

Interceptors were at one time mandatory in all English boroughs. However, it is now generally accepted that the use of interceptors should be optional. Not all local authorities have, as yet, adopted these suggestions, probably due to overriding considerations of rat infestation. The disadvantage of an interceptor is that it tends to obstruct the flow and retain foul matter in the seal which can cause a blockage. If an interceptor has been fitted check that the rodding eye is fitted with a stopper to prevent foul air passing into the drainage system. If a chain is not provided the stopper may well be found lodged in the trap.

It is advisable to recommend that a galvanised iron chain is provided and secured to the chamber wall as shown in Fig. 15.8. Where an interceptor is provided, a fresh air inlet with a mica-flap is theoretically necessary as shown in Fig. 15.8. It should be located away from windows and doors and sited where the front will not be accidentally damaged. The mica-flaps should be closely examined. They may cause a nuisance by acting as a foul air outlet if the flaps are damaged.

15.15   Access covers and frames

Although cast-iron is a fairly rigid material, cast-iron frames are not themselves completely rigid. It is essential that the frames are properly bedded and levelled to prevent rocking and also lateral movement of the frame. All solid cast-iron covers are manufactured in three grades, light, medium and heavy duty, and finished with a non-slip surface to provide adequate grip. An elementary, though frequently overlooked, requirement is that of ensuring that the access covers have not been cracked by vehicles and need replacing with a heavier duty cover. The writer has found that external inspection chambers to carriageways around commercial or industrial properties are often fitted with light or medium duty covers which have been damaged by passing vehicles and obviously require a heavy duty cover.

Covers used for internal inspection chambers should be double seal airtight fitted with stainless steel locking bolts. Many of these covers are recessed for filling to match the surrounding surface finish. There are a number of special covers and frames. A typical example is the medium or heavy duty cover in a non-slip plate hinged to a single seal frame fitted with a patent locking device with a self-locating stay. This type of cover is usually found in footpaths and carriageways.

Steel access covers and frames are now in common use and are zinc sprayed after manufacture for protection against corrosion. A combined lifting and locking device provides a smooth flush-to-floor surface with a minimum area in which dirt collects. The covers are produced in the same range and patterns as previously described for cast-iron.

Covers and frames should be closely examined and recommendations made for removing rust, treating with a rust inhibitor and rebedding the covers in grease. In many cases the hand lift recesses or key holes will be found to be solid with earth or other debris, and the lifting bar corroded which should be noted for replacement.

Some drainage regulations require internal inspection chambers to be sealed with bolted down cast-iron covers at channel level in addition to the double seal cover at floor level. As a general rule, these bolted covers need not be removed unless a blockage is suspected.

GULLEYS AND GREASE TRAPS

15.16   Gulleys

Gulleys are designed to disconnect the foul drain from the waste or surface water inlet. Present day gulleys are to BS 539 for salt glazed ware and BS 1130 for cast-iron. The various types in common use are described below.

Open gulleys (round or square top)

This type of gulley is used outside a building and is fitted with a vertical or horizontal inlet as shown in Fig. 15.9b. For use within a building this type of gulley must be fitted with a sealing cover and frame secured with screws and washers. The inlet should be connected below the grating, but above the water seal; be self-cleansing; have a seal of at least 50 mm and a flat base for bedding in concrete. During his examination the surveyor will find that many of the older type gulleys will not comply with these requirements and are often fitted with insanitary connecting channels to receive the discharge from waste pipes or, alternatively, the waste pipes discharge over the top of the gulley grating.

Access gulleys

This type of gulley is similar to the above but is fitted with a rodding arm and sealed plug. Access gulleys sometimes permit a waste water branch drain to be connected to the main drain using a 'Y' junction and thereby avoiding the construction of an additional inspection chamber. In such cases it is important to see that the inspection chamber on the main drainage run is within 'roddable distance' of the 'Y' junction.

Grease traps and gulleys

In normal domestic drainage grease traps need not be provided, but where substantial quantities of grease are discharged into a drainage system from large kitchens and wash-up rooms a grease trap should be provided. The intention is that the cooling effect of the water in the trap solidifies the grease which rises to the surface and collects in a solid mass. The heavier matter collects in a galvanised sediment pan situated at the base of the grease trap. The grease trap consists of a cast iron or galvanished steel tank fitted with a single seal cover at ground level. A vent grating at ground level is fitted close to the inlet and the connection to the main drain is fitted with a 'P' or 'S' outlet. The disadvantages of these traps are that they are rarely cleaned and the removal of grease is usually an offensive operation. However, the surveyor will often find that the increased use of detergents has lessened the need for providing grease traps or gulleys (see Fig. 15.9c).

15.17   Inspecting and checking gulley defects

The following points concerning gulleys should be checked:

- Firstly, the covers or access plates should be removed. Check that the waste pipes are properly connected to the vertical or horizontal inlets. Gulley gratings tend to trap small particles of waste matter and leaves from adjacent trees and should be cleared.
- The water should be removed from the trap to enable an inspection of the trap to be carried out. The material may have cracked or the traps may have silted up, so causing blockages.
- In older drainage systems where the waste pipes discharge some distance above the gulley grating, they tend to splash the external wall if not protected by a slate or waterproof rendering. However, this defect is not found frequently but it can cause dampness in the external wall, and if found it should be noted in the report.
- Surrounding brick or concrete kerbs should be checked for cracks or other defects. They are often damaged by accident.
- Concrete dishing or channels around the gulley may be cracked or unsound causing leakage.
- Where properties have been vacated for some time, there may be a smell from the system due to evaporation of the seal. This defect is easily remedied with a bucket of water or by turning on a tap. Some old gulleys have very shallow seals and are likely to smell.
- The surrounding ground should be checked for wet areas, which may be due to leaking joints between gulley and branch drain.
- Grease traps and gulleys should be periodically cleaned out, but in the writer's experience this operation is seldom carried out. If the trap is

found to be choked through the deposition of grease and silt then this must be reported to the client with a recommendation for all the offensive matter to be removed and the trap thoroughly flushed out with clean water. In such cases a second inspection may be necessary, care being taken to examine the inlets, outlets and vent gratings etc.

ANTI-FLOOD DEVICES, PETROL INTERCEPTORS AND DRAINAGE CHANNELS

15.18   Anti-flood devices

Drainage systems which are liable to backflooding from a surcharged sewer will be found to have an antiflooding device installed. A simple device which has been commonly used consists of a fitting similar to an interceptor and fitted with a copper or plastic ball which floats on the water seal. As the water level rises the ball is guided by a specially shaped underside of the main cover plate, into a rubber seating, thus closing the drain until the flood water subsides.

15.19   Petrol interceptors

Where there is a risk of pollution from petrol or oil spillage from garage wash downs entering a drain communicating with a sewer, a suitable interceptor should be incorporated in the system. For a single domestic garage a deep gulley trap fitted with a perforated lifting tray is usually sufficient. For a large number of garages or a filling station, a brick or concrete structure consisting of three chambers is necessary, as shown in Fig. 15.10. The chambers allow for any petrol that may be present to rise to the surface and evaporate into the atmosphere through the vent pipes. The construction details are similar to brick inspection chambers, but many petrol interceptors in modern buildings are constructed of precast concrete or pre-fabricated in glass reinforced polyester (GRP). GRP interceptors are factory manufactured; rot proof, impact resistant and are usually installed in a concrete surround. The condition and construction of the following parts of the interceptor should be noted.

- Inspection chamber covers and frames should be checked as described in Section 15.15. The majority of covers will consist of the heavy duty type.
- The vent pipe should be taken up to a height of not less than 2.4 m so as not to cause a nuisance. If the vent pipes are secured to a building, the outlet should not be less than 900 mm above the head of any window or other opening within a horizontal distance of 3 m.

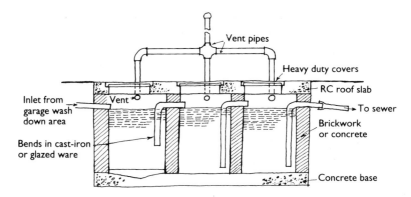

Figure 15.10 Petrol interceptor.

- The first chamber should contain a shorter bend as shown in Fig. 15.10. The purpose of the shorter bend is to allow for the accumulation of grit etc. without blocking the end of the pipe.

15.20 Drainage channels and gratings

The type of floor channel shown in Fig. 15.9d is usually found in industrial buildings. It consists of a cast-iron or precast concrete half-round channel fitted with a light, medium or heavy duty slotted grating. Some channels have a built-in fall with an outlet at one end connected to the drainage system.

The condition and construction of the following parts should be noted:

- Floor finishings around the channels may be fractured or unsound and leak water.
- Remove some of the gratings to ensure that they sit firmly into the channel rebates and do not rock.
- The channel joints must be intact. A water test may reveal leakages which may be due to poor jointing.
- Check outlets for blockages.

OTHER MEANS OF SOIL AND WASTE DISPOSAL

15.21 Cesspools/cesspits

Where there is no public sewer available, foul water drains are connected to a cesspool or a septic tank. The use of a cesspool is often considered to be the 'last resort' method of removing soil and waste water from a building, but it is recognised that in rural areas a cesspool may be the only practicable method of dealing with sewage especially where the disposal of the final effluent after treatment is impracticable.

In the past cesspools have been constructed of brick, *in situ* concrete or precast concrete and required to be impervious so that the contents cannot come into contact with the surrounding ground. In some modern developments or where defective cesspools have been replaced they are usually constructed of GRP which has eliminated many of the waterproofing problems, and are far superior to the traditional brick and concrete methods.

When dealing with cesspools the surveyor will obviously have some difficulty in checking the construction below ground and the general watertightness of the storage compartment. There are several methods of dealing with this problem. Firstly, before attempting an examination of a cesspool the surveyor should be fully acquainted with all the precautions listed in the Building Regulations which apply to both cesspools and septic tanks. These regulations are most vital to prevent pollution of the ground, buildings and water supplies and are designed to protect the health of the inhabitants living nearby. Secondly, old cesspools frequently leak and permit foul water to discharge into the surrounding ground and occasionally into nearby streams and wells. If the ground is suspect it is advisable to dig two or three trial holes around the perimeter of the cesspool and examine the subsoil. If the ground is found to be polluted, a serious nuisance could arise involving very costly repair work.

Thirdly, if the waterproofing is suspect, it is advisable to arrange for the chamber to be emptied. This could be an expensive operation and the client should be informed before action is taken. The local authority usually undertakes this service at an agreed charge. Alternatively, the owners of the property may have a private contract arrangement. If the emptying arrangements are successful, the chamber can be filled with water and allowed to stand for 24 hours, and then topped up if necessary and left for a further 48 hours. The fall in water level should not exceed 25 mm. Allowance must be made for absorption but if the cesspool is constructed of engineering bricks and is properly rendered, the absorption will be minimal. Apart from making notes of the construction it should be possible to check the capacity of the chamber during this part of the examination. The internal defects causing leakage are usually confined to cracked or loose rendering on the internal face. The surveyor should also bear in mind that if the ground is waterlogged there could be water penetration from the outside particularly if the outer face of the walls have not been properly filled with puddled clay or concrete.

If the surveyor still considers that further information is essential then a visit to the local authority surveyor's department would no doubt be helpful. A sight of the plans and any past history of defects arising from cesspools in the area would be useful. Details of the charges for periodic emptying which could be a financial burden on the client should also be noted and included in the surveyor's report.

One of the essential points to check is the distance of the cesspool from the nearest building. Under the old Building Bye-laws it varied from 15 to 18 m away from the building. These distances have been removed from the current Building Regulations. However, this does not mean that it is safe to install a cesspool nearer to a building. Ideally, the cesspool should be at least 15 m away on the downward side of the building and in ground that slopes away from the building. The main drain should be separated from the cesspool by an inspection chamber fitted with an interceptor trap. The condition of the ventilation system should be carefully noted. Both the cesspool and the interceptor chamber must be adequately ventilated as described in Section 15.14. The condition of the access covers and frames to both cesspool and interceptor chamber should be carefully noted as explained in Section 15.15.

When carrying out an examination of the inspection chamber check that the interceptor is not blocked with sewage or other debris which will indicate that the cesspool is full and requires emptying. The surveyor should also ensure that the cesspool is not provided with an overflow, discharging into a soakaway. This arrangement is, in fact, insanitary and may contaminate the underground water supplies. The cesspool must also have vehicular access to within about 9 m to enable the local authority emptying vehicle to approach within a reasonable distance, although some authorities will extend this distance in difficult situations. The local authority will be able to advise on any special circumstances which affect a particular situation.

To carry out repairs and make a traditional cesspool completely watertight can be a long and costly operation. It is, therefore, necessary to carefully consider the cost of all the defects as it might prove cheaper to install a prefabricated GRP cesspool or septic tank than to carry out extensive repairs to an old brick structure. In this connection the drainage system would be out of action while the repairs or a new installation is being carried out. A note concerning this possibility should be included in the drainage report.

15.22   Septic tanks

This type of soil and waste disposal is preferable to a cesspool. Septic tanks work on the principle of breaking down the solids by anaerobic bacteria in an enclosed chamber and purifying the liquid in a filter bed. The effluent is sometimes discharged into a humus chamber which allows the unstable material from the filter to settle. The water resulting from the process usually passes to a stream or soakaway, although the surveyor may well find that in some districts this discharge may not be permitted. Size and length of run-off will vary with porosity of the subsoil and location conditions. The general construction of the walls, floors and pipework etc. is similar to that described for cesspools in Section 15.21.

The design of a septic tank allows for supervision and maintenance to be reduced to a minimum. As mentioned in Section 15.21 certain sections of the Building Regulations also apply to septic tanks.

Although most surveyors are aware of the principles of septic tanks, there are many different systems, some of which were installed many years ago and have not always been properly maintained. It must also be remembered that sewage disposal at the present time is often more complicated to deal with due to the use of disinfectants and detergents. If the surveyor has not made a special study of sewage treatment, and is unfamiliar with local conditions, then in the writer's view it is advisable to obtain the client's authority to instruct a sanitary engineer or a firm who specialises in the design of this type of installation to evaluate the condition of the system and submit a report.

15.23   Pumping stations

In some cases the relative levels of site and sewer, or where the sanitary appliances are installed in basements, render lifting pumps essential. Apart from pumps, pneumatic ejectors are often used where small quantities of sewage have to be pumped against relatively small heads. However, pumps are to be preferred over pneumatic ejectors where the rate of flows is substantial. There are many different systems working on various principles, and as mentioned earlier there are excellent reasons to employ a professional firm of consultants to test and report on the various items of equipment. As with other mechanical equipment there may be a contracting firm responsible for the maintenance of the equipment. Many of these installations are housed in brick and concrete structures which must be examined by the surveyor as described in Chapters 5 to 13.

SURFACE WATER

15.24   Disposal systems from roofs

At this stage of the survey it will be necessary to decide whether the surface water is taken separately to a surface water sewer, combined with the foul sewer or to soakaways. If no access is provided to a separate system it may be necessary to check the drainage plans at the offices of the local authority.

Where surface water is discharged into a combined or separate sewer, trapped gulleys are invariably found at the feet of rainwater pipes as shown in Figs 15.3 and 15.4. In low density areas, where the discharge is to soakaways on the site, rainwater shoes, sometimes referred to as trap-

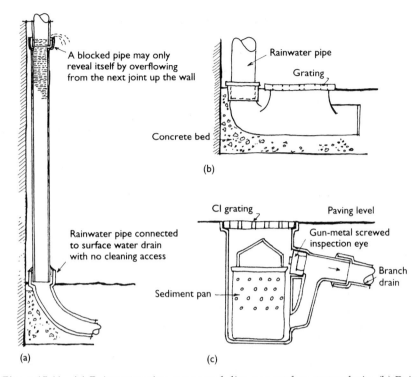

Figure 15.11 (a) Rainwater pipe connected direct to surface water drain. (b) Rainwater shoe. (c) Yard gulley.

less gulleys, may be used. However, in many domestic properties the bottom of the rainwater pipe is jointed direct to the socket of the surface water drain as shown in Fig. 15.11a. This method has several disadvantages because the state of the drain is unknown, and having no cleaning access the silt will gradually build up in the rainwater pipe and drain. This build up will not be noticed until the rainwater overflows from the next joint up the wall.

The writer has found this problem in many of the older domestic and commercial properties. The usual signs are damp patches on the external wall surface adjacent to the pipe joint. It is essential to draw the client's attention to this defect which if allowed to continue may well cause damp problems internally. To rectify this defect usually necessitates the removal of the complete length of rainwater pipe in order to clear the pipe and drain. The dictum here, as in many aspects of drainage work, must be to remove the problem and recommend the installation of a rainwater shoe to enable easy access in the future as shown in Fig. 15.11b. If the surface water is on a 'separate' or 'combined' system then the appropriate tests described in Sections 15.7 and 15.8 can be carried out.

Rainwater gutters and down pipes may be made of cast-iron, pressed steel, asbestos, lead, zinc, aluminium, or uPVC. Pressed steel should be carefully examined. It is very vulnerable to corrosion and it is necessary to ensure that adequate protection is applied.

Precast concrete gutters which have been used in domestic work should be lined with bitumen or bitumen felt, and should be examined for faults. Owing to the small sections used the gutters have a considerable number of joints which often shrink causing leakage. Reinforcement is usually placed so near the surface of the concrete that it will tend to rust resulting in expansion and spalling of the units.

The main troubles associated with rainwater pipes are impact damage and leakage at the joints. Defective joints invariably lead to dampness in the building. Due to the difficulty of access for painting, corrosion attacks the portion of the pipe nearest the wall. Rust in cast-iron pipes will expand and crack the collars, causing dampness in the wall. Less obvious causes of dampness are slight weeping at the joints of rainwater and waste pipes and small pinhole leaks in pipes.

An important point to check during the survey is whether the gutters are adequate to cope with heavy rainstorms and whether the gutters are properly inclined to the outlets. Rainwater pipes should also be checked to ascertain that they are sufficient in number and are of adequate bore. The writer has found that in many properties the gutters or rainwater heads are blocked at the outlets or are poorly fixed and the metal brackets corroded. Localised leakage may occur as a result of defective gutter joints causing damp patches on the fascia, soffit boards and walls. In such cases water spillage can saturate walls and eventually lead to internal dampness.

15.25   Disposal of surface water from paved areas

The adequate drainage of paved areas must always be considered. Only the smallest of paths can be expected to drain into a flower bed. The surveyor will often find that the gulleys receiving rainwater pipes are used to drain small paved areas, although this does necessitate grading the paving towards rather than away from the building. The use of yard gulleys solely to drain paved areas is a little unsatisfactory, if unavoidable at times, as the trap tends to dry out during hot weather.

The type of gulley illustrated in Fig. 15.11c is designed to take surface water from yards, pavings and garage wash down areas. There are several variations some of which are fitted with a perforated sediment pan to facilitate the removal of grit and mud. The pans should be periodically emptied, but in the writer's experience this operation is seldom carried out. If there is a build up of silt or other matter, this should be noted. In such cases it is advisable to recommend that the branch drain from the yard gulley is cleared by rodding.

15.26   Soakaways

The design of a soakaway system will vary according to the nature of the locality. In certain districts where there is a tendency to flooding during heavy storms, it is often advantageous to install a system of soakaways having capacities in excess of the normal requirements.

There are basically two types of soakaway, 'filled' and 'dry wall lined', both of which consist essentially of a pit dug in permeable ground. The 'filled type' consists of suitable hardcore or clinker up to the level of the drain inlet from which surface water percolates through into the surrounding ground. The 'dry wall lined type' resemble large inspection chambers constructed of brick or stone dry walling with open bottoms and perforated sides to support the surrounding earth.

The 'dry wall lined type' has far more storage capacity than a 'filled' type and therefore is the type most designers prefer. Both types have a concrete or stone slab top to prevent soil being carried down by percolation of rainwater from above.

On some building sites the soakaways are often constructed of perforated concrete rings or segmental units. This type of soakaway is usually fitted with a concrete or cast iron cover at ground level which makes an internal examination much easier.

If no access is provided to a soakaway system it is difficult to ascertain the condition of the drain or if the soakaway is functioning effectively. As for soil drains without access, it would be unwise for the surveyor to pass an opinion on their condition.

Failures are often indicated by flooding, waterlogged conditions or by subsidence if the structure of the soakaway fails. It must also be remembered that if soakaways are situated too near a building it may be a possible cause of settlement in the external walls. Soakaways should always be at least 5 m from any building. The only method of checking this distance is to instruct the builder to pass rods through the drain until they reach the soakaway. The rods can then be withdrawn and measured, and so establish the location of the soakaway and at the same time clear any obstruction. Any soakaway of the 'filled' type constructed over fifty years ago should be suspect.

In the writer's experience many of these soakaways serving domestic properties simply consisted of a pit filled with debris from the original building site and are now silted up, collapsed or are defective due to root penetration. The surveyor will obviously have some difficulty in checking these defects below ground, but the problem mentioned in Section 15.24 concerning the lack of access and blocked rainwater pipes will no doubt give some indication of what has happened in the soakaway system.

REPAIR PROCESSES

15.27   Recommendations

After carrying out the various tests described earlier, the surveyor may find that some of the pipe runs have failed or that there are fundamental signs of poor construction, such as incorrect falls or the drains being out of true alignment. Under such circumstances the report must include clear details of the problem and prescribe a remedy together with an estimated cost of the work.

Serious consideration must now be given to the type of repair work to be undertaken. If the failure consists of relatively minor defects in the pipe joints due to slight ground movements or old age, then it may well be advisable to recommend one of the patent internal repair processes. These processes involve the repair of a drain run either by a grouting process or by forming an impervious liner inside the pipes. However, before recommending such a process, the surveyor must ensure that the drainage repair company would guarantee an effective repair. The advantage of an internal repair system is a worthwhile saving in expenditure, time and upheaval, but at the same time the limitations should also be apparent. In cases where drains have serious defects and the cause can be attributed to one of the defects mentioned in Section 15.6 then a repair by one of the internal systems will not solve the problem as the drains will still be subject to the same pressure as before. In these circumstances the surveyor will need to give careful consideration to renewing by traditional methods.

In calculating the cost of re-laying a seriously defective drainage system, an allowance should be made for renewing the concrete bed and providing flexible joints to ensure that the system remains stable.

PAVING AND CARRIAGEWAYS

15.28   The function of carriageways

This section covers external carriageways, footpaths and parking areas. The main function of any carriageway is to provide a hard, dry, non-slippery surface which will carry the load of traffic asked of it. In domestic work, garden paths and surrounds to houses are being vastly improved with coloured paving slabs and blocks. Defective carriageways and footpaths are expensive to repair and are, therefore, of considerable importance when carrying out a structural survey.

15.29   Flexible paving

Flexible pavings are easily laid to varying falls and irregular shapes and are desirable in areas liable to subsidence, but sometimes require periodic surface dressing to seal the surface. The pavings usually consist of tarmacadam, bitumen macadam, mastic asphalt, hot rolled asphalt or fine cold asphalt. Pre-mixed bituminous materials are generally used for the surface of vehicle parks.

The important properties of such surfaces are that they shall not deform under standing loads and that the surface shall be resistant to the softening effect of oil droppings. Slow moving and stationary vehicles create greater stresses in the carriageway structure than vehicles travelling at high speeds on a normal highway and an allowance is usually made for this by increasing the thickness of the base and surface material.

15.30 Concrete paving

Concrete can be used for both vehicular access, parking areas and footpaths. If adequately compacted and cured, with ample provision for expansion and contraction concrete provides a hard-wearing surface.

15.31 Blocks and slabs

When carrying out a survey of more modern structures the surveyor will no doubt find a greater variety of attractive concrete paving slabs. Blocks and slabs of high density concrete which are laid on sand in an interlocking pattern, form a strong surface capable of carrying heavy vehicles. Precast concrete flags either hydraulically pressed or cast in moulds are made in many subtle colours and a variety of textures.

15.32 Tiles and setts

Smaller units such as square tiles and setts are often used where a demarkation is required to some special purpose, such as parking areas or for borders and surrounds. Cobbles set in concrete will often be found in special pedestrian areas used as hazards to keep people away from grass or plants and at the same time contrast with smooth areas of paving.

15.33 Gravel and hoggin

This type of material is not suitable for heavy loads. It is also expensive to maintain as it will require periodic rolling and making.

15.34 Examination of pavings

The condition of the paving should be examined with regard to the following aspects:

(1)   The construction and finish of all types of paving should be described in the report and any areas of settlement, potholes or cracking should be carefully noted. The question of settlement particularly in clay soils was discussed in Chapter 5. Settlement and cracking warrant close examination as the defective areas may require reconstructing. Any settlement around inspection chambers and gulleys should also be noted. Depressions in the surface causing 'ponding' can become dangerous in cold weather.

(2)   Precast concrete kerbs are now extremely popular and are usually placed at the edges of flexible pavings in order to prevent the lateral spread of the paving material. In most cases a flush kerb is all that is necessary but where it is required to channel surface water at the edge of the paved area, the top of the kerb should project above the surface. All kerbs should be checked for cracks and settlement.

(3)   The importance of the provision of an adequate drainage system cannot be over-emphasised. The falls, slopes and ramps should not be too steep or unexpected. The different types of surfaces require the following cross-falls:
   • Concrete pavings: 300 mm in 18 m.
   • Bituminous or tarmacadam surfaces: 300 mm in 12.1 m.
   • Paving slabs: 300 mm in 22 m.
   • Gravel: 300 mm in 9.1 m.

(4)   If considerable cracking is found over a large area it is usually due to insufficient thickness and consistency to withstand the various temperature variations. In some cases it may be that the top surface was not adequately finished and allows rainwater to penetrate.

(5)   It is not uncommon for paths and carriageways to be built up over the years and ultimately finish up level with the damp-proof course. Paths and carriageways are often found sloping towards external walls with the result that water will pond against the wall. The paving levels should be carefully checked and any defects noted. (This problem is also dealt with in Section 8.9.)

(6)   Concrete carriageways provide a hardwearing surface, although irregularities sometimes occur at the joints. They are, however, vulnerable to chloride attack from de-icing salts or sea-spray effects. Construction joints are made by forming a groove in the top of the slab to a depth of about 25 mm. The groove is filled with a pliable sealing material, but often requires periodic re-sealing. Where slabs meet adjoining walls or inspection cover frames, an expansion joint should be formed finished with a joint filler. The edges of all slabs should be arrised to prevent spalling. The condition of all joints should be carefully noted.

(7)   Damage by vehicles mounting footpaths, construction damage, and tree roots often results in broken paving and should be noted and described in the report.

(8)   The provision of turf up to the face of a building can result in damage to the face of the building by grass cutting machines. Rainwater running down the face of cladding may result in the turf adjoining the cladding becoming a surface-water drainage area. It is sometimes advisable to recommend the installation of a small paved area around the building with a slight fall away from the external wall.

(9)   Weeds have a way of thriving in the most inhospitable areas including cracked and damaged paving. Weed growth encourages dampness and this increases corrosion of metal and decay in timber. It is also a serious fire hazard and impairs drainage. If large areas are found it is advisable to recommend some form of weed control treatment.

(10) Steps should be examined to ensure that they have sufficient fall to throw water clear and do not become dangerous during wet or frosty weather.

BOUNDARY WALLS, FENCES AND GATES

15.35   Introduction

Difficulties may sometimes arise as to the ownership of boundary walls, fences and hedges between adjoining plots. Before dealing with the physical examination the surveyor should make himself acquainted with the elementary law of boundaries (Anstey and Powell 2004). The main points are dealt with briefly in the following and could be retained for reference if the occasion arises.

15.36   Ownership of fences and walls

A boundary is an imaginary line which delineates the limit of the property. At common law there is no obligation upon a landowner to fence or erect a wall around his land unless there is an express covenant requiring him to do so. When dealing with leasehold properties the surveyor should examine the section of the lease concerning boundaries. Covenants to erect and maintain fences are often found in leases, and they may be by either party. However, there are some exceptions to this rule. The Highways Act 1959 section 144, imposes an obligation upon an owner of land adjoining a public highway, footpath or lane to adequately fence anything thereon which is a source of danger to persons using the highway. The Occupiers' Liability Act 1957 requires an occupier of property to make it reasonably safe for visitors, and it may, therefore, be necessary to erect a fence to discharge this duty.

In some cases it can be difficult to tell precisely where the boundary between two properties runs, particularly in old properties situated in rural areas. The most obvious way of finding out about the boundaries is to inspect the title deeds of the property, but all too often they fail to specify the position of the boundary line in relation to the fence or wall. In some cases the boundary on the deed plan or lease will be marked with a letter 'T' which shows who is the owner of the fence or wall. If the title deeds fail to provide an answer, the surveyor should inform his client's solicitor so that he may make inquiries in order to clarify the matter. It is highly desirable that the client should be clear upon this issue as this may avoid misunderstanding at a later date. It must also be remembered that where a fence has been erected even though there is no obligation to do so, and it is in a dangerous condition its owner may be liable for damage resulting from its condition. However, it is the usual practice for two properties to be separated by a physical division not only to define the boundary, but to keep out intruders. The surveyor can then ascertain who is responsible for its upkeep.

Brick, stone or concrete block walls separating the properties of different owners, when built on the line of junction of two properties are usually deemed to be shared by the respective owners and are known as 'party fence walls'. A considerable number of terraced houses built before 1900 were provided with this type of boundary wall which is usually from 1.5 to 2 m high. Section 38 of the Law of Property Act 1925 provides that party fence walls shall be treated as being severed vertically, but that each owner shall have a right to have his part of the wall supported by the other (Fig. 15.12a). This means that neither can remove his part of the wall if this would mean that the other part collapses. Occasionally, projecting piers indicate ownership since they are invariably built on the land of the owner of the wall. In such cases the actual boundary line may be the far face of the wall (Fig. 15.12b).

Unlike walls, fences generally belong wholly to one owner. The facts that have a distinct bearing on the ownership of fencing are still valid. An owner must erect a fence entirely on his land and therefore places the posts and rails on his ground with the boarding fixed on the outside (Fig. 15.12c). Thus, the face of the fence defines the boundary line even though the owner of the fence must sometimes enter upon the neighbouring ground to drive the nails into the woodwork.

Artificial boundaries formed by man consisting of hedges and ditches are described below although contrary evidence may exclude these assumptions.

- Where two properties or fields are separated by a hedge or ditch there is a presumption that the boundary is the centre of the hedge or ditch (Fig. 15.12d).
- If the properties are separated by an artificial ditch alongside a bank with a hedge on it there is a presumption that the boundary is along the edge of the ditch furthest from the hedge (Fig. 15.12e).

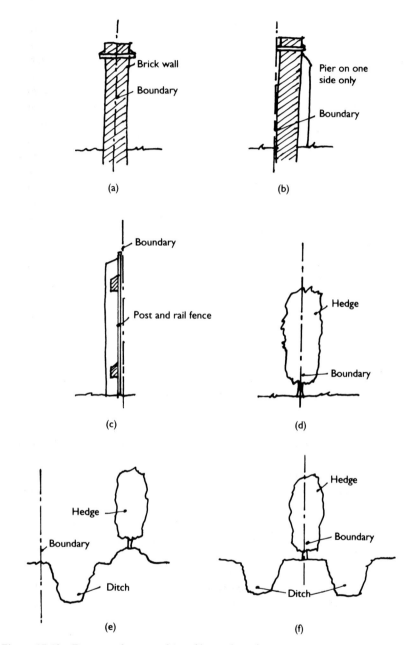

Figure 15.12 Presumed ownership of boundary features.

- If there is a ditch on either side of the hedge the ownership of the hedge must be proved by acts of ownership. No presumption operates, because the original landowner's actions in digging the ditch or planting the hedge cannot be logically deduced. However, Ordnance Survey practice is to take the centre line of the hedge as the boundary feature (Fig. 15.12f).

There are many ways in which boundary walls and fences can deteriorate and their condition and probable life should be carefully noted as described in the following paragraphs.

15.37   Brick and stone boundary walls

Maintenance is a constant problem with fencing and walling. Boundary walls have to withstand the effects of the weather on both sides, including movements due to temperature changes, moisture penetration and settlement. Many brick boundary walls were built in long lengths without stiffening piers or expansion joints causing the brickwork to bulge or lean. Old brick walls were often built with insufficient foundations and no DPC. In such cases the brickwork will be badly decayed at ground level and settlement may have occurred. The surveyor must give careful consideration to walls in poor condition and decide whether the brickwork can be repaired or it is more economical to rebuild using the old bricks.

When rebuilding a whole boundary wall it must not be forgotten that the existing foundation may be inadequate and a trial hole may be necessary in order to ascertain the depth and quality of the foundation. If it is decided to rebuild it is a comparatively simple matter to recommend the inclusion of a vertical movement joint at intervals not exceeding 10 m to accommodate movements. A movement joint should also be provided where the wall meets an existing building.

Care should be taken to examine the coping. If the top course consists of brick on edge and the joints are defective, water will soak down causing two or three of the top courses to be come green. If the coping brick is decayed it is advisable to recommend replacement with an engineering brick to provide resistance to damp and frost. A projecting tile creasing weathered with a cement mortar fillet is often introduced to throw rainwater clear of the wall. Unfortunately, the fillet nearly always shrinks away from the brick and the crack forms a trap for rainwater. A stone or precast concrete projecting coping should have a weathered top and be throated on the underside. Unless the coping stones are adequately secured to the wall they may become loose, eroding the pointing and bedding mortars. The incorporation of a DPC in the joint below the coping is to be strongly recommended.

Repointing is often a major item in building maintenance, much of which is made necessary by the faulty methods used in the original work.

Repointing large areas can be an expensive item and all areas where the pointing is defective should be carefully recorded.

Damp penetration through stone boundary walls is unlikely except in situations where there are inadequate foundations and no DPC. Penetration through joints is more likely to be a problem and these should be carefully examined.

15.38 Retaining walls

When a boundary wall acts as a retaining wall it is usually built of brick or concrete blocks, but where there is considerable side pressure, of reinforced concrete. The following points should be checked and noted:

- Retaining walls are likely to be wet for long periods and are particularly vulnerable to frost attack which may cause spalling.
- The walls must permit drainage of moisture. This is usually achieved by forming perpendicular open joints at approximately 900 mm intervals in the course above the finished ground level. On wet sites clay field drains will be found built into the wall. It is advisble to examine the pipe and insert a probe as they often become clogged with soil.
- Unless the side of the wall facing the earth is protected by a waterproof material there is an increased risk of sulphate action as described in Section 6.6.
- Walls extending above the higher level of retained ground should have a properly designed overhanging coping of stone, precast concrete or other durable material. It should be weathered and throated to discharge water to the lower ground level as described in Section 15.37.
- A retaining wall must withstand soil pressure. A bulged wall is obviously suspect and could be dangerous. Bulges are usually caused through inadequate thickness in relation to height or through a lack of lateral restraint. The details should be carefully noted and if there is any doubt concerning rebuilding then it is advisable to recommend the services of a structural engineer who will investigate the structure and make the necessary calculations.

15.39 Timber and metal fencing

The majority of timber and metal fencing has the advantage of durability and easy maintenance particularly if the timber and metal has been treated with a preservative. The most durable and commonly used timbers are oak, sweet chestnut and larch. However, it must be remembered that few, if any, species are immune to decay. The life of timber fencing will depend on the natural durability of the timber and the type of preservative treatment before and after erection.

Fencing is divided into two basic types – 'open' or 'closed' and the principal types are described in Parts 1 to 13 in BS 1722.

15.40   Types of fencing and their defects

The most common defect in timber fencing is that the wooden posts tend to rot where they enter the ground as the portion sunk in the earth is kept damp by the moisture in the soil, while the portion above ground is alternately wet and dry, according to the weather conditions. Defective posts are easily tested by rocking the fence and if the bases are weak they can be strengthened by the provision of concrete spurs bolted to the old parts. This method of repair must be carefully considered in relation to the condition of the whole fence. Some very old close-boarded fences are often on the point of collapse.

Precast concrete posts are now extensively used and should last many years without attention. They are grooved on two opposite sides and the timber panels are dropped into the slots. The advantage of this system is that any defective panels are easily replaced. Tops of posts should be rounded or shaped to throw off rainwater.

Paled fences become loose because the nails if not galvanised will rust and break under the swaying movement caused by high winds. Defects in the pales are usually due to rising damp where the boards touch the ground or earth or earth is banked up against them. It is very desirable, in pale fencing, to have a gravel board running from post to post so that the pales rest on top of the board and can be replaced if required without much difficulty. If a replacement is necessary the surveyor may consider the provision of a precast concrete gravel panel which will no doubt extend the life of the fence.

Arris rails are usually cast from rectangular lengths of timber approximately 75 mm square by cutting them diagonally down their length. Alternatively, cant rails are generally formed by canting or chamfering one face of a rectangular piece of timber. The shape of both types is such that rainwater will run off quickly. The most common defects in arris rails that call for remedy are sagging and wet rot causing the pales to become loose.

Two other types of fence that have become quite common are the woven board and waney cut panel fencing. The boards next to the posts are kept in place by vertical fillets nailed to the posts on each side of the boards. The top of the fence should be finished with a weathered capping. If some of the boards are worn or split, new boards can be inserted.

Precast reinforced concrete posts and panels will often be found in modern commercial and industrial sites. The posts are grooved on two opposite sides and precast concrete panels approximately 40 mm thick are dropped into the slots. As the panels are comparatively heavy, it is necessary to give the posts a firm seating in the ground by embedding them in concrete. The surveyor should check that the panels and posts are not chipped exposing the reinforcement. Defects in the concrete bases supporting the posts are usually found where the fence is out of alignment and the posts can be moved.

Metal railings were popular prior to 1914 but are seldom seen today. They were often used to fence off the pavement from the front gardens of houses and were frequently fixed on top of a dwarf wall. Corrosion is a defect most likely to occur especially where there is a lodgment of moisture; for instance, where the upright bars pass through the horizontal flat sections or where moisture penetrates the sockets at the base. If the vertical bars are inserted into a coping stone or concrete kerb and are not 'run in' lead, the rust will expand and cause the stone or concrete to fracture. The surveyor will need to examine the metalwork carefully by scraping the paint at the various joints which are suspect and also test the stability of the railings by hand pressure.

Fences of galvanised chain link secured to timber or steel angle posts have often been considered as being of a temporary nature, although standards of galvanising have considerably improved and plastic coated chain link appears to have a much longer life. The most common complaint with fences of this type is that the timber posts have become rotten at their bases or the steel angles and chain links have corroded. Care should be taken to check all metal and timber parts as previously stated.

Cleft chestnut is seldom used as a permanent fencing material although it is moderately cheap. It is made with small chestnut branches split into two and kept upright at intervals of about 150 mm with two rows of horizontal galvanised wire. An advantage of this type of fencing is that it can easily be rolled up and re-erected elsewhere. As the upright pales are passed through loops in the wires, defective pales can easily be replaced, but the wires should be checked for corrosion.

Barbed wire fences are erected at the owner's risk and he or she will be liable for any damage or injury caused to a neighbour or to passers by on the highway. The erection of a barbed wire fence is generally inadvisable and apart from the legal liabilities much injury to animals may be incurred. The surveyor should take care in reporting this type of fence and should consider recommending a replacement such as chain link.

15.41 Gates

Wood is the material of which the great majority of gates are constructed and the defects that may arise are similar to those that occur in other wooden structures. Wet rot and shrinkage are the principal defects that should be investigated. Shrinkage will affect the mortice joints which may open and allow damp penetration.

The essential point for the surveyor to remember is that a gate is a framework supported by the hanging stile alone and if the constructional design is wrong the gate will be unable to support its own weight and will droop at the shutting stile end, thus opening the joints between rails and hanging stile. The position of the brace is important and should be checked. Its purpose is to prevent the top rail dropping at the shutting stile end.

Metal gates are either wrought iron or steel and the previous remarks on defects in metal railings likewise apply to iron gates. Metal gates are usually hung to brick or stone piers or to metal standards. Metal hinges can be affected by weather conditions and poor maintenance and should be examined for corrosion and other defects.

OUTBUILDINGS AND OTHER MISCELLANEOUS ITEMS

15.42   Introduction

The final task of the surveyor is to examine the condition of the various outbuildings following the same order as previously described for the main building. Outbuildings vary from small garden sheds to large blocks of garages and are seldom maintained with the same degree of care given to the main building. They are especially prone to deterioration caused by damp penetration and beetle infestation.

15.43   Inspection and checking defects

The main defects to look for are as follows:

Garages

Garages and storage buildings of traditional construction often consist of the same materials as the main building. Roofs and walls should be checked for rising damp, water penetration and timber decay as previously described. The examination should include checking that the doors operate correctly as described in Section 13.12. This is particularly essential where sliding door tracks or 'Up-and-Over' doors are provided. Lack of maintenance, i.e. clearing the channels of dirt and grease, often causes the doors to jam. Where an inspection pit is provided, the covers should be in sound condition and the pit free from water penetration. Gutters and rainwater pipes require close attention, particularly the brackets and holder bats which often corrode due to neglect.

The surveyor should also note the condition and level of the garage floor and outside paving. It may be found that the outside paving does not fall away from the door opening. Provision of properly designed drainage such as a channel across the door opening or some form of threshold would avoid the flooding that inevitably occurs during heavy rainstorms.

Greenhouses

Older type greenhouses were usually constructed in timber framing set on a low brick wall and are especially prone to deterioration. The difference in temperature between the outside and the inside is considerable in cold weather. This will cause the warm humid air inside to condense on the inner surface of the cold glass, and unless this condensation is dealt with it may find its way into the framework and set up decay. The rebate in the glazing bars should be reasonably wide so as to give adequate support. The glass is exposed to considerable variation in temperature and will expand and contract accordingly. For this reason the glass should not fit tight between the bars. Note cracks in the putty which will allow water to collect in the space and may cause decay in the timber.

Sheds and storage buildings

Garden sheds and storage buildings are often in poor condition and beyond repair. If the floors are supported on timber joists in contact with the earth or supported on brick piers with no damp proof course, they are usually affected by wet rot. Roofs usually consist of one layer of roofing felt and may be found to be in poor condition. If the sheds are in reasonable condition the client should be advised to use a timber preservative treatment which may extend the life of the structure by a few years.

REFUSE COLLECTION

15.44 Small domestic dwellings

The disposal of refuse has become a serious problem in many large towns and cities and the volume is increasing year by year. For the average house storage in individual dustbins is an easy and cheap method, and is not insanitary if reasonable precautions are taken. The surveyor will often find that in some modern domestic dwellings such as small blocks of maisonettes and flats, a dustbin enclosure will be provided. Siting will be within easy reach of the kitchen entrance and will have access to the open air. The enclosure must also be accessible for vehicular collection points with a carrying distance not exceeding 25 m. Enclosures are normally constructed with a single skin of brickwork on three sides and weatherproofed with a concrete slab roof. The brickwork should be built on a slab of concrete to a fall of approximately 12 mm to assist shedding of water when being washed down. The surveyor's report should describe the construction and finish with any defects listed.

15.45   Large blocks of flats

The problem of refuse disposal in large blocks of flats and in some commercial buildings is usually overcome by refuse chutes. Figure 15.13 shows many of the standard features of a refuse chute and storage chamber which are required by the Building Regulations. The chute consists of a vertical pipe with a minimum diameter of 457 mm although some older types will be approximately 375 mm. The pipes must be constructed of non-combustible impervious materials such as salt glazed ware or vitreous enamelled iron, though in modern buildings spun concrete is used. The chute runs from top to bottom of the building and has an inlet hopper on each floor. At ground floor level the chutes discharge into a wheeled container in a specially constructed chamber consisting of a brick structure with a concrete roof slab and floor fitted with a fire-resisting door. Adequate ventilation is a necessity and both chamber and chute must be ventilated as shown in Fig. 15.13.

If the surveyor is not fully acquainted with this system of refuse disposal, then it would be advisable to consult the following relevant British Standards:

- BS 1703 Part I, details materials, finish and design of hoppers for admitting household refuse to chutes.
- BS 1703 Part 2, refers to the dimensions and design of the chute, shutters and ventilating flues.

Although the chute is of vital importance it is generally impossible to examine its condition and effectiveness visually. Unless there is evidence that the chute is defective it must be accepted that it is in reasonable condition. The construction and condition of the following parts should be noted.

- Hoppers for access to the chute at each floor level are usually galvanised mild steel. When open an inner flap closes thus shutting off the main chute. When the hopper is in the closed position the inner flap drops allowing the contents to fall into the chute. Hinges and fastenings should be checked for ease of working. The hopper should have a surround at least 300 mm wide on the external face consisting of glazed tiling or other impervious material.
- The chute must be ventilated at the top and so positioned as not to transmit foul air in such a manner as to become prejudicial to health.
- A hinged cover for access at the top of the chute for cleaning is desirable, as shown in Fig. 15.13.
- The chamber containing the wheeled container must have a fire-resistance of one hour and be impervious to moisture. The floor of the chamber should be laid to a fall towards a trapped gulley situated immediately outside the chamber to enable the floor to be

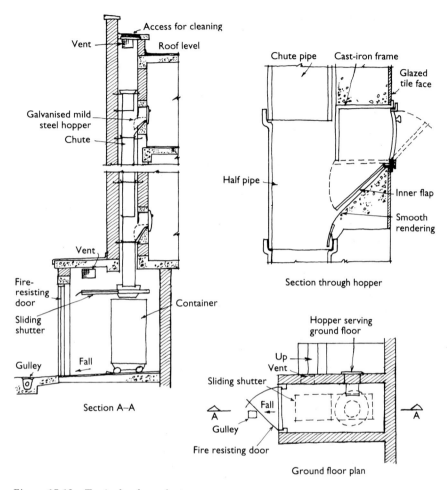

Figure 15.13 Typical refuse chute.

washed down. The construction of the walls, floor and roof should be checked as described in Chapters 5, 6 and 7.

- The lower extremity of the chute must be fitted with a metal sliding shutter capable of closing the outlet of the chute when changing over containers. It is important to check that the shutter is in sound working order.
- The entrance door to the chamber must have a fire-resistance of half-an-hour. It is important to check the condition of the door and frame. In the writer's experience the chamber doors are often damaged by the removal of the wheeled container.
- The inner surface of the chamber walls should be lined with a smooth impervious material. Many chambers are often lined with quarry tiles,

which are easily broken or may have become loose due to impact damage.

- Galvanised metal containers are usually supplied by the local authority. They must be capable of easy removal by means of a jacked trolley or other device.

15.46   Lightning conductors

An important aspect when carrying out a survey of a large building, and especially high buildings in isolated positions, is to ensure that the building is properly protected with a lightning conductor (see BRE Digest 428). Conductors on high and other inaccessible positions should be closely examined, particularly the contact between the tape and the air terminals. If the tapes show signs of pulling away from the building or breaking their fixings the expense of repair can be very high. The bottom of the tape is sometimes open to vandalism and if it is damaged or its electrical bond is loosened the efficiency of the system is in question. In such cases it is advisable to recommend that the tape be protected by a pipe covering.

A test button should be provided just above ground for periodic testing of the earth qualities of the ground rods and the continuity of the system above ground. If the surveyor is unable to obtain a record of recent test results then it is advisable to recommend to the client that the conductor should be tested by a competent electrical engineer in accordance with BS 6651 and a record of the results can then be inserted in the report.

15.47   Trees

Some general notes to describe the condition of planted trees and lawns is all that it is necessary for a report, but more attention is required where trees are present especially if there are large varieties. Retention of trees is obligatory where tree preservation orders have been imposed unless consent for removal is granted by the local planning authority. However, if a tree is in poor condition it may be necessary to advise on its removal.

The danger with trees is that their roots may travel long distances and absorb a considerable amount of moisture from the subsoil as described in Section 5.7. The surveyor should pay particular attention to this point and try to gauge the effects that any trees may have on the building foundations. It is sometimes difficult for a surveyor without training in forestry work to know whether a tree is sound or not. In such cases it is advisable to consult a forestry expert. It is as bad to fell a tree unnecessarily as to allow one to stand when it has become dangerous.

16 Fire and Flood Damage

16.1 Introduction

Another duty the surveyor may be called upon to perform is to act on behalf of an owner in cases of damage or destruction by fire or flood in a property that has been insured, and negotiate a settlement with the insurers' assessor. The insurers must be informed immediately that a fire has occurred letting them know the approximate extent of the damage, and if possible the cause. In fact, the surveyor will usually find that the insurance company has been notified by the owner before he arrives on site.

Initial steps to be taken by the surveyor:

- Examine the insurance policy.
- Carry out a preliminary investigation of the damage in accordance with the terms required by the insurance policy.

It is important to understand the term 'reinstatement'. In the writer's view the definition is as follows:

- Where a property is destroyed, the rebuilding or replacement by a similar property in a condition equal to, but not more extensive than the original building.
- Where a property is damaged, the repair of the damaged portion to a condition substantially the same, but not better than the original portion of the building.

In the case of a serious fire or flood, the insurers will often employ an independent firm of loss adjusters to investigate the claim on their behalf. A representative of the firm will contact the owner's surveyor and from then on all negotiations will be between the surveyor and loss adjuster. In the case of serious damage to a large building the surveyor should arrange a meeting with the loss adjuster at the earliest possible date and if necessary, obtain his agreement to have the building cleared of fire debris and demolish or shore dangerous roofs, floors, walls and chimney stacks etc. Until this work has been carried out it is difficult to assess the real damage. If the roof is badly damaged then it may be necessary to propose to the loss adjuster that a tarpaulin or heavy duty plastic sheeting be securely fixed over the roof to keep the structure watertight until permanent reconstruction can be undertaken.

One of the conditions of most insurance policies is that the insured is under an obligation to submit full particulars of the damaged property to the insurers. In the majority of insurance policies a period is named within which the claim must be delivered; usually thirty days, but in serious cases this period may be extended by permission of the insurers.

In view of the big rise in values during the past fifty years it has been advisable for building owners to raise the insurance figure on their property including the contents. If the building and its contents are significantly under-insured the loss adjuster will, no doubt, inform the surveyor that the claim could be subjected to 'average'. This means that if the insurers stipulate that the sum insured is less than the amount properly insurable, then the claim settlement will be reduced to the proportion that the sum insured actually bears to the correct figure. For example, if the value of the building and its contents is £200,000 and the owner has insured it for only £100,000, should fire damage occur to the extent of £100,000, the insurance company may then apply the average clause and pay only one half i.e. £50,000. It will be assumed that insurance for an amount less than the full value means that the owner carries the proportionate balance risk.

The insurers are only bound by the conditions of the policy and it is, therefore, important to check whether or not a comprehensive form of policy is used. This form of policy includes cover for the buildings, contents, loss of rent, claims by the public, theft and larceny etc.

In some cases the owners will have a complete inventory made of the contents prepared by a qualified valuer. In the event of a fire for example, whether a partial or total loss results, both the surveyor and loss adjuster would find time would be saved if the losses were checked against a priced inventory. The advantages of a mutual check are that the surveyor will be able to agree with the loss adjuster at the outset the valuation of the contents as a basis for a claim. If the owner is unable to produce such a document then it will be necessary to compile a list of all the items damaged or destroyed and the owner should be asked to produce accounts or receipts as these would be of great assistance.

When carrying out negotiations it will be necessary to convince the loss adjuster that the owner was in actual possession of the furniture etc. and that the amount claimed is the correct value at the time of the loss. In an industrial situation the building contents, plant or machinery may be separately insured. When preparing a claim, damage by smoke and water can be included, as the insurers are liable for all damage caused by fire, and the question of whether water was used to extinguish the fire does not affect the matter.

If the damage is extensive and a rebuild is necessary, then a complete specification containing quantities which can be priced out in detail will probably be required to satisfy the loss adjuster. Each case must be treated upon its merits, but where a portion of the building has been demolished it may be necessary to attach plans to the specification showing the building in its original state. After the surveyor has agreed the cost with the loss adjuster he will receive a form to sign to confirm that the total figure is the full amount which has been accepted in settlement of the claim. It is important, therefore, to ensure that nothing has been omitted as the insurance company is unlikely to agree a claim for further costs on completion of the work.

The surveyor will certainly be faced with the question of his fees and expenses and this must be agreed with the loss adjuster during the negotiations. It is now customary to insure for the full value of the work plus architects', surveyors' and legal fees, cost of debris removal and local authority requirements. The surveyor will find that settling a claim is very time consuming. The duties covered could include the following.

- Making arrangements with contractors to clear debris and silt etc.
- Arranging for shores and struts to support dangerous portions of the building.
- Examination of the damaged portions of the building.
- Making careful notes in sufficient detail to enable a full specification and quantities to be prepared.
- Obtaining the necessary measurements for producing plans for the portions of the building destroyed.
- Negotiations with the loss adjuster.
- Negotiations with the local authority under the building regulations in respect of any rebuilding work required.
- Negotiations with the supply authorities in respect of damage to gas, electrical and water supplies.

The negotiations are often prolonged, particularly where a serious fire has occurred to a commercial or industrial building. It is, therefore, advisable to let the owner of the property know that a considerable period of time may elapse before the settlement of the claim and reinstatement of the damage is completed.

On occurrence of a serious fire or flood it is important that the owner or the surveyor should take precautions to see that nothing is lost by theft during or after the fire or flood. The police should be informed at once and will no doubt assist in the matter.

EFFECTS OF FIRE

16.2 Preliminary investigation

Fire damage varies from the superficial to complete disintegration where a complete rebuild becomes a necessity. If arson is suspected the police will be involved before the surveyor begins his investigation. Before dealing with the various defects there is the problem of fire debris. All burnt rubbish should be removed and the damged areas cleaned before an examination can commence. At the same time there is always the possibility that some items may be salvaged and re-used. Precautions should be taken to protect any delicate features such as carved or moulded stonework, ornamental plaster and woodwork which may have only been slightly damaged. Plastic sheeting is useful in such circumstances provided it can be properly secured.

When dealing with badly damaged buildings, measures must be taken to prevent further movement and to ascertain that no part of the building is in danger of collapse. Failure to take adequate precautions might risk the lives of the occupants and members of the public and may also extend the damage. It is the duty of the surveyor to advise on this matter and arrange for temporary supports such as shores, struts and needles. The propping of damaged buildings is difficult to describe in writing as so much is governed by sound common sense. As mentioned in previous chapters, the owners should be informed of any emergency action being taken and a careful record made of the costs.

In the following sections individual materials will be considered.

16.3   Brickwork

Although bricks are non-flammable they can be damaged by heat according to the severity of the fire, and the thickness of the walling. The factor of safety of brickwork is so great that minor cracking can generally be ignored and the joints made good by repointing. If the wall has bulged to a major extent or if there are serious fractures it may well be advisable to demolish the wall and rebuild. Alternatively, remedial work may also be carried out by inserting tie rods to stabilise the wall. If the walls are constructed of sand lime bricks or concrete panels they may be seriously damaged and the structural stability of the wall may be affected which necessitates demolition and rebuilding. If cavity walls are subjected to severe fire they may act as flues exposing the walls to excessive temperatures, and if prolonged will cause vitrification of the brickwork and destruction of the mortar. In such cases demolition of the wall and rebuilding may be necessary.

Water from hoses used during the fire fighting can saturate right through to the internal face of a brick wall and the sudden quenching after being hot may cause cracking and spalling of the brickwork.

One of the most important factors that the surveyor must consider when dealing with fire damage is that the drying out process is very slow and may take several months. In the meantime the remedial work may have been completed. It is, therefore, extremely important to ensure that all embedded timbers such as plates, fixing blocks and the like are carefully removed and replaced with sound materials.

16.4 Concrete structures

The behaviour of concrete during a fire depends largely upon the type of aggregate used. Crushed clay, brick and slag do not cause spalling. Flint gravel expands considerably and causes spalling, and most other stones behave similarly to a lesser degree. Fire damage to concrete will normally affect the outer face of the concrete to a depth of 50 to 100 mm which will cause cracking followed by surface spalling. Exposed reinforcing bars absorb heat and expand, and will tend to twist. The surveyor must note that steel loses strength above a temperature of 570°F and at temperatures of between 800° and 900°F there is a loss of strength of up to 80%. High temperatures may cause concrete walls and floors to fall away and leave the fire free to spread into other parts of the building which may not have been affected when the fire first occurred. In cases where concrete floors and walls have suffered a loss of strength it will be necessary to demolish and rebuild the complete structure.

16.5 Stonework

The effect of fire can cause very considerable damage to stonework. Where the rise of temperature is very rapid stone is subjected to considerable stresses, but these will vary according to the thickness of the stonework, the intensity of the fire and the type of stone used. When water from fire hoses is poured on, considerable spalling of the stone is to be expected due to the sudden cooling. It is a fact that when limestone is heated to a high temperature (about 700–800°C) and carbon dioxide is driven off it becomes calcium oxide or quicklime. This process is known as calcination. The Building Research Establishment has carried out heat tests on limestone and has found that calcination usually occurred on arrises and edges of mouldings. Limestone walling is not significantly affected by fire calcination and the strength of the stone is not normally affected to any serious degree.

An important point for the surveyor to recognise is the colour changes which occur in most building stones when subjected to heat. Limestone and sandstone may turn brown, pink or purple. Limestone that is free of iron oxide, such as Huddlestone or Portland, usually turns a greyish colour.

Stonework will need careful examination in order to detect any hidden defects. Projecting features such as mouldings, string courses and cornices will require close examination. If the stability of the stonework has been

seriously affected by spalling, consideration must be given to the replacement of the defective stones or alternatively, whether they can be repaired
if only superficial damage has occurred.

16.6   Steel beams, columns and roof trusses

Steel beams built into a wall can suffer damage by distortion and expansion. These defects are likely to cause severe cracking or displacement of
the walling. When the steel beam is encased in a fire-resistant material the
rise in temperature will be delayed, but when the heat finally penetrates
through to the steelwork the temperature will rise quickly causing expansion and distortion. At 400°C the movement can lead to a total collapse of
the beam. Steel roof trusses consisting of small angles will tend to become
misshapen or twist with loss of strength. This movement will affect the
connecting joints by shearing or loosening the bolts or rivets. The surveyor
must, therefore, consider all metal as suspect if it has been subjected to
prolonged heat and has possibly lost strength. Badly burnt steel should be
rejected.

16.7   Timber

Although timber is combustible it will often function satisfactorily for a
longer period than metal provided it is of adequate size. Several species
such as iroko and teak are highly resistant to fire. However, small sections
will quickly burn and disintegrate, but large sections will burn on the
outer surface only.

16.8   Roof structure

A description of the common features of timber roof construction has been
given in Sections 10.1 to 10.5. It is necessary to consider the nature and
extent of the fire damage to be examined. If all but a few members have
survived the fire, the repair may be confined to the joints and members
which have failed. Longitudinal splits should be looked for in members
which show signs of deflection or have lost their support. Where purlins
are in reasonable condition it is well to see that their bearing is still adequate even though the walls have minor cracks. Rafters, struts and collars
should be carefully examined as many may be found to be badly charred
or split.

Damaged struts and collars are usually readily accessible and can be
easily replaced. This is not always possible with rafters or hips. For
instance, a rafter secured at the ridge and birdsmouthed at the wall plate
could not be withdrawn without disturbing the existing tiling, battens and

felt etc.; and ceiling joists cannot be replaced without disturbing the ceilings. A careful note should be made of these defects to enable a full specification and estimate to be prepared. All joints should be examined and a note made of any weaknesses that require strengthening. Badly burnt timbers must be replaced; however, large sections may be cleaned of charcoal and if sound they can be faced with boarding and left in position.

16.9   Pitched roof coverings

Slates and tiles are normally durable but are nevertheless brittle and will often crack or completely disintegrate under intense heat. It is, therefore, essential for the surveyor to make a close examination of outer and inner surfaces. It often happens that the condition of the roof is so bad that 'patching' would be totally ineffective. When dealing with older buildings it will be found difficult to obtain a similar tile or slate to match the colour and texture of the existing ones. It has often been said when a patch of smooth red tiles has been used to make good a patch in a handmade clay tile roof that 'they will weather down in time!' This is not true: the patch of red will remain a patch as long as the roof lasts. In addition the renewal of the battens and felt etc. may be required.

16.10   Flat roof coverings

The efficiency of flat roof coverings of all types is dependent to a large extent on how much damage was done to the supporting structure. It is difficult to classify the types of damage in a simple manner. In most cases damage may amount to minor cracks, charring of the surface, small holes, joints may have opened, skirtings may have pulled away from the walls, flashings become disturbed and general surface deterioration due to the effects of fire. Many of these defects can be made good which will give a worthwhile extension of life, but where the roof covering has been badly damaged total replacement may be the only remedy. Metal and plastic gutters may exhibit superficial defects if subjected to a minor fire, but intense heat particularly if applied to the joints will cause structural failure with possible collapse.

16.11   Floors

Floor joists usually run in the direction of the shortest span resting on trimmers or beams over openings and around staircases. Being protected by the ceiling and flooring it is unlikely that the joists will suffer any serious damage. Serious defects in flooring and ceilings are to be expected as a result of fire and water damage and will often require complete renewal. Serious fire damage to the walls may leave the floor joists without proper support. As long as the joists are in reasonable condition the floor can be temporarily carried on struts until the walling forming the permanent support has been repaired or replaced. It is unlikely that a building so severely damaged as to let the floor drop could, in any case, be worth repairing.

16.12   Internal and external finishes

During the course of a fire, most of the internal and external finishes are damaged or completely destroyed by fire, smoke or water. Damage to internal plastering or external rendering may have arisen from the fire, or from water from fire hoses, or from exposure of the building to the weather before external damage was repaired and the structure made weatherproof. Fire will cause a looseness from the background and will be readily detected as described in Sections 13.5 and 13.6. In the case of gypsum plasters a high temperature will cause the plaster to burn into anhydrous substances and in the case of lime plasters it will change into calcium oxide. The strength of gypsum plasterboards and insulation boards is often permanently impaired where they have been subjected to prolonged soaking or after repeated cycles of wetting and drying. Boards which have been damaged in this way do not constitute a suitable base for plastering or decorating and should be replaced. However, if the boards are only slightly damp their strength is largely regained.

Fire damage and water penetration may have caused serious damage to the woodwork and if the members are badly charred they should be completely removed. Fire resisting partitions and doors will often withstand a considerable amount of heat, and if slightly charred may be faced with plywood, if the strength of the frame structure is adequate. Woodwork affected by smoke stains can usually be cleaned and repainted. With some types of wood heat may cause a resinous exudation from knots and flaws in the timber, but the source can easily be removed and the timbers made good with fillers.

Plain glass may crack and splinter when subjected to heat and often forms into a fused condition at high temperatures. Roof lights, fire resisting doors and glazed partitions are usually fitted with wired cast glass which usually withstands a certain amount of heat before splintering.

Many defects in paintwork and decorations generally are to be expected as a result of fire damage and excessive water from fire hoses. Badly blistered or charred paintwork should be completely removed. If the surface of the wood has been slightly damaged, local repairs such as making good with a wood filler may be necessary.

16.13 Services

Electrical, gas, plumbing and heating services should be examined and tested by the various specialists. Testing of the electrical installation is particularly important. Electrical faults are responsible for many fires. Loose connections and damaged insulation can introduce a serious fire risk. It is advisable for the surveyor to obtain an 'inspection certificate' stating that the work has been tested and found efficient.

16.14 Recording the defects

After the defects have been identified the next step is to record every item systematically and mark on the drawing the defective areas with any necessary dimensions in order that a complete specification and estimate can be prepared. When fire damage repairs are to be carried out on a large complex structure, photographs of the damaged areas will be found most useful and will ensure that no details are omitted. It is also important to identify all damaged windows, doors and other fittings. Sketches of all the internal wall surfaces and partitions showing the position of the openings should be produced together with notes describing the defects. The openings should be numbered so that they are easily identified. It is of little consequence whether the numbering commences at the top of the building or in the basement so long as it is clearly understandable. In addition to these items large scale sketches should be made of any special features such as cornices, mouldings or recessed panels and reveals together with any necessary notes such as 'badly damaged cornice to be removed and rebuilt'.

FLOOD DAMAGE

16.15 Causes

Flood damage is usually due to the following causes:

- Excessive rains or melting snows which cause rivers to rise.
- 'Tidal waves' occurring when high winds drive the sea on to the land at period of high tides.
- Burst storage tanks or water mains causing local flooding in the ground floors or basements.

16.16 Preliminary examination

In the majority of cases the surveyor will find that when called to a flood damaged property the owners have already contacted the local fire services and have had the building drained or pumped out. Trapped water is a problem and the first step is to check that underfloor areas, ducts, cavities in cavity walls and basement areas have been properly cleared. Cavity walls are the worst problem and if water and mud are not properly cleared this could lead to permanent rising damp. The cavity walls should be cleared by removing some of the bricks at regular intervals along the base of the wall to form outlets. Particular attention must be paid to ducts and conduits containing electrical and telephone cables. Water must be removed by opening the inspection boxes and elbows.

Before commencing the examination of the flood damage it is important to test the water, gas and electrical services and isolate them if necessary.

If mud and silt have accumulated under floors or have piled up above the damp-proof course and blocked the ventilation, then arrangements should be made to have these areas cleared as soon as possible and the spaces sprayed with disinfectant. After clearing, the damp areas should be dried out and ventilated (BRE Digests 152 and 163). Heaters are useful for this purpose but all windows and doors should be kept open as long as possible to obtain a good draught. The traditional timber suspended ground floor should be well ventilated, with an occasional floor board removed to increase the draught. Thorough drying out cannot be too strongly emphasised. This operation will minimise the risk of fungal decay.

16.17   General effects of flooding

Brick, stone and concrete in general are unlikely to be seriously damaged by flood water, although brickwork will absorb fairly large quantities of water and will often take months to dry out. The surveyor should check the following points in respect of the main wall structure:

- Scouring and erosion may occur in soft brickwork or sand lime bricks, especially where flooding has been caused by sea-water.
- Cracking often occurs in lightweight structures such as timber-framed buildings due to expansion and contraction when the flood water rises and falls.
- Steel and ironwork which are built into walls will corrode, particularly in flood water containing a high proportion of salt or sea water.
- Damage by efflorescence as described in Section 6.6.

16.18   Foundations

In case of serious flooding the foundations of a building are most likely to be damaged especially where they are situated on shrinkable clay. When saturated, the clay is likely to swell and cause movement in the main walls and possible cracking. Erosion from the foundations may also occur, in which case the ground should be allowed to dry out and trial holes dug at intervals along the wall adjacent to the suspected position of the failure. The foundation and base of the wall must be carefully examined to ascertain its condition before a decision on remedial work is taken.

16.19 Ground floors

Solid concrete floors normally will withstand flooding, although defects may occur as a result of damage to the surrounding walls. However, some concrete floors may already be weakened by defects in the original construction or base such as chalk or burnt colliery shale which may cause expansion and serious cracking (see Section 8.9). It is advisable to remove small areas of the floor covering and examine the screed or slab.

16.20 Suspended ground floors

Suspended ground floors consisting of joists on wall plates will require careful examination, and where possible a few boards should be lifted to enable a thorough examination to be made. This operation, however, may well have been carried out previously as described in Section 16.16 above.

Particular attention should be paid to joists embedded in damp walls, skirting boards, panelling and bottoms of door frames and if badly affected by damp it is advisable to recommend their replacement. The ends of defective joists should be cut off and new brick sleeper walls built to support the ends of the joists. No joist should be in contact with the outer walls. A moisture meter with deep probes should be used to measure the moisture content of the timber. If the reading is more than 20% then further drying out is essential.

16.21 Floor finishes

Wood blocks and boarding tend to swell and lift. Plastic and ceramic tiles, linoleum and carpets secured to a screed will often deteriorate when water penetrates through to the screed causing loss of adhesion. As in the case of suspended timber floors it is advisable to lift a few tiles or blocks to ascertain the condition of the base before any remedial work is recommended.

16.22 Wall finishes

Undercoats of cement and sand or cement, lime and sand are not usually affected by flood water, but the old type of lime plasters and calcium sulphate plasters may soften when saturated. These defects are often followed by expansion and a rippling of the plaster face. Plasterboard, insulation board and fibre board become soft when wet and tend to warp or sag.

16.23   Metal finishes and fastenings

Metals are unlikely to suffer much damage if the flood waters are cleared as quickly as possible. However, metals are liable to corrode if they have been submerged for long periods particularly if the flood water was contaminated with sea-water or salts. The metal surfaces should be cleaned where possible as a precaution against corrosion. Door furniture and locks should be cleaned and oiled.

16.24   Drainage systems

Inspection chamber covers should be removed and the interiors checked for defects. All main drains and branches should be rodded to remove mud or silt that may have built up causing partial or complete blockage. In cases of serious flooding it is advisable to dig two or three trial holes at the side of the drain to ascertain the condition of the concrete bed. The soil supporting the bed may be washed away by underwashing currents of flood water leaving the base unsupported.

16.25   Pavings

Flood waters that penetrate through to the subsoil causing clay to swell may often cause the pavings to break up. Hardcore bases will require careful examination and if washed away will require complete reinstatement. If the area is suspect, one or two small trial holes should be dug around the edges of the paving or slabs lifted to ascertain the condition of the base.

16.26   Recording defects

As with fire damage, sketch plans of the various rooms and external works should be prepared showing all relevant details, and if large areas are affected, photographs will no doubt be found useful. A full description of all defects and dimensions should be included on the drawing sheet. If the flooding is extensive the surveyor will require a builder's 'attendance' to deal with cavity wall blockages, removing floor coverings and test drains etc. It is the surveyor's responsibility to explain the service he requires.

The builder will usually charge for this service on a 'daywork basis' and the client should be made aware of the costs.

The surveyor should always bear in mind that flooding can cause serious deterioration and decay if debris and silt are left behind when the water subsides. It is, therefore, necessary to allow the structure to dry out completely (per BRE Digest 163) before recommending any remedial work to take place. A useful 'check list' describing the various defects that may occur when dealing with flood damaged buildings is contained in BRE Digest 152.

17   Report Writing

17.1   Introduction

Report writing is, perhaps, the most important form of communication between the surveyor and the client. It can form a lasting impression of the surveyor's quality of work in the mind of the client. The majority of building surveys are carried out on behalf of clients who propose to purchase a property or take one on lease. A typical example is given in Appendix VII. Other types of reports are described in Section 1.3. An example of a report on a specific defect with some typical recommendations is given in Appendix VI.

17.2   Presentation

When the survey is complete and all information collated the next stage is to prepare a report on the structural condition of the building together with a clear and definite recommendation and assessment of costs. It is suggested that the report be prepared immediately following the inspection while the details are fresh in the mind. If reports are required from electrical and heating engineers or other consultants, they should be included in the surveyor's report.

As previously mentioned in Chapter 1, it is important that the purpose of the report is given careful consideration. Not only should it set out the information the client requires, but must also deal with matters that a technical person notices, but are sometimes overlooked by the layman. On the other hand, the report must aim at giving concise details to the layman and not be couched in technical language and long involved sentences that he finds difficult to understand. In certain circumstances technical terms are unavoidable, but in such cases they should be simply defined. As far as the client is concerned, he has instructed the surveyor to study and diagnose the structure and prepare a lucid report stating whether or not the building is basically sound. The surveyor should always endeavour to diagnose the cause of a defect and estimate its prognosis. He should also give some consideration to prescribing a remedy and detailing its costs. This will enable the client to assess how much money he will need to restore the property to a sound structural condition.

Some clients, such as a large company with a board of directors or an amenity society, may require an inspection of a property for a number of purposes. They may be trying to decide whether the property is suitable for conversion and will require the surveyor to report on the options. In such circumstances the report could be in two parts – firstly, describing the development possibilities and secondly the condition of the building. How much detail should be included will depend on the client's requirements and the skill of the surveyor, and on his ability to set down his observations in a carefully written report.

17.3   Report writing criteria

Primary criteria

Apart from good presentation, the primary criteria for the contents of any professionally or scholarly written piece of work are that they must be:

- *Accurate*, i.e. no false/misleading assertions or incorrect figures, and minimal grammatical or spelling mistakes in the script.
- *Coherent*, i.e. they make sense, are not contradictory, and have a clear and unified structure.
- *Critical*, i.e. they provide an analytical or objective view that is balanced or impartial, and any distinct bias or vested interest is clearly stated.
- *Rigorous*, i.e. they do not contain any fallacious arguments, *non sequiturs* or any other errors in logic.
- *Substantiated*, i.e. any assertion or claim must be evidence-based.

Secondary criteria

The secondary criteria are, by implication, not as critical but nonetheless are as desirable as the primary ones. Ideally, therefore, the contents must be:

- *Confidential*, i.e. never divulge a client's name or identify a property without prior permission. If necessary use pseudonyms.
- *Concise*, i.e. brief enough to cover the main points and non-repetitive – any padding or excessive detail should be removed.
- *Detached and dispassionate*, i.e. avoid being too subjective.
- *Original*, i.e. wherever possible be innovative or original in your work.
- *Poignant*, i.e. incisive and to the point.
- *Readable*, i.e. short sentences, maximising the use of 'active' verbs rather than 'passive' ones.
- *Scrupulous*, i.e. be scrupulously honest in writing; be prepared to accept the outcome of an experiment or investigation, even if it disagrees with your cherished theory.

17.4   Arrangement of information

No matter how experienced a surveyor may be he is bound to have some difficulty in preparing a report for a large building containing major faults. As indicated in Chapter 2, it is always advisable on arriving on the site to adopt a definite procedure that can be used as a framework for the main body of the report. It is not an easy matter to prepare a report so that the principal defects are the first items to make an impression on the client's mind. The main items a client will require from a report are as follows:

- A general description of the property.
- Condition of the structure with details of any faults and failures and their cause.
- Condition of the services – electrical, gas, heating, water supply, lifts (if any), and telephones.
- Condition of the fittings and finishes.
- Garages and outbuildings should be dealt with in the same manner as the main building.
- Condition of the boundaries, evidence of ownership and any details of rights-of-way and way-leaves.
- A few paragraphs at the end of the report headed 'Conclusions and recommendations' are of vital importance to a client. It is here that the surveyor must explain the significance of what has been observed and also why it was not possible to examine inaccessible parts of the building. An indication of actions and further investigations needed is important here.
- An approximate cost of the repairs where this is relevant.

It should always be remembered that if insufficient information is obtained on site a second visit may be needed. This makes the surveyor appear incompetent and makes explanations to the client difficult. Recommendations should be listed in order of importance, particularly with regard to urgent work such as a dangerous parapet wall or badly leaning chimney stack.

Distinguish clearly between structural defects and normal maintenance work, which could be deferred, together with an approximate estimate of costs. Reports describing extensive repairs or reconstruction work should be illustrated and any sketches or photographs attached to the report.

Many clients seem to believe that their surveyor has a duty to remove floorboards, bath panels and roof insulation without authority. No reasonable vendor will object to a surveyor using a ladder in order to examine roofs and gutters etc., but rarely will he agree to floorboards or roof tiles being removed which may result in damage to his property. The surveyor's report should state precisely the permissible limits and his inability to comment upon any part of the structure which is inaccessible or concealed (see Chapter 18 'Legal Aspects').

On the other hand, the report should not contain a large number of 'escape clauses' which will reduce the value of the survey. Moreover, it must be clearly understood that the onus is on the surveyor if he fails to investigate any suspicious areas. Escape clauses will not exonerate him if he has been negligent. Some of the legal pitfalls are dealt with in Chapter 18. It is wise to remember that an aggrieved client will certainly study the report for evidence of any omissions in order to claim against the surveyor. The surveyor's safety is obviously important and he cannot be expected to risk life and limb. The final decision whether or not to climb on to a pitched roof must lie with him. He must, therefore, report which parts of a building were not examined on the grounds of safety.

17.5   Format

Every technical report should have a clear beginning, middle and end – in that order! Most 'good' reports contain at the very least three main well-structured sections: introduction, investigation and analysis (main body – in big reports this is often more than one section), and conclusions and recommendations. For example, the following structure should be used when compiling a diagnostic survey report on a specific building defect (e.g. dampness problem):

1.0	**INTRODUCTION** (*main headings should be in upper case*)
1.1	**Instructions/Remit** (*consider bold headings and subheadings for emphasis*)
1.2	**Description of property**
1.3	**Survey/investigation details**
2.0	**INVESTIGATION & ANALYSIS**
2.1	**Condition** (*use lower case for sub-headings*)
2.1.1	History of defect/s
2.1.2	Anomalies/symptoms
2.2	**Diagnosis**
2.2.1	Hypothesis (e.g. condensation dampness)
2.2.2	Evidence (e.g. moisture meter readings, salts analysis, etc.)
2.2.3	Cause (i.e. confirm/confute hypothesis)
2.3	**Prognosis**
2.3.1	Significance
2.3.2	Likely consequences
3.0	**CONCLUSIONS & RECOMMENDATIONS** (or if no recommendations are needed: SUMMARY & CONCLUSIONS)
3.1	**Conclusions**
3.1.1	Main findings

APPENDICES (containing supporting material, e.g. graphics, etc.)
Appendix 1:   Location plan of property
Appendix 2:   Photographs of defects*
Appendix 3:   Schedule of costs

These may be inserted within the body of the report if recorded on a digital camera.

17.6   Valuations

As regards the value of the property the client may express a wish that the surveyor includes at the end of his report a few comments on whether or not he thinks the purchase price is fair. This obviously amounts to a valuation and unless the surveyor is sufficiently knowledgeable to give a valuation he would be unwise to attempt to do so. Today many factors have to be considered when assessing the value of property, such as rising values in a particular locality, the position of local shops and schools, the proximity of transport and the future development of the area. A valuer must make searching enquiries before he can express an opinion, especially when dealing with commercial or industrial properties.

Naturally, the client will hold the surveyor responsible for such advice, so unless he is qualified to carry out this service he should suggest to his client that he obtains the services of a competent valuer, preferably one who has some knowledge of the district.

18 Legal Aspects

18.1 Introduction

Liability for negligence has become a matter of great importance to all members of the building professions in recent years. Until about 1960 the law of tort (or delict as it is known in Scotland) remained fairly static, but during the past thirty-five years the various categories of negligence have been greatly extended by the courts.

This chapter is written to provide surveyors with some guidance on points of law concerning professional negligence. The following items are not a substitute for professional legal advice when complex problems arise, but are written to make the surveyor aware of the legal ramifications. We now live in a state of rapid change, not only in building techniques, but also in legislation. Building faults are continually being reported and journalists in the national press frequently report the public awareness of these problems. This is partly because of the increase in the number of property owners and also the government trend towards greater consumer protection. The public is entitled to protection, and if necessary recourse to law. The purchaser of a property will naturally seek legal advice and will wish to sue the surveyor if he is negligent and fails to detect structural defects or other significant problems (e.g. dampness). This usually means financial compensation.

The conduct of a surveyor may result in negligence if his performance falls below the required standard of care. However, a surveyor, like any other professional person, is neither expected to be an expert nor perfect. The standard required is that of an ordinary skilled surveyor exercising reasonable care – i.e. the norm in the surveying profession.

18.2 Negligence defined

Negligence is a tort or civil wrong of particular importance to surveyors. It is the failure to exercise the degree of care that the circumstances demand and is separate and distinct from the law of contract (see Section 1.6 'Contracts and fees'). What amounts to negligence will depend upon the facts of each particular case and can be defined as a breach of legal duty to take care which results in damage to the plaintiff. The duty of care arises independently of any contractual relationship. In order to decide whether a defendant has been negligent the plaintiff must prove that:

(1)   The defendant owed to the plaintiff a 'duty of care'. This is an area of particular importance in building disputes. The courts now have many cases to guide them in deciding whether there is a duty of care (see Section 18.11).

(2)   The defendant was in breach of that duty. This is of central importance in all negligence litigation.

(3)   The plaintiff has suffered loss or damage as a result of a breach of duty by the defendant. The plaintiff will usually be entitled to recover damages, the object being to provide recompense by money for the wrong that has been done.

18.3   Duty of care

The duty to take care has been defined by Lord Atkin in the celebrated case of the snail in the ginger beer bottle, *Donoghue* v. *Stevenson* (1932):

'You must take reasonable care to avoid acts or omissions which you reasonably foresee would be likely to injure your neighbour. Who in law is my neighbour? The answer seems to be the persons who are so closely and directly affected by my act that I ought reasonably to have them in contemplation as being so affected when I am directing my mind to acts or omissions which are called into question.' [1932] AC 605–623.

What is reasonable care will vary with the circumstances. Clearly the seriousness of the potential damage that a mistake may cause must be taken into account. The likelihood of injury must also be taken into account if there is a probability of something going wrong, and great care must be taken to avoid any injury to a person or persons.

However, if a surveyor can show that he acted in accordance with the generally approved practices of his profession he will be likely to have discharged his duty of care. The important issue is not what the surveyor actually foresaw but what he ought reasonably to have foreseen. A typical case concerning negligence, *Daisley* v. *B.S. Hall & Co* (1972) is quoted below:

'The surveyor was held to have been negligent in failing to report to his client (the plaintiff) the risk created by the combination of poplar trees growing nearby and the clay sub-soil, where certain signs and cracks indicating shrinkage were already discernible in the house that the plaintiff was intending to purchase.' (1972) 225 EG 1553.

The court has to decide in each individual case whether or not there has been professional negligence and the principle that the courts usually adopt is that, if there are suspicious signs or visual evidence of any defects, then it is the surveyor's duty to have parts of the structure opened up for examination.

18.4   Breach of duty

As mentioned in Section 18.2 the issue of 'breach of duty' is of great importance, because the question that must be asked is whether the defendant was in fact negligent and therefore liable in tort. The text of what constitutes a breach of duty is illustrated by the often cited dictum of Baron Alderson in *Blyth* v. *Birmingham Waterworks Co.* (1856).

> 'Negligence is the omission to do something which a reasonable man, guided upon those considerations which ordinarily regulate the conduct of human affairs, would do, or doing something which a prudent and reasonable man would not do. Thus the standard is objective and impersonal. Furthermore, where anyone is engaged in a trade or profession and holds himself out as having professional skill he will be expected to attain the normal competent professional standard. It is no defence to say that he acted to the best of his skill and ability if that falls short of the professional standard. It is not enough to do one's "incompetent best".' (1856) 11 Ex 781.

Where a person holds himself out to be a competent professional surveyor, whether he possesses qualifications or not, he is implying that he is reasonably competent to carry out a building survey. This was clearly stated in *Freeman* v. *Marshall & Co.* (1966) 200 EG 777 High Court. The defendant was an unqualified estate agent, valuer and surveyor who failed to detect rising damp and wood rot in a basement. In his defence, he stated that he only had a basic understanding of building technology and structures involved in routine estate agency work. Mr Justice Lawton said (at p. 777):

> 'He had no organized course of training as a surveyor and had never passed a professional examination in surveying. He was a member of the Valuers Institution through election, not by examination. In fairness to him, he claimed only to have a working knowledge of structures from the point of view of buying and selling, but if he held himself out in practice as a surveyor he must be deemed to have the skills of a surveyor and be adjudged upon them.'

In summary: the plaintiff engaged the surveyor on account of his special technical skill, and it is implicit in law that he will display such skill. It is no defence to say that he was very inexperienced.

In all cases of negligence where a breach of duty is concerned it is for the plaintiff to establish what actually happened, and what can reasonably be expected from the defendant. Proving what actually happened is not an easy task when bringing an action against a professional adviser.

The plaintiff will undoubtedly take legal advice and consult other experts in building matters to discover whether the original survey fell below the recognised professional standard. The onus lies with the plaintiff to prove that the defects were in existence at the time of the original survey.

It must also be remembered that a surveyor is expected to keep up-to-date with current research. The professional standards to be applied are those currently prevailing, not those when the surveyor qualified.

18.5 Damages

The third element in the tort of negligence is damages. This simply means financial compensation for the wrong that has been done. It is essential for the plaintiff to establish not just the fact of damage, but to prove that, but for the breach of duty, the damage would not have occurred.

As a general rule the courts will award damages on the basis of the cost of the repair work necessary to put the plaintiff back in the position he would be in if the tort of breach of contract had not been committed. In some cases the loss is purely financial and is known as 'economic loss' but this rarely occurs in connection with building surveys. The considered view is that economic loss which is not due to damage to persons or property is not usually recoverable. In the case of *SCM (United Kingdom) Limited v. W. J. Whittall & Son Limited* (1970) Lord Denning MR stated:

> 'In actions of negligence, when the plaintiff has suffered no damage to his person or property, but has only sustained economic loss, the law does not usually permit him to recover that loss. Although the defendants owed the plaintiffs a duty of care, that did not mean the additional economic loss which was not consequent on the material damage suffered by the plaintiffs would also be recoverable.'

18.6 Accuracy of estimates

In certain circumstances there is a duty not to make false statements which may cause financial damage. This will be particularly important to a surveyor when he presents an estimate to his client in respect of repairs required to a property. It must be clearly understood that the term 'approximate estimate' implies that the estimate is not exact but is the probable cost of carrying out the repairs. The estimate must give the client some reasonable idea of the expenditure to which he will be committed if he decides to purchase the property.

If the surveyor's estimate is too far out he could be liable for negligence. Occasionally, the client will require a 'firm quotation'. If the surveyor is unable to carry out this service with complete confidence it is advisable to employ a competent estimating surveyor, and pay the fee for his service.

18.7   Brief reports

The surveyor must be particularly vigilant when the client says he only requires a general opinion and not a detailed examination. This happens when a client is anxious to purchase the property and expects his surveyor to give a 'quick answer'. This may lead to wrong conclusions and provide subsequent trouble to all parties concerned. In such cases the surveyor still has a responsibility to discover serious defects and can be sued for negligence if the defects are not correctly reported. When submitting his report he should clearly state that he has not had an opportunity to examine the building in detail and cannot be absolutely sure that there are no hidden defects. Thus in the case of *Sincock* v. *Bangs, Reading* [1952] CPL 562, the architect was instructed to inspect a farm and to give a general opinion and not to make a detailed survey. The architect was held to be negligent in not discovering dry rot, woodworm and settlement.

18.8   Parties in tort

All partners in a professional firm are liable for a tort committed by a partner or an employee during the course of employment. A partner cannot say that a mistake was made by one of his assistants and exonerate himself from blame. This is known as 'vicarious liability'. Thus it follows that in the case of an employee surveyor carrying out a building survey by a professional partnership, the partnership will be responsible for the torts committed by their surveyor provided the torts were committed during the course of the surveyor's employment. Furthermore, where an employee is instructed to carry out a survey and he is negligent in his work, then the employer and employee will be liable in respect of that negligence. The plaintiff in such circumstances can sue the employee, or both.

In the case of *Lister* v. *Romford Ice and Cold Storage Co. Ltd* (1957) the House of Lords held that an employee has a duty to perform his work with reasonable care and skill, but if he is negligent and causes his employer to be 'vicariously liable' for his tort he can be liable in damages to his employer for breach of contract. The surveyor is also liable for his actions in recommending any specialist. If he has not taken reasonable steps to check their competence and the client suffers damage as a result, the surveyor is liable for his own negligence in failing to discover that the firm was incompetent. However, if the surveyor recommends a specialist whom he considers to be fully competent, but who subsequently carries out the work badly, the surveyor is not liable.

18.9   Type of survey required

At this stage it is necessary to emphasise the difference between a building survey and a valuation survey. As previously stated, the building surveyor in his examination looks to see if the structure is sound and, if not, to report faults and suggest remedies.

The valuation of a property involves a fairly superficial inspection in order to estimate what it is likely to be worth, taking into consideration its location, amenities and facilities, etc. The fee paid relates to the relatively shorter time spent on this assessment. There is, however, still uncertainty among the public as to what such an inspection should involve. Many homebuyers, even some solicitors, believe when they have received a mortgage valuation that the property must be in reasonable condition to warrant the building society's offer, and therefore a building survey is not necessary. From the evidence of reported cases it has been proved that a homebuyers survey and valuation (Scheme 2 survey), if not a full building survey (Scheme 3 survey), is necessary, as is demonstrated by the cases described in section 18.11.

18.10   Professional negligence relating to surveying buildings

As indicated above, the basic principle of negligence rests upon breach of duty of care. Negligence in relation to inspecting and surveying buildings usually relates to one or more of the following deficiencies on the surveyor's part.

Failure to carry out instructions

For example, this could relate to not addressing the client's requirements as articulated in the letter confirming instructions.

Failure to make sufficient enquiries

Not consulting local authority records or omitting other checks on the property as regards ownership, easements, etc. – particularly at the desktop stage.

Inadequate level of knowledge, skill or experience

Recognising the limits of one's own competency is a skill in itself.

Failure to inspect properly

The survey was undertaken in a hasty, superficial or partial manner – as evidenced by the incomplete or incorrect guidance to the client contained in the report. Typical examples of this are:

- Failure to inspect particular parts (e.g. roof space, subfloor void).
- Failure to uncover or open up (e.g. hatches providing access to ducts and other hidden spaces).
- Failure to observe (e.g. to note anomalies or other adverse changes in construction).
- Limited survey (e.g. restricted to exterior).
- Failure to recognise/diagnose (e.g. mistaking condensation for rising dampness).

Inadequate report

Poorly presented report or, even worse, badly written or inadequately formatted and failing to comply with the criteria described in the previous chapter.

18.11   Recent negligence cases

Many of the negligence cases that have come before the courts during the past twenty years are concerned with valuation surveys for mortgage purposes. The following cases questioned the previously accepted view of the valuation reports prepared by surveyors acting for building societies.

(1)   *Yianni* v. *Edwin Evans & Sons* [1982] 1 QB 438.

This case concerned a report prepared by a valuer acting for a building society when a prospective purchaser applied for a loan. The mortgagee asked the defendant to value the property. The property was valued at £15,000 and the plaintiff accepted the building society's offer. Some years later cracks appeared in the structure due to settlement and the cost to put the building into a good state of repair was £18,000. The plaintiff claimed damages against the defendants on the basis that they had been negligent. The matter at issue was whether the surveyor making a valuation for a building society was under a 'duty of care' to the purchaser. The High Court decided that they were, even though there was not a contract between them and the plaintiff.

(2)   *Smith* v. *Eric S. Bush* and *Harris* v. *Wyre Forest District Council* (1989) 17 EG 68.

In April 1989 these cases both came before the House of Lords. Both cases concerned valuation surveys where the valuer had failed to detect and advise on the extent of the structural defects. The two appeals, although different in detail, gave rise to three questions, namely:

- Whether the valuer who was asked to value the property for mortgage purposes owed the prospective purchaser a duty to exercise reasonable care and skill.
- Whether the disclaimer notice which seeks to remove all liability was valid under the terms of the Unfair Contract Terms Act 1977. This Act is of great importance and considers disclaimers in considerable detail and operates in tort as well as in contract. The surveyor should make himself familiar with its contents.
- Whether the disclaimer satisfied the requirements of reasonableness.

The decisions handed down by the House of Lords in 1989 have clarified the principles first laid down in *Yianni* v. *Edwin Evans & Sons*, namely that a surveyor carrying out a valuation of a property can be sued for negligence by both the mortgagee and the purchaser if he negligently made significant errors in his report. In the *Smith* and *Harris* cases it was held that the disclaimers were invalid because of the Unfair Contract Terms Act 1977. As can be seen from the above mentioned cases surveyors are very much at risk from the ongoing and undoubted liability that exists for careless statements that can lead to law suits.

(3)   *Merrett* v. *Babb* (2001) EGCS 20

This was a controversial case, which was funded by the RICS when it went to the Court of Appeal. It concerned a claim of alleged professional negligence relating to a mortgage valuation survey. The outcome was that although the defendant surveyor was an employee, he personally owed a duty to the claimant (Glover 2003).

(4)   Other important cases

It is beyond the scope of this book to consider in detail those who could be a party to an action in tort. However, if the reader requires further information on the development of tort in respect of property surveys, the history of the following cases, listed in chronological order, should help:

- *Morgan* v. *Perry* (1973) 229 EG 1737 (High Court)
- *Fryer* v. *Bunney* (1982) 263 EG 158 (High Court)
- *Perry* v. *Sydney Phillips & Son* (1982) 1 WLR 1297 (Court of Appeal)
- *Bolton* v. *Puley* (1982) 267 EG 1160 (High Court)
- *Howard* v. *Horne & Sons* (1989) 5 PN 136 (High Court)
- *Watts and Another* v. *Morrow* (1991) 14 EG 111 (High Court)

For further details of these and other cases see the works by Maltz (2001), Murdoch (2002) and Powell and Stewart (2002).

18.12   Disclaimers and limitation periods

During recent years it has become the custom to insert a disclaimer notice in both building and valuation surveys stating the inability of the surveyor to report on any part of the structure which is inaccessible, covered or unexposed. Generally speaking, the courts do not like disclaimers, particularly the sweeping type used by some building societies and valuers which seek to remove all liability.

In the cases of *Smith* v. *Eric S. Bush* and *Harris* v. *Wyre Forest District Council* their Lordships criticised valuers, building societies and other lending institutions for seeking to avoid liability by sweeping disclaimers. However, it would appear that limitation clauses which inform the purchaser of the extent of the examination are acceptable, and the surveyor should make this clear in his report. The House of Lords also stated that a valuation surveyor owes the purchaser a 'duty of care'. It is likely that no disclaimer clause can completely absolve the duty of care and any clause which limits that duty or liability can only be applied to the extent that it is fair and reasonable. Obviously this is rather a grey area with wide ranging implications for liability and the circumstances in each will determine the validity of a disclaimer notice.

Different types of surveys will obviously serve different functions, and will in some cases affect the standard of care. In the writer's view the surveyor should make clear at the outset what the purpose of the survey has been and the extent of the responsibility which he is prepared to accept. It is clear that the profession needs some guidelines concerning valuation surveys stating what level of inspection the survey should provide, and what can be done to ensure that the purchaser of the property knows the extent of the inspection. It is advantageous to both parties in a litigious situation if they have both recorded their rights and liabilities. Each party should know precisely what is required of him and any limitations should be stated so that it would be clearly understood whether or not obligations have been complied with.

The period after which the injured party is prevented from pursuing a surveyor for breach of contract or negligence has now been determined by statute. The Limitation Act 1980 specifies a limitation of actions as follows:

- A claim under an oral or written contract will be statute barred six years after the time when the cause of action accrued (section 5).
- A claim in tort will be statute barred six years from the accrual of the cause of action (section 2).

- A claim in contract under seal will be statute barred 12 years from the time when the cause of action accrues (section 8).

The legal view is that the defendant who has committed a negligent act should not have the possibility of legal action hanging over him forever, and at the same time the plaintiff who has been injured by a wrong should not delay in presenting his case.

On the 29 November 1984, the Law Reform Committee published a report on latent damage. This report concluded that there were certain sections in the Limitation Act 1980 which might give rise to uncertainty. This could cause injustice to both defendants and plaintiffs where the issue is when the damage occurs and this could give rise to a problem. The Committee's recommendations resulted in the Latent Damage Act 1986. Its main provisions are:

(1)   The basic claim period is still six years from the date on which the cause accrued.

or

(2)   Three years from the 'starting date' if that is later. The 'starting date' is the earliest date on which the plaintiff had or could reasonably have had knowledge of the facts about the damage.

(3)   The Act created a 15-year 'long stop' which runs from the date of the breach of duty on which the negligent act occurred. (This is also in section 14B of the Limitation Act 1980.) The 'long stop' provision protects the defendant in the long term. It is important to remember that the 'long stop' provision can expire before the other two periods mentioned above, and that all three periods run from different starting dates.

This is an area of law where the fixing of dates can be a complex matter. Even the simplest case can give rise to difficulties. Capper (1987) discusses many of the problems of latent damages.

18.13   Trespass

Trespass is a long established tort and occurs where there is direct interference with the plaintiff's goods or land by another. As a general rule it is not a criminal offence (as opposed to mass trespass, which is) and only gives rise to civil action. As far as the surveyor is concerned it will arise from entering on to the land owned by the vendor. The mere entry on to another person's land, without actual or implied consent, is technically a trespass. When a surveyor receives instructions from his client to carry out a building survey this normally includes permission to enter the property. Despite this, it is advisable for the surveyor to obtain permission from both owner and tenant and adjoining owners and tenants. This permission is particularly important when carrying out a drain test or the inspection of boundary walls which may involve walking over the property of others.

18.14   Party structures

Party structures are those wall, fence and floor elements that separate properties of different owners (Watt 1999). The Party Wall etc. Act 1996 controls the procedures relating to party structures in England and Wales (see Party Wall Guidance Note Working Group 2003 and Chynoweth 2003). Common law applies to party wall matters in Scotland (Melville and Gordon 1992).

Surveyors inspecting a dwelling will have to consider any party structure implications for their clients. This is particularly the case with flats, semi-detached or terraced dwellings (see Anstey 1998, 2003). Common repairs, conversions and other adaptations to party structures affecting neighbours are typical issues that surveyors need to consider in such instances (Melville and Gordon 1992).

18.15   Indemnity insurance

As indicated earlier, it is a mandatory requirement of being a chartered surveyor to have an appropriate level of professional indemnity cover. This form of insurance covers the surveyor against clients' claims (legitimate or otherwise) of negligence. Undertaking building survey work is one of the most onerous activities that any construction professional can undertake. Compared to many other professional commissions, the risks are high whilst the financial rewards are low with this type of work.

It is advisable for all surveyors to protect themselves against claims based on alleged neglect, omission, or error when acting in a professional capacity. Damages for all types of claims have increased over the past twenty years and will no doubt go on rising. Indemnity policies are mandatory with some professional bodies and partly so in others and cover the damages and legal costs incurred in cases of negligence.

It is advisable for the surveyor to obtain the services of a reputable insurancee broker specialising in such insurance who will advise him with regard to the various types of policies available. The usual procedure is for the surveyor to complete a proposal form which is the basis of the insurance contract. If all the material facts are not disclosed, the surveyor may find that the insurer can lawfully refuse to settle a claim because of this non-disclosure. This applies to facts which he knows, even though the insurer may not have asked the relevant questions. It follows that the surveyor will only be covered for liability that falls within the description of the cover in the policy. It is well to remember that insurers can refuse to deal with a claim that is not properly presented in accordance with the terms of the policy.

Considering the claims that may arise against surveyors for some omission in their reports, they must not disregard the inflationary effects when arranging adequate insurance cover and the fact that this could increase their overheads. It must also be remembered that insurance cover should be continued when a surveyor retires since claims may still be made in respect of negligence during the period he was in practice. Claims may be made for as long as the law of limitation of action allows (see Section 18.11 above). Legal advice should be sought as to the length of time for which insurance cover should be maintained.

Appendix I
Definitions of Inspections and Surveys of Buildings (CIC Explanatory Leaflet)

Introduction

Professionals carry out many different types of inspections, each for a particular reason and each resulting in a report with specific information produced for a particular purpose. The professional undertaking any of the services described in this leaflet should be suitably qualified, and may be a member of one or more professional bodies. The professional undertaking any particular service must ensure the client is clear as to which type of inspection is required, and must work to the agreed brief.

It is in the interests of both the professional undertaking the service and the client that the conditions of engagement, incorporating a definition of the brief and the extent of the inspection to be undertaken, are agreed in writing at the outset.

This document has been prepared by organisations representing the various professionals whose work includes inspecting existing buildings or those in the course of construction. It is intended to define clearly the various types of inspection normally undertaken, with a view to avoiding different interpretations between the professionals involved and their clients.

The term 'structural survey' has come into common use to describe a survey which covers all visible and accessible parts of a building, including those which are not part of the structure, such as the roof covering, windows and drains. The term should be avoided since misunderstandings with clients have arisen when, for example, a 'structural survey' has concentrated on the structure alone. Similarly, some clients have been confused when a 'structural survey' report has included a recommendation to commission a 'structural investigation'.

In view of this the CIC and the contributing organisations have agreed to urge their members not to use the term 'structural survey' to describe any of the services defined in this leaflet.

When a structural engineer is engaged to investigate the structure of a building, it is recommended that the work should not be referred to as a 'structural survey'; the terms 'inspection', 'appraisal', 'investigation' or 'assessment' are more appropriate.

(1)   Valuation

The primary purpose of a valuation is to provide an opinion of the price a property might achieve if it were sold, or of the rent if it were let. Whilst other types of survey and inspection focus on condition, constructional issues and associated remedies, a valuation inspection normally precedes a valuation, which advises on the value of a legal interest in property.

Valuations may be undertaken for a variety of reasons and the following issues need to be understood and agreed between the parties before the valuation is carried out:

- The purpose and date for which the valuation will be required.
- The interest to be valued.
- The assumptions upon which the valuation will be made.

The extent of the inspection will usually be defined by reference to guidance published by the body to which the professional belongs. The valuation inspection will take account of relevant factors affecting condition, location and aspects of construction that are readily observable on a walk-round inspection.

It will not be usual for the professional to be under any contractual duty to advise in detail as to defects or as to their remedy, neither will he or she be obliged to uncover, open-up or investigate individual features of a property.

A valuation should not be interpreted as, or used in substitution for, a survey report. Generally, the scope of a valuation inspection, and any measurements taken, will be confined to what is necessary to establish aspects or factors material to value.

In some cases the person undertaking the valuation may need more detailed advice on condition before completing the valuation.

(2)   Property purchase survey and valuation

A property purchase survey and valuation will advise on value as well as giving factual information on significant aspects of the condition of the property, usually a home, to the prospective purchaser.

The survey will be carried out under conditions (or terms) of engagement agreed with the professional undertaking the work. For straightforward properties where the client requires no out-of-the-ordinary advice, the report may be in a standard format and under the relevant standard conditions of engagement, such as those produced by the organisation of which the professional undertaking the work is a member.

Minor items of disrepair that do not materially affect the value will not normally be reported. Whether services will be tested or not will be agreed between the parties beforehand. Whilst a property survey and valuation may be undertaken for the prospective purchaser or lessee of a commercial property, it may be more appropriate for a valuation (see (1)) together with a building survey (see (3)) to be carried out in respect of larger, older or non-standard buildings, or buildings of a commercial or industrial nature. Prospective lessees may often need to arrange for the preparation of a Schedule of Condition (see (8)) before the commencement of the lease.

(3) Building survey

A building survey is an investigation and assessment of the construction and condition of a building and will not normally include advice on value. A member of any of the contributing organisations who has appropriate experience may carry this out. The survey will generally include the structure, fabric, finishes and grounds. The exposure and testing of services are not usually covered.

The extent of the survey will be subject to specific agreement between the professional and the client and advice on costs of repair will be subject to such agreement. The report will include reference to visible defects, and guidance as appropriate on maintenance and remedial measures.

The report may recommend that elemental or specialist investigations (see (4)) are undertaken or other specialist advice obtained relating to specific issues.

(4) Elemental or specialist investigation

An elemental or specialist investigation will be carried out when concern exists over specific parts of, or defects in, a property. Examples of this specialist work are the detailed study of movement, cracking, bulging, timber decay or the testing of building services such as wiring, heating, drainage, etc.

The scope of this type of investigation will be specific to each individual job and, if necessary, the brief may be altered as findings on site dictate.

(5) Investigation prior to alteration

An investigation prior to alteration will sometimes be required when alteration work is to be undertaken on a building. Subject to the necessary authority, it may involve opening up, measuring, calculations to check the adequacy of structural and service elements and detailed tests.

This investigation can be more detailed than the building survey referred to earlier (see (3)) and is likely to involve the services of a number of specialists. The scope of this type of survey will be specific to each individual job and, if necessary, the brief may be altered as findings on site dictate.

(6) Reinstatement cost assessment for insurance

A reinstatement cost assessment for insurance will advise on the anticipated cost of reconstructing a building in the event of damage by an insured risk and will have no direct relationship to the market value of the property. Usually a Reinstatement Cost Assessment will be made by reference to figures published by the Royal Institution of Chartered Surveyors/Building Cost Information Service (RICS/BCIS).

When this is not appropriate, for example in the case of a listed or nonstandard building, it may only be possible to assess reconstruction costs by the preparation of approximate building quantities and costing them. In such instances the method of assessment is to be agreed in writing by the parties beforehand.

(7) Stock condition survey

A stock condition survey will be commissioned to assess the state of repair or condition of an organisation's current building stock in connection with the preparation of a maintenance programme. The results of stock condition surveys will often be stored on a database.

(8) Schedule of condition

A schedule of condition will record the condition of the building at that particular time and may be supported by photographs, sketches and drawings. It may be prepared at the beginning of a lease or prior to commencement of adjacent construction work or in connection with litigation.

(9) Schedule of dilapidations

A schedule of dilapidations will identify the wants of repair of tenanted premises in relation to the repairing obligations under the terms of the lease. It may be prepared for service on the landlord or tenant, depending on their respective lease obligations.

(10) Measured survey

A measured survey, sometimes referred to as a *Dimensional Survey*, will involve taking measurements of a building and/or its site in order to prepare accurate drawings to scale. Such a survey may include taking levels. The purpose for which the drawings will be required, their scale, detail and acceptable tolerances, need to be agreed between the parties beforehand.

(11) Inspection of buildings under construction

Inspections of buildings under construction are required for a variety of purposes, including statutory requirements, which will dictate the frequency and scope of inspection and the reporting format, and these must be agreed between the professional and the client.

Appendix II
Sample Extracts of Survey Schedules

Sample extract of schedule of condition

Schedule of Condition
describing the state of repair of the domestic premises at
62 Acme Crescent, Anytown, UK

by

J. Bloggs, Chartered Surveyors
33 Acme Terrace, Anytown, UK
recorded on
30 July 2004

Item No.	Element	Description	Condition	Rating (a) U, D, N (b)	Approx. (a) Cost
1.0	ROOF				
1.1	Roof coverings	'Rosemary' clay tiles	Several tiles loose/missing (10 No.); some broken (5 No.)	U	£1,000
1.2	Gutters	Cast iron	Clogged in places; some joints leaking (c)	D	£500
1.3	Flashings	Zinc	Some rust pitting noted in places	D	£200
2.0	WALLS				
2.1	Front elevation	Cavity facing brick	Some organic staining near eaves; generally in good condition otherwise.	N	N/A
3.0	OTHER etc. (d)				
				Subtotal c/f	

Notes:
(a) These columns are optional.
(b) The ratings relate to the urgency of the attention required:
- U = urgent (within 1 week or 1 month – depending on the priority required for the facility).
- D = desirable (within 1 year – so that further problems or repairs are not triggered).
- N = non-urgent (within 5 years – or within the current planned maintenance programme cycle).

(c) Note that no attempt is made in a schedule of condition to diagnose the cause of any defects indicated. In other words it is descriptive not analytical in style.
(d) The remaining elements (such as windows, doors, services, etc.) would of course be included within most Schedules of Condition.

Sample extract of schedule of dilapidations

Schedule of Dilapidations
and wants of repair required to make good the commercial premises at
12 Acme Road, Anytown, UK
in accordance with the covenants of the lease dated 1 January 1994

by

J. Bloggs, Chartered Surveyors
33 Acme Terrace, Anytown, UK
recorded on
30 July 2004

Item No.	Covenant (a)	Location	Breach	Remedy	Cost (b)
EXTERNALLY					
1	2.7 To keep in repair	Main roof	Failure to keep watertight owing to roof tiles missing/broken	Take off and renew 10 tiles	£200
				Refix 5 slipped tiles (c)	£50
2	2.7 To keep in repair	Walls	Failure to repair defective mortar joints	Rake out joints in brickwork and repoint in a 2 m² area	£200
3	2.8 Redecoration clause	External elevations	Failure to rectify defective paintwork to external windows and doors	Repaint as required during the last year of the occupancy	£5,000
INTERNALLY					
4	2.7 To keep in repair	Reception area	Failure to repair defective floor covering	Take up and renew 20 m² of worn carpeting	£600
5	2.8 Redecoration clause	Reception area	Failure to redecorate internally	Repaint in accordance with the redecoration cycle	£3,000
6	2.8 Redecoration clause	Office areas	Failure to redecorate internally	Repaint in accordance with the redecoration cycle	£12,000
				Subtotal c/f	

Notes:
(a)   As listed in the lease.
(b)   Typical costs can be determined by using the cost data handbook published annually by the Building Cost Information Service run by the RICS's Building Maintenance Information (BMI2002).
(c)   Note that a schedule of dilapidations is not a specification. No attempt is made to prescribe the method or standard of work. It is presumed that the repair work will be done on a like-for-like basis.

Appendix III
Checklist for Property and Site Surveys

Item	Remarks
General particulars	
(1) Client's name, address and telephone number.	
(2) Address of property to be surveyed (if different from above).	
(3) Date of survey.	
(4) Occupied/unoccupied.	
(5) Freehold/leasehold.	
(6) Any special instructions received from the client.	
Survey of site and buildings	
(7) Prepare sketch of site including all existing buildings. All basic dimensions to be taken including ties where necessary in sufficient detail for plotting to scale.	
(8) Details of adjacent properties and roads. Rights or easements affecting the site or adjoining properties.	Names and addresses of the owners or occupants of the surrounding properties if necessary.
(9) Note position of trees and shrubs etc.	
(10) Show correct aspect of site.	
(11) Details of existing vehicular and pedestrian access adjoining footpath.	If new cross-overs are necessary the matter should be discussed with the Local Authority.
(12) Particulars of boundary walls, fences and gates etc. Obtain ownership of fences.	Structural details will be required if alterations to the access or rebuilding is necessary.
Levels	
(13) Levels on the site and surrounding areas to be taken are referred to Ordnance Datum where possible and checked against existing bench marks. If an extension is contemplated extra levels should be taken close to the area of the proposed extension and where any special features occur.	If datum is not known then some permanent feature must be selected such as a step or inspection chamber cover.
Sub-soil	
(14) Describe the general character of the site or landscape. Dig trial holes over the site of the proposed extension and mark the positions on the plan, giving details of the sub-soil and likely bearing capacity. State the depth at which a suitable foundation can be obtained. State water table and liability to flooding.	

Item	Remarks
Existing buildings	
(15) Obtain particulars of the existing walls, floors and roofs including all openings adjacent to the proposed area of the extension or alteration work.	The area should be fully photographed.
(16) General condition of materials and finishes to floors, walls, ceilings, stairs and any other structural member.	
(17) Details of any structural failures to the existing buildings or material failures peculiar to the district.	
(18) Evidence of woodworm, wet rot, staining or dampness.	
Services	
(19) State if gas and electricity is obtainable and if possible show on plan the position of the mains and meters.	
(20) If company's water is obtainable show on the sketch the position of the main.	
(21) Plot soil and surface water drainage system. Inspection chamber cover levels and inverts.	
(22) Show the position and direction of the sewer (if required) and if possible state the depth.	This information can usually be obtained from the Local Authority's surveyors' department.
(23) Check if the authorities require a separate system of drains for soil and surface water.	
(24) Description of existing heating and ventilation systems, including details of any flues or ducts.	Only required if these services are being altered or the equipment is being repositioned.
(25) Details of sanitary fittings and waste pipes.	Only required if these services are being altered or the equipment is being repositioned.
(26) Obtain the address and telephone number of the Local Authority's surveyors' department.	

Appendix IV
Checklist for Building Surveys

Detailed condition of the several parts of the fabric design below should be carefully noted.

Item	Remarks
(1) *General particulars.*	
(a) Client's name, address and telephone number.	
(b) Address of property (if different from above).	
(c) Date of inspection.	
(d) Occupied/unoccupied.	
(e) Freehold/leasehold.	If freehold ascertain if there are any restrictions. If leasehold find out term of years and when term commenced. What restrictions are imposed.
(f) Approximate age of the building.	
(g) Prepare a rough sketch plan of the various rooms with sizes and heights starting with the ground floor (or basement).	
(h) Prepare rough sketch of the site showing outbuildings, boundaries, entrances and orientation.	
Internal	
(2) *Rooms and offices etc.*	
(a) Partitions.	Check if load bearing.
(b) Skirting and picture rails.	
(c) Wall finishes.	
(d) Steel columns and beams.	Note fire resistance.
(e) Floor finishes.	
(f) Ceilings and cornices.	
(g) Doors, frames and architraves.	Check for timber decay.
(h) Ironmongery.	
(i) Windows and glazing including fittings.	
(j) Fireplaces (if any).	
(k) Decorative condition of all walls, ceilings and joinery etc.	
(l) Any special features.	
(m) Condensation problems.	
(n) Prepare list of all fixtures in the various rooms.	
(3) *Rooms below ground*	
Cellar or basement (walls and floors).	Check for rising damp in walls and floors.
Stairs and balustrades.	

Item	Remarks
(4) *Floors*	
(a) Floor construction.	Check for stability and deflection.
(b) Staircases, balustrades and loft ladders etc.	Carefully check for wet and dry rot.
(5) *Roof (interior)*	
(a) Note construction of roof.	
(b) Method of insulation (if any).	
(c) Cold water storage cisterns, ball valves and pipework.	Mainly domestic work. For other systems see 'Services' in part 6 below.
(d) Flues in roof space.	
(e) Party or gable walls.	
(f) Rooms in roof space.	
(g) Open steel or reinforced concrete roof trusses.	Mainly found in industrial properties.
(6) *Services*	
(a) Cold water supply pipes including provision and location of stop valves.	Note position of rising main
(b) Electrical installation.	
(c) Gas installation.	Note position of meters,
(d) Hot water and space heating (gas, electrical, oil fired or solid fuel central heating). Type of insulation	The checks described in Sections 14.5 to 14.15 should be carried out.
(e) Lift or hoist equipment.	
(f) Fire appliances and sprinkler systems.	
(g) Telephone system.	
(h) Burglar alarm system.	
(i) Sanitary fittings: Baths, lavatory basins, WCs, urinals, sinks, showers and bidets.	
(j) Waste pipes and traps etc.	Items b, c, d, e, f, g, h and k may require a special report.
(k) Mechanical ventilation systems (if any).	
External	
(7) *Roofs (exterior)*	
(a) Roof coverings.	
(b) Eaves.	
(c) Gutters and rainwater pipes.	
(d) Chimney stacks.	
(e) Flashings etc.	
(f) Parapet walls and copings.	
(g) Dormer windows including the external coverings.	
(h) Ventilators.	
(i) Tank rooms – Lift motor rooms.	
(8) *External walls and cladding*	
(a) External walls.	Examine walls for signs of settlement.
(b) Lintels and arches.	Evidence of deflection.
(c) Reinforced concrete frames and cladding materials.	
(d) Rendered or rough cast surfaces.	
(e) Expose foundations and examine sub-soil (if necessary).	
(f) Presence of large trees in the vicinity may cause fractures.	
(g) Damp-proof course.	

Item	Remarks
(h) Balconies.	
(i) Steps.	
(j) Air vents.	
(k) Fire escape stairs and ladders.	
(l) External decorative condition.	
(m) Special note of any unusual features.	
(9) *Drainage systems*	
(a) Prepare sketch plan of drainage system. Show positions of inspection chambers and gulleys.	
(b) Soil drains (Disposal – sewer or cesspool etc.)	
(c) Method of surface water disposal.	
(d) Soil and vent pipes – waste pipes (externally).	
(e) Take up covers and gratings and examine inspection chambers and gulleys.	The simple tests described in Section 15.7 should be carried out. If the system is suspect then recommend that the drains are properly tested.
(f) Note if the system has been properly ventilated.	
(10) *External work*	
(a) Boundary fences, walls and gates.	Check ownership.
(b) Site frontage.	
(c) Ground condition.	Trial holes if necessary.
(d) Paved areas and ramps.	
(e) Natural features (trees, shrubs etc.)	Mark position on sketch plan.
(f) Outbuildings – Garages, workshops, sheds, greenhouses, conservatories etc.	
(g) Liability to flooding.	Wet or dry area.
(11) *Local authority enquiries*	
(a) Specific restrictions or preservation orders in respect of development.	
(b) Building improvement lines.	
(c) Any future development in the area such as road widening or drainage work.	
(12) *Estimated cost of repairs* or *decorations*.	

Appendix V
Surveying Safely
(Based on RICS 2004)

Protecting yourself

Some of us take better care of other people than we do of ourselves.

Please don't overwork. Don't squeeze in an extra job at the end of the day. Don't overwork in great heat, and keep warm in cold weather. Don't take unnecessary risks on site, and perhaps most importantly, don't drink alcohol, as many deaths on site are related to alcohol.

Also use extreme caution whilst taking medication. In addition:

- Make sure that business vehicles are well maintained.
- Don't try to drive long distances when you are tired.
- Keep a copy of the Highway Code in the office.
- Have you had anti-tetanus vaccinations?

The case of Suzy Lamplugh, an estate agent who disappeared while doing her job, brought home the need to be wary of escorting strangers around empty premises. Carry personal alarms and telephones. Make sure people know where you are going and with whom you are going. Work in pairs if visiting particularly dangerous areas.

Surveying can be a dangerous business and members are asked to take particular care in matters of personal safety.

However well a survey or pre-inspection is 'pre-planned', the individual surveyor needs to be alert to potential hazards during the inspection. The following hazards could be encountered when surveying buildings or sites.

Structures

The chance of partial or total collapse of:

- Chimney stacks, gable walls or parapets.
- Leaning, bulged and unrestrained walls (including boundary walls).
- Rotten or corroded beams and columns.
- Roofs and floors.

Timbers

- Rotten and broken floors and staircases. Flimsy cellar flaps and broken pavement lights.
- Floorboards, joists and buried timbers weakened by age, decay or attack.
- Projecting nails and screws. Broken glass.
- Glazing in windows and partitions may be loose, hinges and sash-cords weak or broken. Glass panels in doors and winglights may be painted over.

Roofs

- Fragile asbestos cement and plastic coverings.
- Fragile rooflights (often obscured by dirt or temporary coverings).
- Low parapets or unguarded roof edges. Loose copings.
- Rusted, rotten or moss-covered fire escapes, access ladders and guard rails.
- Rotten roof decking and joists.
- Slippery roof coverings (slates, moss or algae-covered slopes).
- Broken access hatches.
- Mineral wool dust, mortar droppings and birds' nesting material and excrement in roof voids.
- Cornered birds and vermin.
- Insects, bugs and lice. Bee and wasp colonies.
- Water cooling plant may harbour legionella.
- Unguarded flat roofs.
- Broken, loose, rotten and slippery crawling boards and escape ladders.
- Weak flat roofs and dust covered rooflights.
- Slippery roof surfaces.
- High winds during roof inspection.
- Ill-secured or flimsy, collapsible, sectional or fixed loft ladders.
- Concealed ceiling joists and low purlins.
- Ill-lit roof voids.

Unsafe atmospheres

- Confined spaces with insufficient oxygen, including manholes, roof voids, cellars, vaults, ducts and sealed rooms.
- Rotting vegetation which may consume oxygen and give off poisonous fumes.
- Accumulation of poisonous or flammable gases in buildings on contaminated land.

- Stores containing flammable materials, such as paint, adhesives, fuel and cleaning fluids.
- Hazardous substances, including toxic insecticides and fungicides.
- Gas build-up in subfloor voids.

Danger from live and unsecured services

- Electricity, gas, water and steam supplies.
- Awkward entrances into sub-stations and fuel stores.
- Temporary lighting installations: mains connections and generators.
- Buried cables and pipes.

Hidden traps, ducts and openings

- Lift and services shafts, stairwells and other unguarded openings.
- Manholes, including those obscured by flimsy coverings. Cesspools, wells and septic tanks.

Intruders and others

- Physical dangers from squatters and vagrants. Guard dogs.
- Health risks (including AIDS) from discarded syringes and condoms.
- Structures weakened by vandalism or arson.
- Aggressive tenants and property owners.

Contamination

- Asbestos, lead and other substances hazardous to health.
- PCB and PCN chemicals in electrical transformers and capacitors in fluorescent lighting fittings.
- Overhead electrical cables.
- Contaminated water supplies.
- Contaminated air conditioning systems (legionella).

Vermin and birds

- Rats and mice: Weil's and other diseases.
- Bird droppings.
- Lice may be present in bedding, soft furniture and carpets.

Travelling to and from site

- Driving too fast, for too long or when tired.
- Use of hand-held car telephones on the move – now illegal.

After a survey, the property should be left secure. You should make sure at all times that someone in the office or at home knows where you will be, especially when working in remote places or empty buildings, by establishing an open diary regime.

The Health and Safety at Work etc. Act 1974 places a duty on employers to take reasonable measures to ensure the safety of their employees. Employees have similar responsibilities to take care of their own safety. Discharging these responsibilities involves the process of risk assessment in which hazards, or events likely to lead to harm, are identified and then assessed in terms of the likelihood of the event occurring and the severity of the harm that would result.

Having identified a hazard and assessed the risk involved there are two options: to remove the hazard by replanning the work process or activity, or to accept that the hazard will remain but replan the work process or activity in order to reduce the likelihood of the hazard occurring or to reduce the severity of the consequences if it does. Either option will involve consideration of working methods and, if necessary, documenting a safe method of work or method statement for the activity.

No amount of training can deal with every eventuality but these notes should provide a useful checklist and prompt for deciding whether a hazard exists and how best to reduce risk.

Premises

When instructed to inspect premises, the basic factors that need to be considered will include the following.

Type

- What type of site/building is it?
- If a construction site, what stage has been reached?
- If an existing building, what is the form of construction and age?

High structures

- Are any towers, masts or tall chimneys involved?
- How will they be accessed for inspection?
- Is special access equipment needed and who is to provide/manage it?

Condition

- Are the premises known to be derelict or in poor condition? What is the nature of the damage?
- Are security measures in force and how is access to be gained?
- Is protective clothing required?
- Is any special access equipment needed?

Occupation

- Who are you likely to encounter on site?
- Are the occupants or neighbours likely to be aggressive or disaffected?

Activity

- If a building/site is occupied, what is the nature of that occupation? What might you encounter, e.g. noise, fumes, vehicle movements?

Special risks

- Is the nature of the building or site such that it presents special hazards?

Special access

- Will special access arrangements be required and who will provide and manage them?
- Does the surveyor need special training?
- If a scaffold exists, is it safe to use? When was it last inspected by a competent person?

Site rules

- Does the premises manager have house rules?
- Are there permit to work/enter procedures?
- If a site, is there a construction phase health and safety plan including induction procedures?

Dangerous substances

- Is the inspection likely to bring you into contact with noxious atmospheres or explosives?

Vermin or diseases

- Is the nature of the site such that it could be contaminated with any form of clinical waste?

- Is the surveyor likely to encounter used syringes/needles, condoms, razor blades?
- Could the site be a source of anthrax which, for example, could be present in haired plaster?
- Could legionella be present?
- Might the surveyor encounter vermin, e.g. rats, on the site?

Having considered the physical hazards that might exist, these need to be considered in the light of personal and environmental issues, as follows.

Environmental

- Will weather conditions and/or light levels increase risk? Will temperature extremes present a hazard?

Personal

- Does the surveyor's gender or level of fitness have any bearing on the hazards which have been identified? Would lack of fitness present a hazard?
- Are special skills required and does the surveyor concerned have those skills?
- Does the surveyor have phobias/suffer from vertigo or claustrophobia?

Welfare and security

- Is lone working a safe option? What provisions are made for communications in an emergency?
- How would access for rescue be achieved?
- Does a lone worker suffer from any medical condition which could affect personal safety?
- Are toilet, wash and first-aid facilities required?

Special equipment

In certain circumstances any of the following equipment may be necessary: gloves, respirators, safety helmet, ear defenders or temporary lighting.

The above lists are not exhaustive and the extent to which any of the items might be relevant will vary.

Appendix VI
Report on Roof Defects (Village Hall)

Preamble

We were asked to examine the defective roofs, parapet walls and surface water drainage system and report upon their condition. For obvious reasons the address of the property is fictional.

Roof Defects: The Village Hall, Hall Lane, Blankton, Kent

1.0 INTRODUCTION

1.1 Background
Further to our meeting on (insert date) and in accordance with your instructions contained in your letter dated (insert date) we have examined the following portions of the above property in order to advise you as to their condition:

(a) Main slate roof.
(b) Roof trusses and other supports internally.
(c) Small asphalt flat roofs to the north and south sides of the hall.
(d) Gutters and rainwater pipes.
(e) Parapet walls and copings to all roofs.
(f) Surface water drainage system.

As authorised in your letter, we have obtained the services of a builder to provide attendance to enable us to gain access to the roofs and to test the drainage system.

1.2 Description
The hall consists of a single storey brick structure built about 1895. The main roof to the hall is slated and there are two brick gable walls to the north and south ends of the hall finished with brick parapets and stone copings. There are two small asphalt covered flat roofs at the north and south ends of the building.

2.0 ROOF DEFECTS

2.1 Main roof (internally)
There are four timber roof trusses supporting the purlins and rafters. Access to each truss was obtained by ladder. The sizes of the truss members and purlins are adequate for the span of the roof and the general condition of the timbers is sound having regard to the age and character of the building. The rafters are lined internally with tongued and grooved matchboarding. It was, therefore, impossible to examine the rafters internally. However, a close inspection of the boarded slopes internally revealed three small damp stains. The problem is no doubt due to damp penetration in the past caused by some defective slates. The boarding was dry at the time of our inspection, but the condition will no doubt deteriorate unless the slate problem is dealt with (see item 2.2 below).

2.2   Main roof (externally)

The main roof is covered with Welsh slates. Both slopes were closely examined from roof ladders. In accordance with our agreement four small areas of slating were removed to enable the battens and rafters to be examined. The areas uncovered coincided with the damp stains on the inner face of the boarding. The general condition of the slates is not satisfactory. The following defects were noted:

(a) The slates are principally affected around the nail holes. This indicates 'nail sickness' and means that a large number of fixing nails have failed. This was confirmed by the large number of slates supported by zinc clips.

(b) Some of the slates near the ridge were soft as a result of atmospheric pollution causing a breakdown of the bond between the laminated layers.

(c) There have been extensive renewals carried out to isolated slates. Where this has occurred the slates are in reasonable condition.

(d) Several slates to both slopes are cracked and chipped. This is often caused by high winds lifting the slates without dislodging the nails.

Due to the condition of the slates the battens show signs of wet rot in several places and it will be necessary to renew them in order to provide a firm fixing. As far as could be ascertained the rafters and boarding have not been affected, the minor penetration having been contained in the battens. The small damp stains visible on the boarding internally have not seriously affected the boarding or rafters and can be dealt with during normal redecoration internally. You will, of course, appreciate that the examination was of a small localised nature and would not have facilitated a complete examination of all the rafters and boarding. The ridge is covered with angular ridge tiles fixed with iron nails. Apart from three or four loose tiles the general condition is satisfactory.

The joints between the roof slopes and parapet walls are formed with cement fillets. The fillets have shrunk and cracked and small pieces have broken away in several places.

Due to the random disrepair as stated above and because of the evidence of damp penetration it is our opinion that total recovering of the two roof slopes including new battens should be carried out as soon as possible. To prevent a recurrence of damp penetration in the roof timbers we recommend the use of roofing felt laid over the rafters. We also recommend the provision of zinc flashings in lieu of cement fillets. Flashings are more efficient than the fillets and have a longer life. They also allow for the slight movement that may take place in the timber roof structure.

With regard to the Welsh slates we recommend that consideration be given to recovering the west slope with new Welsh slates and the east slope be recovered with good salvaged slate, i.e. the best taken from both slopes and the residue cleared away. The existing angular ridge tiles can be refixed.

Having stripped both slopes the opportunity would arise for the timber structure to be inspected before the coverings are reinstated. It would also be our recommendation to include a provisional sum in any future specification. This would allow for any timber repairs it is not possible to anticipate at this survey stage. All obvious repairs will be included in the estimated figure shown in the summary.

2.3   Small asphalt flat roofs to north and south elevations

We are unable to give a report on the structural timbers to the flat roof on the north side as no means of access is provided and inspection is not possible without removing the plaster ceiling below. We understand that you are not agreeable to carrying out such work and to bearing the cost involved. However, 'soundings' were taken at various points on the ceilings and we were able to establish that the supporting structure consisted of timber joists. We were able to calculate the depth of the joists and test for vibration. This investigation revealed that the capacity for roof loading is adequate.

With regard to the structural timbers to the flat roof above the entrance (south side) our builder was able to remove one of the insulation board panels without causing damage. The roof consists of boarding laid on 180×50 mm timber joists. The joists and boarding are satisfactory having regard to the age and character of the building. There was no evidence of damp penetration. Calculation of the size of the joists compared with the span is adequate.

We understand that the two asphalt roofs were renewed about ten years ago and apart from the two minor faults described below the asphalt is generally in good condition.

Ladders appear to have been used for maintenance work or window cleaning to both flat roofs. The ladder feet have caused small depressions in the asphalt but have not caused any serious deterioration of the surface. This defect is fairly common in asphalt roofs. The asphalt upstands to the parapets of both roofs were examined and in both cases a length about 1 m has slightly sagged no doubt due to differential thermal movement. The defects are not serious and can be easily made good. However, this type of roofing repair is usually carried out by a specialist. The falls to the gutters are satisfactory and a metal drip to direct the rainwater into the gutter has been provided in accordance with standard practice.

Two criticisms of a general nature might be made concerning the asphalt. If regular maintenance traffic is anticipated it would be advisable to provide duck boards. This would allow means of access and avoid damage to the roofing material. Secondly, the surface of a flat roof being fully exposed will gain heat by solar radiation. With an existing roof the use of solar reflective paint is generally beneficial. This could be done at a moderate cost and we would be pleased to supply details of a suitable treatment.

2.4 Parapet walls

The brick parapet walls to the main roof and asphalt roofs were closely examined. No damp course had been provided below the coping stones, but there was no evidence of damp penetration. The brickwork and coping stones are in a satisfactory condition. The condition of the brickwork and coping stones at present does not justify the insertion of a damp course, but you must be advised of the possibility arising in the future. Small areas of loose pointing to the joints in the coping stones and the inside faces of the brick parapets require some attention.

2.5 Gutters and rainwater pipes

The gutters and rainwater pipes are of cast iron and are probably the original. They were examined from ladders and found to be in poor condition. The supporting gutter brackets were badly corroded and there is evidence of leakages from the joints. The gutter interiors show signs of corrosion. This was particularly noticeable in the gutters serving the main roof. Several lengths of gutter are out of alignment, mainly due to the defective brackets. Rainwater seepage from the gutters has affected the fascia board and there was evidence of wet rot. All the soffit boards were satisfactory, but require repainting.

The rainwater pipes were not properly jointed and in some cases the backs of the pipes were split due to corrosion. This is not an unusual problem where pipes are close to the wall. Access for painting is obviously restricted. There were clear signs that rainwater percolates through the back of the pipes and runs down the walls. Accordingly, we advise the following replacements as a matter of urgency:

(a) That all rainwater gutters are replaced with 125 mm half round PVC system. All gutter outlets to be fitted with plastic balloons to prevent possible blocking by leaves and other debris.

(b) That all rainwater pipes are replaced with 76 mm diameter PVC pipes with all necessary offsets and fixing clips.

(c) The softwood fascia board to both sides of the main roof to be replaced and treated with a preservative before fixing.

Plastic rainwater fittings are relatively trouble free and require little, if any, maintenance.

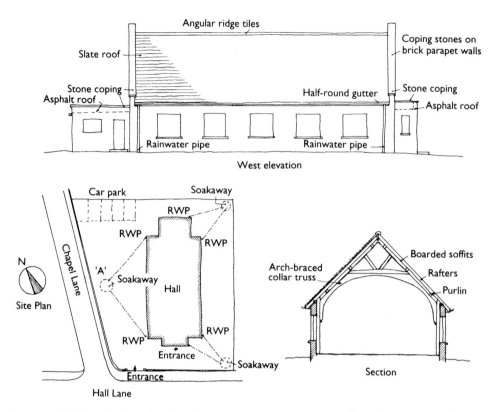

Figure AVI   Details of roof and surface water drainage system referred to in Appendix VI.

2.6   Surface water drains and soakaways

As with many cases of soakaway drainage, the rainwater pipes have been connected direct to the drain with no access. This method has several disadvantages as the condition of the drain is unknown and having no clearing access the silt will gradually build up in the drain and rainwater pipe. This defect is not usually noticed until the rainwater overflows from the next pipe joint up the wall. The usual signs are damp patches on the external wall surfaces. Unfortunately this defect has occurred in three of the rainwater pipes. As agreed the builder exposed the lower section of three of the rainwater pipes and a portion of the drain on the west and east sides of the building. In all cases the drains were completely blocked and out of alignment. The jointing material to the drain pipes has shrunk and cracked and in some places has completely broken away.

We located the approximate position of the soakaways and judging by the subsidence in the ground in that area the soakaways are now ineffective. We removed the top soil to the soakaway marked 'A' on the attached plan and it was found to be in a very poor condition and completely silted up. Experience has shown that any soakaway constructed over fifty years ago is either (a) silted up, (b) collapsed, (c) defective due to root penetration. Taking into consideration the age of the system and defects already found it would seem that the whole of the surface water system is in the same deplorable condition. We, therefore, advise that a completely new surface water system be installed. The system we propose will consist of the following:

(i) PVC rainwater shoe fitted with a metal sealing plate for access at the foot of each rain-water pipe. The rainwater pipe being connected to a vertical back inlet.

(ii) New 100 mm diameter drains will be required and it is suggested that PVC is used in long lengths to minimise the number of joints (this will reduce labour costs).

(iii) Replace the soakaways. It will be necessary to contact the local authority regarding the method of construction and siting in relation to the building.

3.0 Summary

We trust the foregoing descriptions are clear. As you will have observed there are a number of faults in the roofing and drainage system which require attention as soon as possible. To carry out the roof recovering work, repairs to asphalt, repointing parapet, renewal of rain-water systems, renewing fascia boards and surface water drainage indicated within this report would necessitate a budget in the region of £......... This figure is approximate and does not include VAT or professional charges. As agreed we have included a provisional sum of £......... to cover any unforeseen works to the main roof timbers. This sum would be expended in whole or in part during the course of the works (see item 2.2 above).

In terms of programming the works it is suggested that upon receipt of your instructions to proceed a period of approximately five weeks would elapse for preparation of a specification, tendering and for analysing the results and submitting a report.

We will now await your further instructions. Should you wish to discuss this mater further, please make an appointment to see the writer.

Signed: .... (Surveyor)

Appendix VII
Report on Property to be Purchased

Preamble

We were asked to carry out a building survey on behalf of a client who had visited the property but wanted to have an expert opinion before he actually committed himself. It was agreed that we are not required to remove or take down any part of the structure for detailed examination. We were asked to examine the services in as much detail as possible but not to employ specialists, except to test the soil drainage system. If any of the other services are suspect then we will recommend that they should be tested by a competent specialist.

Property survey: No. 10 Orchard Road, Blankton, Kent

1.0 INTRODUCTION

1.1 Instructions
In accordance with your instructions we have examined the above property in as much details as possible in order to advise you as to its condition.

1.2 Description and situation
The property consists of a brick built detached bungalow with a tiled roof built about 1937. A detached garage is situated on the south side of the bungalow and a greenhouse in the garden on the west side. All elevations to the bungalow and garage are faced with brown sand-faced bricks. There is a bay window to the front elevation and a loggia overlooking the garden at the rear covered by the main roof. The property stands on a plot of land at the junction of Orchard Road and Mill Lane and is approximately rectangular with a frontage to Orchard Road of 19 m and a total depth of 48.7 m. Vehicular and pedestrian access is provided at the south east corner of the plot facing Orchard Road, with a brick boundary wall and gates. Access from the gates to the garage and front entrance is by means of concrete paving. The details concerning shopping, schools and local transport are all contained in the estate agent's particulars.

1.3 Accommodation
In the following accommodation list the first dimension in each case is parallel to the frontage (see sketch plan attached).

Entrance Porch	(front elevation)
Entrance Hall	1.370 m wide
Sitting Room	3.500 × 4.200
Dining Room	3.500 × 3.450
Front Bedroom 1	3.200 × 3.650

Back Bedroom 2	2.280 × 2.600
Kitchen	2.440 × 2.600
Bathroom	2.400 × 1.524
WC	1.520 × 914 mm

Outbuildings

Garage	2.600 × 6.095
Greenhouse	1.900 × 1.600
Garden shed	1.800 × 1.600

2.0 EXTERNAL CONDITION

2.1 External walls

The wall thicknesses were measured and were found to be 280 mm thick. This suggests cavity wall construction which was confirmed by the outer leaf of brickwork being laid in what is called stretcher bond, i.e. all end-to-end. Cavity construction consists of two structural leaves with a 50 mm space between the two leaves being connected together with metal ties. This type of construction improves the damp resistance and thermal insulation qualities. The pier to the west corner of the loggia is 457 mm square in facing bricks matching the main walls. The garage is constructed with the same facing bricks.

We were able to identify a damp-proof course in the main walls which was situated approximately 150 mm above ground level in order to prevent dampness from rising in the brickwork. Some time was spent in examining for evidence of rising damp that might cause problems in the future. There were no signs of damp penetration in any of the external walls. However, on the north flank wall earth has covered the damp-proof course for a height of about 50 mm. It was evident that this earth was deposited comparatively recently and has not yet caused any serious deterioration in the brickwork. This soil should be removed as soon as possible.

We were given permission by the present owner to excavate a small trial hole close to the east flank wall and expose the concrete foundation. Our examination showed that the subsoil consisted of shrinkable clay which necessitates the foundations being taken down to a sufficient depth to avoid atmospheric changes and the effects of vegetation. We are satisfied that the foundations are adequate and at a suitable depth so as not to be disturbed by any movement in the clay subsoil. Trees close to buildings can cause unequal settlement when active roots dry out. There are no mature trees close to the building to cause any problems. The four small trees at the NW corner of the garden are well away from the building and are unlikely to cause problems in relation to the building.

There were no signs of any settlement cracks in the main walls, and the condition of the facing bricks is generally good. The brick pier at the SW corner of the loggia was closely examined. This pier carries the ends of two oak beams supporting the rear part of the roof. We were satisfied with the design of the pier and beams having regard to the loading and the subsoil. However, we did observe slight signs of movement at the two corners of the rear wall to the garage. This consisted of two fine vertical cracks at the junction between the brick pier and the half-brick rear wall extending from the floor to the eaves. The crack also extends across the concrete floor between the two flank walls. It was apparent from the signs internally that some settlement had taken place in the concrete floor slab. This is not unusual where the concrete floor also supports the walls. However, there is a certain amount of restraint provided by the roof structure mainly from the rafters and hips. If this type of foundation is not carried down sufficiently deeply seasonal movement will take place in the slab close to the surface causing a vertical fracture. We are unable to state whether this movement is likely to continue. An examination of this nature can only take place over a period of time in order to establish with any degree of certainty the exact cause of the failure. In our opinion the defect is not sufficient to warrant repair work at the present time but if you decide to purchase the property and the movement continues underpinning of the concrete slab and rebuilding the rear wall and piers may be needed.

The condition of the pointing to the whole of the property was examined and apart from a small area on the west wall below the windows is in reasonable condition. Defective pointing can sometimes cause deterioration of the brickwork due to frost action. If you purchase the property we suggest that the defective areas are repointed.

The brickwork to the bay window was closely examined. In properties of this age it is often found that the brickwork supporting bay windows tends to separate from the main structure due to the fact that it is lightly loaded and the foundation concrete is arranged at a different level to the main structure. There were no signs of movement in the brickwork, and the woodwork to the bay window is in sound condition.

2.2 Windows and doors

The windows throughout the property consist of timber casements and frames and are obviously the original. All the casements were opened and closed and found to be in serviceable condition. The paintwork is in fairly good condition but there are signs of minor outbreaks of wet rot in the lower members of the sashes on the west elevation. Some of the putties are cracked and loose. The repairs required are of a minor nature but we recommend that they are dealt with in the near future. The metal window furniture i.e. stays and catches have recently been renewed and are in sound working order. The sills are properly constructed and have an adequate 'run-off'. The joints between the frames and adjoining brickwork are sound and there is no evidence of damp penetration.

The external doors and frames are of timber construction. The lower portion of the doors is panelled and the upper portion glazed with glazing bead fixings. The front door is in reasonable condition. The kitchen door has been neglected over the years. The problem is due to the entry of moisture through the glazing beads and open joints. The absence of a weatherboard on the external face means that rainwater is not thrown clear of the gap under the door and moisture has penetrated the joints between the bottom rail and stiles. The only satisfactory method of dealing with this problem is to renew the door completely.

The glazed doors between the dining room and loggia are in a satisfactory condition, but are difficult to open and require easing.

The double timber doors to the garage are well painted, but have several faults. The external face consists of tongued and grooved boarding secured to rails, stiles and braces. The problem here is that the strap hinges are inadequate and cannot support the weight of the doors. The doors have, therefore, dropped towards the centre of the opening and are difficult to open. A loosening of the joints between the rails and stiles has also occurred causing entry of moisture. Moisture has also penetrated the tongued and grooved joints of the match boarding.

Being close to ground level the lower timbers have become saturated with rising damp from the paving. There is also evidence of wet rot at the feet of the door frame. If deterioration is excessive it is sometimes less expensive to replace a complete set of doors and frames than to replace with small pieces of timber. We are of the opinion that this is such a case and accordingly recommend a complete replacement. As an alternative to timber doors you may wish to consider the installation of an 'up and over' door. They are suitable when doors are situated in a confined space and usually consist of aluminium sheeting secured to a mild steel frame operated by springs or balance weights.

The ornamental wrought iron gate between the bungalow and the garage is in reasonable condition and the hinges are firmly fixed into the brickwork.

2.3 Roofs (externally)

The main roof and garage are covered with machine made clay tiles laid on battens. Roofing felt has been provided over the rafters, but not to the garage roof. The ridges and hips are covered with half-round tiles, the hips being secured at the foot with galvanised hip irons. The tiles are generally in sound condition. Approximately six tiles close to the eaves are chipped or cracked and we recommend that they are replaced. We understand that the present owner has several matching tiles which have been kept for this purpose. The jointing material to the hips could be improved in various places to form a better watertight joint.

The galvanising to four of the hip irons (two on the garage and two on the east side of the main roof) has deteriorated and we recommend that these are replaced in the near future. The open valley gutter to the front of the main roof is lined with lead sheeting. We are of the opinion that this is the original lining and shows signs of 'making good' in two places, but after an examination internally and externally we are satisfied that the lead is sufficiently sound for a further period.

Apart from the minor defects mentioned above there are no signs of any serious problems to the roof coverings.

2.4　Gutters and rainwater pipes

The gutters and rainwater pipes to both roofs are of cast-iron and judging by their design they are the original. The gutters and their supporting brackets were examined from a ladder and were found to be in a reasonable condition. However, the interior of the gutters to the main building have patches of rust in various places. We, therefore, recommend that at the time of the next external painting the gutter interiors be cleaned out and two coats of bituminous paint applied.

With regard to the cast-iron rainwater pipes, two pipes on the west elevation are split at the back due to corrosion. This is not an unusual problem where pipes are close to the wall. Access for painting is obviously restricted. There were no signs that rainwater had caused dampness in the brickwork, but we recommend that these two sections of pipe be renewed and that the remainder are periodically checked. This problem could worsen in future years.

The soffit and fascia boards are sound and the paintwork is generally satisfactory.

2.5　Chimneys

The two chimney stacks on the north and south sides are of brick construction and match the main walls. The brickwork is sound, but some repointing is required to the top sections. The flues are no longer used for solid fuel fires as the open fires have been replaced by gas-fires. We are not able to give an opinion on the condition of the brick flues or make any comments on their efficiency (see Gas Fires). The flue terminals (chimney pots) were examined through binoculars and were found to be satisfactory. The terminals were secured to the top of the stack with a cement fillet known as flaunching. The flaunching to both stacks is cracked. Replacement is suggested in order to prevent dampness penetrating the stack below. This repair work is not unduly expensive and can be carried out when repointing the stack as previously described.

2.6　Soil and surface water drains

The main soil and surface water drains run on the north side of the property being drained on a totally separate system.

The soil drain receives the soil and waste from the various fittings in the bathroom, WC and kitchen and discharges into a public sewer in Orchard Road. The system is provided with three inspection chambers. The covers were lifted and the interiors examined. The brickwork channels, and benching to chambers 2 and 3 were found to be in sound condition. The metal cover to chamber 3 at the NW corner of the property is cracked and should be replaced as soon as possible to avoid the risk of personal injury. The frames to all three covers require cleaning and sealing with a grease compound. The front inspection chamber 1 is provided with an interceptor which prevents foul air from the sewer from entering the drains to the property. The chain attached to the interceptor cap is broken and the cap is stuck fast in the interceptor trap causing a partial blockage. The interceptor is, therefore, ineffective. A new chain and cap should be provided as soon as possible. The cement benching is cracked in places and requires making good. The chamber is ventilated by a fresh air inlet as shown on the attached sketch plan. The fresh air inlet was examined and found to be broken. A new head and mica flap valve is required. The head of the system is properly ventilated by a soil and vent pipe attached to the north wall adjacent to the WC compartment. The cast-iron vent pipe is taken approximately 900 mm above the eaves. The pipe was examined and found to be in sound condition.

From measurements taken between the inspection chambers and the depths of the drains at the three chambers we are of the opinion that the drains have been laid to a satisfactory fall. The main drain and the various branches are of 100 mm diameter glazed earthenware. The sizes are satisfactory for this type of system.

The gulleys receiving the waste water from the kitchen and bathroom were examined. The gulleys were found to be satisfactory but the grids are partially blocked with grease and hair and need cleaning.

In accordance with your instructions a water test was applied to the soil drainage system. Firstly, the main drain between inspection chambers 1 and 2 was tested and found to be in sound condition. Secondly, the main drain between inspection chambers 2 and 3 was tested and found to be in sound condition. The branches leading to inspection chambers 2 and 3 were also tested and found to be in sound condition.

The rainwater pipes receiving the surface water drainage from the roofs discharge over trapped earthenware gulleys. The various branches connect to a main drain on the north side of the property as shown on the sketch plan and discharge into a surface water sewer in Orchard Road. Rodding eyes have been fitted at the various junctions and an inspection chamber provided just inside the front boundary wall. The rodding eye covers and inspection chamber cover were partially covered with earth and weeds and had to be cleared before we could remove the covers. The covers to the inspection chamber and rodding eyes were removed and the interiors examined. The pipes, channels and brickwork were found to be in sound condition. The frames and edges of the covers are covered with rust and should be cleaned and coated with grease.

As agreed we have not tested the surface water drains and therefore cannot report on the condition and workmanship. However, when water was poured down each gulley it ran clearly through the system which seems to indicate that no obstructions have occurred and that drains have been laid to a satisfactory fall. Although this is only a simple test we are reasonably confident that a satisfactory system has been provided.

2.7   Garden outbuildings

The greenhouse on the west side of the property is constructed of cedar wood framing with a hinged glazed door of the same material. The lower members of the frame are bolted to a concrete slab. The timber is in fairly good condition, but we recommend that a coat of preservative is applied in the near future. The glass panels are bedded in putty and secured with sprigs externally. Some of the sprigs are defective and need replacing.

The shed is timber framed and lined externally with weatherboarding. The corners are protected by an angle corner post. The door is ledged and braced and the floor consists of plain edge boarding on joists supported on brick piers. The roof is of lean-to construction covered with felt on boarding. The felt has deteriorated and there are damp stains internally due to rainwater penetration. Apart from the roof covering the timberwork is in sound condition and has recently been treated with a timber preservative.

2.8   Paving

The pavings shown on the attached plan are all constructed in concrete with a slightly textured finish. The pavings appear to have been laid at different periods, the drive to the garage being a recent addition. The paving to the garden adjoining the west boundary is cracked in several places and a section approximately 4 m long on the south side of the garden has sunk. The defective areas are not dangerous but will no doubt have to be relaid in the near future.

2.9   Boundary walls, fences and gates

The boundary walls facing Orchard Road and Mill Lane are constructed of stock brickwork 114 mm thick strengthened with brick piers at approximately 2.700 centres finished with a creasing tile coping. The walls have been provided with a suitable damp course. The wall facing Orchard Road is 600 mm high with brick piers to the garage drive entrance supporting a pair of ornamental steel gates. The brickwork and gates are in a satisfactory condition.

The wall facing Mill Lane is approximately 1.8 m high. However earth from the garden has been allowed to pile up above the damp course at the NW corner of the site causing damp staining in the brickwork. We therefore recommend that the earth is moved. The pointing to both walls has been neglected in the past. An area of approximately 12 square metres requires repointing.

The fence to the west side of the property consists of waney cut larch panels secured to concrete posts. The fencing is a fairly recent improvement and is in sound condition. We understand from the present owner that the walls and fences marked 'T' on the attached plan belong to the property and that you will be responsible for their upkeep. The timber fence on the south side is under the ownership of the adjoining owner. However, we suggest that you check this matter with your solicitors.

3.0   INTERNAL CONDITION

3.1   Floors
The floors throughout the property are of solid concrete construction. This type of floor is normally constructed with a damp-proof membrane incorporated below the floor surface and its purpose is to prevent dampness from the ground reaching the surface. The finish in the sitting and dining rooms consists of oak block flooring, but the remaining rooms are finished with thermoplastic tiling and is obviously a fairly recent improvement.

A problem which sometimes occurs in solid floors arises from damp penetration or when the slab sinks or drops out of level. You will no doubt appreciate that we were not able to investigate the condition of the slab since the floor finishes could not be raised without causing damage. However, tests for dampness were made with a moisture meter on the surfaces throughout the property and the result of our examination was satisfactory. We also examined the floor for any signs of settlement or movement out of level, but there were no signs of settlement or disrepair. The garage floor is of solid concrete construction and apart from the fine hair crack described in item 2.1, is in fairly good condition. There were no signs of damp penetration.

3.2   Partitions
The majority of the internal partitions which support the ceiling and roof structure are of brick construction while the non-load-bearing partitions in the bathroom and WC area are evidently of concrete block construction. The partitions were examined in a number of places and from the roof space above were found to be in sound structural condition. There were fine cracks in the plaster at the junction between the bathroom partition and the external wall on the north side of the property. This is due to the fact that the partitions and external walls are formed of different materials. This is a common fault and we do not consider that it is in any way a serious matter.

3.3   Roofs (internally)
Access to the roof was obtained through a trap door situated in the hall ceiling. Boarded walk-ways were provided from front to rear of the roof space. The structured timbers supporting the roof coverings to all slopes were examined. The rafters to the two main slopes are supported by timber purlins. Each purlin is braced by timber struts and collars. The struts are secured to timber plates fixed to the ceiling joists which in turn are supported on the brick partitions. The feet of the rafters are nailed to the ceiling joists and wall plates to prevent outward spreading. The rafters are secured to a timber ridge piece at the top. The timbers are adequate for the height and span of the roof and we could find no evidence of undue pressure on the walls and partitions from the weight of the roof.

There is one criticism that might be made of the roof space. It is usual today to provide some form of insulation in the main roof against loss of heat. A glass fibre quilt or similar material laid between the ceiling joists would be a great improvement and would help to prevent heat loss.

The garage roof consists of timber rafters and collars, the rafters being secured to wall plates and a ridge piece. A close examination revealed that the timbers are securely fixed and their condition is satisfactory.

4.0   INTERIOR FINISHES

4.1   Wall and ceiling plaster
The ceilings throughout the property consist of plasterboard and plaster. There are minor cracks at the junctions between wall and ceiling plaster in the bathroom and kitchen no doubt due to slight movements in the ceiling joists. The plaster was tested and found to be generally sound. Cove cornices have been provided in the dining and sitting rooms. Slight cracking is noticeable in the cornices on the south wall. This is due to the movement of the structure referred to above and can easily be made good during normal redecoration.

4.2   Wall tiling
The walls to the bathroom and WC compartment are half-tiled with coloured ceramic wall tiles. The tiling appears to be a recent improvement. They have been laid with straight joints in both directions and are in sound condition. The only tiling in the kitchen consists of a white ceramic splash-back behind the sink unit. Two of the tiles were slightly cracked near the corners. This is not a serious matter and could easily be dealt with during normal redecoration.

4.3   Internal doors and other internal joinery
The doors are those provided when the bungalow was built. They are all a four panelled pattern and apart from the few minor defects listed below are all in sound condition:

(a) The door furniture to the bathroom is loose.
(b) The door to bedroom 2 needs easing.
(c) The folding doors between the sitting room and dining room are single panelled and in sound condition. However, the metal fittings and supporting track need adjusting and oiling.

There are fixed fanlights above the kitchen and bedroom 2 doors to enable 'borrowed light' to enter the hall. The doors and other joinery items to the meter and linen cupboards are generally in reasonable condition.

The kitchen contains formica faced floor units and work-tops and these are generally in reasonable condition. The brackets supporting the work-top adjacent to the cooker are loose.

4.4   Skirtings
The skirtings throughout the bungalow are 150 mm deep with rounded tops and are typical of that period. Apart from scratched paintwork the condition is generally fairly good.

4.5   Internal decorations
We understand that if you purchase the property a complete internal redecoration will be carried out and that we are not to report on the interior decorative condition.

4.6   Sanitary fittings and waste pipes etc
The property contains the following sanitary fittings:

(a) Pink enamelled cast-iron bath length 1.67 m fitted with two handgrips. The bath is fitted with enamelled hardboard front and end panels with chromium plated angle strips.
(b) Pink vitreous china pedestal wash basin fitted with pillar valves with starheads.
(c) The WC compartment contains a pink vitreous china low level suite fitted with a 9 litre cistern with side supply, overflow, plastic seat and cover.
(d) The stainless steel sink unit in the kitchen is a fairly recent addition. The sink is fitted with 'Supatap' pillar valves.

The sanitary fittings are generally in sound condition. However, the bath shows slight stain-ing around the outlet and the WC cistern is slow in filling and needs some adjustment.

The majority of the traps and waste pipes were concealed which made inspection impossible. However, the fittings discharged properly when filled and there were no signs of blockages.

5.0   SERVICES

5.1   Cold water supply
Water is laid on from the main in Orchard Road and a 12 mm copper rising main enters the property on the north side and rises up in an internal angle of the kitchen partitions. The supply pipe is taken to a 227 litre asbestos cold water storage cistern in the roof space. From the rising main a 12 mm branch is taken to the kitchen sink. A stopcock is fitted where the rising main enters the kitchen. Ideally, there should be a drain cock fitted directly above to enable the system to be drained down should it be necessary to carry out repairs.

The cold water storage cistern is positioned over the brick partition between the kitchen and hall and is in a suitable position to supply the bathroom, WC and hot water system. The cistern is properly supported on timber bearers and is of adequate capacity for the number of fittings served. The entry of the mains supply to the cistern is controlled by a copper ball valve. The valve was examined and found to be in serviceable condition. A 25 mm overflow pipe is taken through the roof space to discharge just below the eaves on the north wall of the bungalow. The cistern is covered with an insulation board cover. Down services from the cistern consist of a 19 mm copper pipe taken across the ceiling joists into the bathroom and from there 12 mm branches are taken to the various sanitary fittings and to the hot water cylinder in the linen cupboard for the hot water supply. The main down service is fitted with a stopcock close to the cistern. The branches feeding the sanitary appliances are also fitted with stopcocks close to the appliances.

No defects in the pipework or layout were observed, but two of the stopcocks in the bath-room are stiff and should be freed in order to make sure that they can be turned off easily when required. All the taps ran satisfactorily, but the kitchen tap needs a new washer. Nev-ertheless, in the event of your purchasing the property, we advise that the pipework and cistern in the roof space are properly lagged. There could be a danger of freezing during very cold weather.

5.2   Hot water system
Hot water is provided by an electric immersion heater installed in a 180 litre copper cylinder with a sectional insulated jacket held in position with draw strings. The hot water cylinder is situated in the airing cupboard as shown on the attached plan. From the head of the cylin-der a 19 mm supply pipe is taken through the roof space to the bath with 12 mm branches to the wash basin and kitchen sink unit. An expansion pipe is taken to a point above the could water storage cistern. When hot water is drawn off the cylinder it is replenished from the cold water storage cistern. The hot water system was not tested, but the pipework, stop-cocks and joints were examined. We are satisfied that the system is in sound condition and that there is an adequate supply to the sanitary fittings. The pipework in the roof space is well insulated with hessian wrappings.

6.0   SPACE HEATING

6.1   Gas fires
A gas supply is laid on from the main in Orchard Road and enters the property on the east side and is taken to a meter in the hall cupboard. From the meter, pipe runs are taken to the gas fires in the sitting room and bedroom 1. We understand that the gas fires were installed some five years ago and when switched on were found to be working efficiently. The fires are connected to the original fireplace flues. Gas connections have been correctly made and a closure plate has been provided at the back to seal the void behind the appliance. The

plate has a slot at the base to allow air to enter. We understand that the fires have been regularly serviced by the Gas Board and are in serviceable condition.

Permanent ventilation has been provided for introducing air to the rooms. The main supply
to the meter is hidden and therefore, we cannot answer for its condition. The majority of the
pipe runs internally are on the surface secured to the skirtings and well painted. We were
only able to make a superficial examination of the pipes but from their appearance we are
satisfied that they are in a satisfactory condition. If you decide to purchase the property we
suggest that you request the Gas Board to advise you as to the condition of the main intake
and supply runs.

6.2   Electric heaters

An off-peak electric storage heater is installed in the hall. This type of heater has a core of
special bricks which are heated up during the night. The heat is then dissipated by the hot
bricks during the day. The unit is fitted with an independent electric circuit as required by the
current electrical regulations. The heater is working satisfactorily and is no doubt adequate
for the present arrangement.

The kitchen and bathroom are heated by wall mounted infrared heaters with pull cord
switching. The heat output is 1.8 kW. The dining room and bedroom 2 are heated by electric panel fires secured to the wall in the positions shown on the sketch. Both these appliances are fairly modern and are in serviceable condition.

6.3   Electrical installation

Electrical supply is laid on from Orchard Road and enters the property on the east side and
thence to a meter in the hall cupboard. A single consumer unit, made of metal, holding a row
of plug-in rewireable fuses and a single main switch is fixed on the left hand partition. This
type was first installed about 25 years ago and this unit is obviously a replacement.

In accordance with your instructions we have not had the installation tested. The 13 A
square holed socket outlets, cooker point and flush mounted light switches are modern. The
following is a list of socket outlets fitted in the principal rooms:

Sitting Room	3
Dining Room	3
Bedroom 1	3
Bedroom 2	2
Kitchen	4
Hall	1
Garage	1

The electric storage heater and infrared heaters are connected to the consumer unit by PVC
insulated cables which is the most commonly used system of wiring at the present time.
Judging from appearances we are reasonably confident that the fittings are in serviceable
condition. We examined the cables in the roof space and other places where they were
exposed and found that they were rubber insulated and above 40 years old. This type of
wiring is now obsolete and often dangerous due to the age of the insulation material and
mechanical damage. We are, therefore, of the opinion that the cables are a mixture of various
types and if you decide to purchase the property we strongly advise that the system is examined and tested by a competent electrical engineer or the supply authority.

7.0   SUMMARY

As can be seen from the foregoing report, the property is in a fairly reasonable condition.
However, there are a number of matters which require attention at the present time or in the
foreseeable future. The defects which require immediate attention can be summarised as
follows:

(1) Remove soil above the damp-course (north flank wall).
(2) Wet rot in timber casements.
(3) Renew external door to kitchen.
(4) Renew garage doors.
(5) Renew two damaged rainwater pipe sections.
(6) Repoint brickwork to the top of the two chimney stacks and make good fractured flaunching.
(7) Renew cover to inspection chamber 3.
(8) Renew chain and cap to interceptor and make good cracked benching (inspection chamber 1).
(9) Renew broken mica flap to fresh air inlet.
(10) Clean all gulley gratings.
(11) Renew roofing felt to garden shed.
(12) Minor repairs to internal doors.
(13) Release stop-cocks in bathroom.
(14) Adjust ball valve to WC cistern.
(15) Lag pipes and provide insulation to cold water storage cistern in roof space.

In order to give you an approximate idea of what expenditure might be involved, we consider that a sum of £ (insert amount) might be required to carry out the repairs and improvements to the foregoing items.

The following items are not serious problems and therefore need not be carried out immediately, but are pointers to repairs that will be needed in the foreseeable future:

(1) Replace cracked roof tiles and hip irons.
(2) Paint interior of gutters.
(3) Clean and grease inspection chamber frames and covers to both soil and surface water drains.
(4) Renew glazing sprigs to greenhouse.
(5) Renew broken and sunken paving.
(6) Repoint boundary walls.
(7) Renew cracked kitchen wall tiles.

We would expect the above items to cost approximately £ (insert amount). With regard to the fractures in the rear wall and floor of the garage, these are likely to need remedial work in the future depending on the circumstances as detailed in the report (see item 3). In order to give you an approximate idea of the cost involved we consider that a sum of £ (insert amount) might be required to rebuild the rear wall and foundation. This figure is based on present day prices.

In addition to the foregoing items, your attention is drawn to the recommendations concerning the need for the electric and gas installation to be tested by appropriate specialists. The cost of these items can be obtained from the specialists concerned.

We understand from the present owner that the following kitchen fittings will be removed when he vacates the property:

(i) Refrigerator
(ii) Washing machine
(iii) Electric cooker.

The survey of the property has been carried out on the basis of the following conditions:

(1) The examination of the property was visual and was made from ground level, ladders and from roofs where accessible.
(2) That it was not possible to inspect the interior of the two chimney flues.

(3)   That no inspection of the structural concrete slab to the ground floor was possible.

(4)   We have not examined any parts of the structure which are inaccessible or covered, and we are, therefore, unable to report that such parts of the property are free of defects.

We trust that this report give you all the information you need but if there is any further information you may require please contact us.

Signed: .... (Surveyor)

Date: ....

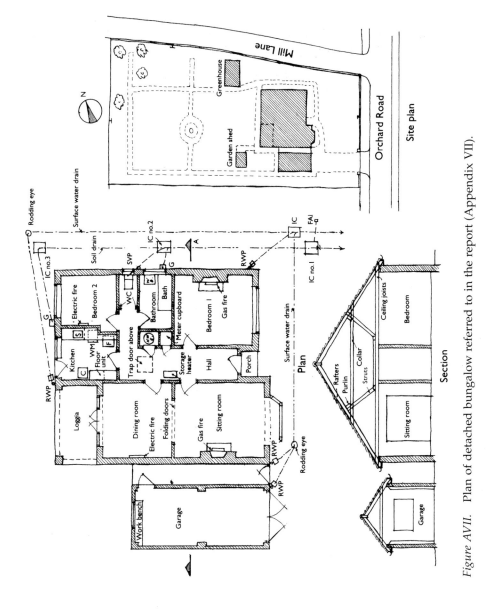

Figure AVII. Plan of detached bungalow referred to in the report (Appendix VII).

Appendix VIII
Glossary of Building Inspection and Related Terms

adaptation: Any work to a building over and above maintenance to change its capacity, function or performance. In general terms adaptation means the process of adjustment and alteration of a structure or building and/or its environment to fit or suit new conditions (Chudley 1983). However, more specifically it is also considered as work accommodating a change in the use, size or performance of a building, which may include alterations, extensions, improvements and other works modifying it in some way (Douglas 2002).

aetiology: The study of causes of things.

ageing: Degradation due to long-term influence of agents related to use (ISO 15686-1:2000).

alteration: Modifying the appearance, layout or structure of a building to meet new requirements (Watt 1999). It often forms part of many adaptation schemes rather than being done on its own.

anomaly: An indication (or symptom) of a possible defect (CIB 1993).

beam-filling: A crude masonry infill about 600 mm high found in traditional pitched roofs of Georgian and Victorian buildings to provide a degree of fire stopping and draughtproofing at the eaves (Douglas 2002).

bressummer: A large timber beam found in older buildings. It spans a wide fireplace or a bay window. The built-in ends of this wooden loadbearing member are highly prone to fungal attack (Douglas 2002).

building pathology: The systematic investigation and treatment of building defects, their causes, their consequences and their remedies (CIB 1993). It involves a holistic approach to understanding the various mechanisms by which the material and environmental conditions within a building can be affected (Watt 1999).

building survey: A Scheme 3 survey that provides an investigation and assessment of the construction and condition of a building, but will not normally include advice on value. The survey will generally include the structure, fabric, finishes and grounds, but the exposure and testing of services are not usually covered. The report will include reference to visible defects and guidance as appropriate on maintenance and remedial measures (RICS 1998).

carbonation: The transformation in concrete or cement mortar of the free alkali and alkali-earth hydroxides in the cement matrix into carbonates, due to attack by carbon dioxide in the atmosphere (CIB 1993).

condition: Level of critical properties of a building or its parts, determining its ability to perform (ISO 15686-1:2000).

condition survey: Inspection of a building or other structure, at a certain date, to determine its state of repair and any requirements for maintenance. It is often carried out on a regular basis for certain types of property, for example quinquennial inspection of church buildings (Dallas 2003; Haverstock 1999). Usually it is presented in the form of a schedule of condition.

conservation: Preserving a building purposefully by accommodating a degree of beneficial change. It includes any 'action to secure the survival or preservation of buildings, cultural artefacts, natural resources, energy or other thing of acknowledged value for the future' (BS 7913: 1998).

consolidation: Basic renewal and maintenance works to ensure a building's ongoing beneficial use (Douglas 2002).

cryptoclimate: The climate created by the building and influencing its fabric (Harper 1978). Adapting an existing building can change its cryptoclimate (Douglas 2002).

dampness: Condition of being slightly wet – usually not so wet that liquid water is evident – such as wetness caused by condensation on a porous substrate or water transmitted up a porous wall by capillary action (CIB 1993). Dampness can be said to occur when an atmosphere or material is wetter than 85% RH (Oliver et al. 1997).

defect: Non-conformity with standard or shortfall in performance (CIB 1993). Also known as a fault, or deviation from the intended performance of a building or its parts (ISO 15686-1:2000).

degradation: Changes over time in the composition, microstructure and properties of a component or material, which reduce its performance (ISO 15686-1:2000). It also usually refers to the deleterious effects of the weather such as sunlight, in particular on organic materials. Strictly speaking degradation is defined as the conversion of a complex molecule into simpler fragments.

degradation agent: Whatever acts on a building or its parts to adversely affect its performance (ISO 15686-1:2000).

degradation mechanism: Chemical, mechanical or physical path of reaction that leads to adverse changes in a critical property of a building product (ISO 15686-1:2000).

design life: The period of time over which a building or a building subsystem or component (e.g. roof, window, plumbing) is designed to provide at least an acceptable minimum level of performance (Iselin and Lemer 1993). It is the period of use as intended by the designer – for example, as stated by the designer to the client to support specification decisions (BS 7543).

deterioration: A reduction in ability to perform up to the anticipated standard (CIB 1993).

dew point: The temperature at which condensation occurs (Trotman et al. 2004).

diachronic survey: A diachronic survey is a way of studying buildings in terms of how they change or evolve over time. This is the way architectural historians (and building maintenance surveyors) appraise buildings (Brand 1993). Maintenance surveys as well as conservation plan inspections and other record surveys are methods of studying buildings diachronically.

diagnosis: An impartial assessment of all the data and evidence available to determine the root cause of a problem. Deciding the nature of a fault from its symptoms (Watt 1999).

dilapidations survey: Inspection of a property to ascertain the damage sustained to it during the period of a lease (Seeley 1985). It lists the defects usually prepared at the end of a lease in relation to the repairing covenants in the said lease. According to Hollis (1988) 'dilapidations' refers to a state of disrepair in a property where there is a legal liability for the condition of disrepair. It is normally presented in the form of a schedule of dilapidations.

durability: The ability of a building or its parts to perform its required functions over a period of time and under the influence of internal and external agencies or mechanisms of deterioration and decay (Watt 1999). It is also a measure of a building's ability to resist deterioration. 'Durability is not an inherent property of a material or component, although the term is sometimes erroneously used as such' (ISO 15686-1:2000).

economic life: The period of time over which costs are incurred and benefits or disbenefits are delivered to an owner; an assumed value sometimes established by tax regulations or other legal requirements or accounting standards not necessarily related to the likely service life of a facility or subsystem (Iselin and Lemer 1993).

extension: Expanding the capacity or volume of a building, whether vertically by increasing the height/depth or laterally by expanding the plan area (Douglas 2002).

failure: The loss of the ability of a building or its parts to perform a specified function (ISO 15686-1:2000). It can also be classed as the consequence or effect of a defect.

fault: State characterised by an inability to perform a required function, excluding the ability during preventative maintenance or other planned actions, or owing to a lack of external sources (Watt 1999). An unexpected deviation from requirements, which would require considered action regarding the degree of acceptability. It is also considered as a departure from good practice.

homebuyers valuation and survey: A Scheme 2 survey that, along with an indication of a property's open market value, provides basic feedback on its condition including advice on actions that should be followed. It is presented in a proforma format that comprises 11 standard sections.

improvement: Bringing a building and/or its facilities up to an acceptable or higher standard as required by the Building Regulations or the occupier (Douglas 2002). Beneficial improvement entails replacing something with a new item on a like-for-like basis. Substantive improvement, on the other hand, involves the replacement of an element or component with a new item having a higher performance rating.

industrialised building: This term is not necessarily synonymous with 'non-traditional' construction (Douglas 2002). Industrialised processes could be used with traditional building, although generally more extensively with more non-traditional forms. Industrialisation is a wider term than non-traditional because it is primarily concerned with the rationalisation of the building process itself. In building operations it implies the use of mechanical plant and the replacement of in-situ work by prefabricated units (Douglas 2002).

life cycle: The sequence of events in planning, design, construction, use and disposal (e.g. through sale, demolition or substantial renovation) during the service life of a facility; may include changes in use and reconstruction (Iselin and Lemer 1993).

life-cycle cost: The present value of all anticipated costs to be incurred during a facility's economic life; the sum total of direct, indirect, recurring, non-recurring and other related costs incurred or estimated to be incurred in the design, development, production, operation, maintenance, support and final disposition of a major system over its anticipated life-span (Iselin and Lemer 1993).

loss: The consequences of a defect or failure, expressed in terms of costs, injuries, loss of life, etc. (CIB 1993).

maintenance: A 'combination of all technical and administrative actions, including supervision actions, intended to retain an item in, or restore it to, a state in which it can perform a required function' (BS 3811: 1993). Maintenance involves routine work necessary to keep the fabric of a building, the moving parts of machinery, etc. in good order (BS 7913: 1992). The word comes from the French verb 'maintenir', which means to hold (Chudley 1983).

modernisation: Bringing a building up to a standard prescribed by occupiers, society and/or statutory requirements (Douglas 2002).

mortgage valuation survey: A Scheme 1 survey that provides an indication of a property's open market value and gives cursory feedback on its condition.

non-traditional building: This form of construction may use unfamiliar products or the same basic traditional materials in new ways, employing new techniques in fixing and erection, which differ, for instance, from the traditional method of laying by hand one brick or concrete block on top of another. It is a narrower concept than 'industrialised building' but also uses prefabricated elements (Douglas 2002).

obsolescence: The condition of being antiquated, old-fashioned, or out of date, resulting from a change in the requirements or expectations regarding the shelter, comfort, profitability or other dimension of performance that a building or building subsystem is expected to provide. Obsolescence may occur because of functional, economic, technical or social and cultural change (Douglas 2002).

performance: The degree to which a building or other facility serves its users and fulfils the purpose for which it was built or acquired; the ability of a facility to provide the shelter and service for which it is intended (Douglas 2002). It is a quantitative expression of behaviour of an asset or product in use (BS 6019).

physical service life: The time it takes for a building, subsystem or component to wear out. It is the time period after which a facility can no longer perform its function because increasing physical deterioration has rendered it useless (Douglas 2002).

preservation: Arresting or retarding the deterioration of a building or monument by using sensitive and sympathetic repair techniques. Preservation means 'the state of survival of a building or artefact, whether by historical accident or through a combination of protection and active conservation' (BS 7913: 1998). It also can be defined as 'the act or process of applying measures necessary to sustain the existing form, integrity, and materials of an historic property' (Weeks and Grimmer 1995). Preservation focuses on the maintenance and repair of existing historic materials and retention of a property's form as it has evolved over time. It includes protection and stabilisation measures (Douglas 2002).

prognosis: An impartial, technical assessment of the probable course and outcome of a building deficiency. Predicting or forecasting the course of a fault from its symptoms (Watt 1999).

protection: The legal use of this term involves the provision of legal restraints or controls on the destruction or damaging of buildings, etc. with a view to ensuring their survival or preservation for the future. Physical protection may be either temporary (e.g. tarpaulins over an exposed roof surface undergoing refurbishment) or permanent (e.g. over-roofing scheme).

redundancy: The consequence for a building when it becomes superfluous or excess to requirements (Douglas 2002). It is often triggered by obsolescence.

refurbishment: Modernising or overhauling a building and bringing it up to a client's requirements (Douglas 2002). It is usually restricted to major improvements primarily of a non-structural nature to commercial or public buildings. However, some refurbishment schemes may involve an extension.

rehabilitation: Extending the life of a building and making it suitable for habitation and use. It is a term that strictly speaking is normally confined to housing (Douglas 2002). Rehabilitation can also be defined as 'the act or process of making possible a compatible use for a property through repair, alteration, and additions while preserving those portions or features which convey its historical, cultural, or architectural values' (Weeks and Grimmer 1995). It acknowledges the need to alter or add to a historic property to meet continuing or changing uses while retaining the property's historic character.

rising dampness: Moisture rising up masonry walling or solid flooring as a result of capillary action (Trotman et al. 2004). True rising dampness is relatively rare – when it does occur it is usually where the soil is persistently damp and the wall and floor at ground level construction lacks a damp-proof course/membrane. Bridging moisture is the main cause of indirect rising dampness.

service (or working) life: Actual period of time during which no excessive expenditure is required on operation, maintenance or repair of a component or construction – as recorded in use (BS 7543).

settlement (contrast with 'subsidence'): A downward movement of a building, caused by above ground factors – such as overloading, differential movement.

stabilisation: Substantial adaptation works to ensure a building's long-term beneficial and safe use. It often includes strengthening works such as stitching and underpinning.

subsidence (contrast with 'settlement'): A downward movement of a building, caused by below ground factors – such as desiccation of clay soils, mining collapse, landslip.

subsystem: Functional part of a system, and often used interchangeably with that term – e.g. heating subsystem being part of HVAC system (Iselin and Lemer 1993).

sustainability: A set of processes aimed at delivering efficiently built assets in the long term (DETR 1998b). 'Eco-renovation' occurs where sustainable issues are deliberately and explicitly incorporated into an adaptation scheme (Harland 1998).

synchronic survey: A synchronic survey is a snapshot assessment of a building and the way it all fits together at a particular moment in time. This usually means the present, but buildings can be studied in terms of how they worked at one time in the past. In other words, it is about studying buildings in terms of immediacy and is the preference of building surveyors as well as 'city planners and architects looking for design ideas' (Brand 1993). Building surveys, condition surveys and dilapidation surveys are typical examples of this kind of appraisal.

system: Collection of subsystems, components or elements that work together to provide some major aspect of shelter or service in a constructed facility (e.g. plumbing system, heating system, electrical system and roofing system). Also, a set of building components specifically designed to work together to facilitate construction, such as an integrated building system (Iselin and Lemer 1993).

system building: This form of construction usually relies on both non-traditional and industrialised methods (Douglas 2002). It takes these a stage further by producing a combination of materials and methods for a design and construction package. A system building provides both a design and technique (often with specially made plans and components), available from one design (and possibly one construction) organisation. System building therefore is synonymous with proprietary forms of construction, such as 'Airey', 'Bison', 'Orlet', 'Skarne', etc.

technology: This can be defined simply as the systematic study and application of how artefacts are made and used (Douglas 2002).

traditional construction: Traditional methods are basically on the principle of an on-site operation where all the materials traditionally required for the building are first gathered together, such as bricks, cement, sand, ballast, timber, tiles and plaster. It predominantly involves the use of relatively small-scale units such as bricks and blocks, which are assembled or installed on site.

Turnerization: The term used to describe a proprietary system of applying a bituminous coating to slated or tiled roofs to enhance their weathertightness.

upgrading: Enhancing the performance characteristics of a building's major elements, components and/or services.

whole life-cycle cost: Generic term for the costs associated with owning and operating a facility from inception to demolition, including both initial capital costs and running costs (Watt 1999).

Bibliography

Preamble

This bibliography aims to provide a comprehensive list of publications dealing with the survey, inspection and investigation of properties. It contains the sources highlighted in this book as well as other references that the reader may wish to consult. Also included are many titles of works not directly mentioned in the text on various aspects of surveying buildings and related technological topics.

Up until the early 1960s, there were few if any British textbooks covering the inspection and surveying of properties. The works by Melville and Gordon (1st edition 1964) Desch (1st edition 1970) and Bowyer (1st edition 1971) were the main pioneer works dealing with this subject.

Thankfully, however, along with this revised edition there are now well over half a dozen books on surveying buildings, on the market in the UK (e.g. Seeley 1985; Mika and Desch 1988; Bowyer 1993; Melville and Gordon 1992; Williams 1993; Hollis 2005; Parnham and Rispin 2001; Hoxley 2002; Glover 2003; and Melville and Gordon 2004). Several texts written for the North American market are also available (e.g. Scaduto 1989; Ventolo 1996; Friedman 2000; Becker 2002). All of these and more are listed below.

As with any textbooks, each one has its strengths and weaknesses. Some, like the work by Hollis (2002), give excellent and well-illustrated guidance on survey procedures and typical building defects. Others, such as Parnham and Rispin (2001) provide good coverage of the appraisal process of residential property surveys, with a chapter dealing with certain types of non-traditional housing.

Chandler (1991), Marshall et al. (1998) and Parnham and Rispin (2001) provide greater coverage of the problems associated specifically with non-traditional housing. For detailed guidance on the identification of and defects in system-built dwellings and other types of non-traditional construction, readers are encouraged to consult the extensive CD-ROM package of reports and other publications produced by Harrison et al. (2004).

Other sources (e.g. Smith 2001; Vegoda 2002; Building Surveying Faculty 2003) cover the legal issues and procedures involved in dealing with these latter surveys in more detail.

It is important therefore that students and practitioners have a wide selection of texts on building investigations and surveys to consult. The type and condition of properties vary enormously and these cannot be wholly covered in one book.

Books and articles

Some of the books listed here, such as the ones on home inspection, were published in the USA. The term 'home inspection' is the American equivalent to the domestic building survey. Despite the environmental, professional and constructional differences between the UK and North America, though, the texts originating from either country contain much that is relevant to all other developed countries.

Several of the titles listed are out of print given their age. However, most should still be available for reference in any good construction library within academic establishments and professional institutions dealing with the built environment.

Addleson, L. (1992) *Building Failures: A Guide to Diagnosis, Remedy and Prevention*, 3rd edition. London: Butterworth-Heinemann.

Addleson, L. and Rice, C. (1992) *Performance of Materials in Building*. Oxford: Butterworth-Heinemann.

ALGAO (1997) *Analysis and Recording for Conservation and Control of Works to Historic Buildings*. London: Association of Local Government Archaeological Officers.

Ambrose, J.E. (1992) *Building Construction and Design*. New York: Van Nostrand Reinhold.

Anon and Building Research Establishment (1959 and 1962) *Principles of Modern Building*, Vols. 1 & 2. London: HMSO.

Anstey, J. (1998) *Party Walls and What to Do with Them*, 5th edition. London: RICS Books.

Anstey, J. (2000) *A Practical Manual for Party Wall Surveyors*. London: RICS Books.

Anstey, J. and Powell, D. (2004) *Anstey's Boundary Disputes*, 5th edition. London: RICS Books.

Asbestos Working Party (2003) *Asbestos – RICS Guidance Note*. London: RICS Books.

Ashurst, J. and Ashurst, N. (1988) *Practical Building Conservation*, Vol. 1: *Stone Masonry*. English Heritage Technical Handbook. Aldershot: Gower Publishing Ltd.

Ashurst, J. and Ashurst, N. (1988) *Practical Building Conservation*, Vol. 2: *Brick, Terracotta, and Earth*. English Heritage Technical Handbook. Aldershot: Gower Publishing Ltd.

Ashurst, J. and Ashurst, N. (1988) *Practical Building Conservation*, Vol. 3: *Mortars, Plasters, Renders*. English Heritage Technical Handbook. Aldershot: Gower Publishing Ltd.

Ashurst, J. and Ashurst, N. (1988) *Practical Building Conservation*, Vol. 4: *Metals*. English Heritage Technical Handbook. Aldershot: Gower Publishing Ltd.

Ashurst, J. and Ashurst, N. (1988) *Practical Building Conservation*, Vol. 5: *Wood, Glass and Resins*. English Heritage Technical Handbook. Aldershot: Gower Publishing Ltd.

Askham, P. and Blake, L. (1993) *The Best of Mainly for Students*, Vol. 1. London: Estates Gazette.

Askham, P. and Blake, L. (1999) *The Best of Mainly for Students*, Vol. 2. London: Estates Gazette.

Atkinson, M.F. (2000) *Structural Defects Reference Manual for Low-Rise Buildings*. London: E & FN Spon.

Baird, G. and Gray, J. (1995) *Building Evaluation Techniques*. London: McGraw-Hill.

BCIS (2002) *Access Audit Price Guide*. London: Building Cost Information Service.

Bech-Andersen, J. (1995) *The Dry Rot Fungus and Other Fungi in Houses*, 5th edition. Vejdammen, Denmark: Hussvamp Laboratoriets Forlag.

Bech-Andersen, J. (2004) *Indoor Climate and Mould*. Vejdammen, Denmark: Hussvamp Laboratoriets Forlag.

Becker, N. (2002) *The Complete Book of Home Inspection*, 3rd edition. New York: McGraw-Hill.

Becker, N. (2003) *Home Inspection Checklists: 111 Illustrated Checklists and Worksheets You Need Before Buying a Home*. New York: McGraw-Hill.

Beckmann, P. and Bowles, R. (eds) (2004) *Structural Aspects of Building Conservation*, 2nd edition. Oxford: Butterworth-Heinemann.

Bereton, C. (1997) *The Repair of Historic Buildings: Advice on Principles and Methods*. English Heritage Conservation Series. London: English Heritage.

Bett, G., Hoehnke, F. and Robison, J. (eds) (2003) *The Scottish Building Regulations Explained and Illustrated*, 3rd edition. Oxford: Blackwell Publishing.

Beukel van den, A. (1991) Linking building pathology and quality assurance input. *Management, Quality and Economics in Building* (P. Brandon, ed.). London: E&FN Spon.

Billington, N.S. and Roberts, B.M. (1982) *Building Services Engineering: A Review of Its Development*. Oxford: Pergamon Press.

Billington, M.J., Simmons, M.W. and Waters, J.R. (2003) *The Building Regulations Explained & Illustrated*, 12th edition. Oxford: Blackwell Publishing.

BMI (2004) *Building Maintenance Price Book 2004*. London: Building Cost Information Service Ltd.

Bonshor, R.B. and Bonshor, L.L. (1996) *Cracking in Buildings*. Construction Research Communications Ltd, London.

Bowyer, J.T. (1993) *Guide to Domestic Building Surveys*, 4th edition. London: Butterworth Architecture.

Brand, S. (1994) *How Buildings Learn*. New York: Viking.

Brantley, L.R. and Brantley, R.T. (1996) *Building Materials Technology (Structural Performance and Environmental Impact)*. New York: McGraw-Hill Inc.

Bricknell, J.M. (1991) Surveying to determine the presence and extent of an attack of dry rot within buildings in the United Kingdom. In: *Serpula lacrymans: Fundamental Biology and Control Strategies* (D.H. Jennings and A.F. Bravery, eds), pp. 117–30. Chichester: John Wiley and Sons Inc.

Bright, K. (2004) *Disability: Making Buildings Accessible – Special Report*, 2nd edition. Cambridge: Workplacelaw Network.

British Steel (1996) *The Prevention of Corrosion on Structural Steelwork*. Cleveland: British Steel Publications.

Brockett, P. and Wright, A. (1986) *The Care and Repair of Thatched Roofs*. London: Society for the Protection of Ancient Buildings.

Broomfield, J.P. (1994) *Assessing Corrosion Damage on Reinforced Concrete Structures Corrosion and Protection of Steel in Concrete* (Narayan Swamy, ed.). Sheffield: Sheffield Academic Press.

Bruntskill, R.W. (1994) *Timber Building in Britain*, 2nd edition. London: Weidenfeld Nicholson.

Bruntskill, R.W. (1997) *Brick Building in Britain*, 2nd edition. London: Weidenfeld Nicholson.

Bruntskill, R.W. (1999) *Traditional Farm Buildings of Britain*. London: Weidenfeld Nicholson.

Bruntskill, R.W. (2000) *Illustrated Handbook of Vernacular Architecture*, 2nd edition. London: Faber & Faber.

Bruntskill, R.W. (2000) *Houses and Cottages of Britain: Origins and Development of Traditional Buildings*. London: Weidenfeld Nicholson.

Bruntskill, R.W. (2004) *Traditional Buildings of Britain: An Introduction to Vernacular Architecture*. London: Cassell.

Bryan, A.J. (1989) *Movements in Buildings*, CIOB Technical Information Service. Englemere, England: Chartered Institute of Building.

BSI (1992) *BS 7543: 1992 Guide to durability of buildings, and building elements, products and components*. London: British Standards Institution.

BSI (2000) *ISO 15686-1:2000 Buildings and constructed assets – Service life planning*, Part 1: General Principles. London: British Standards Institution.

BSI (2000) *ISO 15686-2:2000 Buildings and constructed assets – Service life planning*, Part 2: Service life prediction procedures. London: British Standards Institution.

Building Surveying Faculty (2003) *Dilapidations*, RICS Guidance Note. London: RICS Books.

Building Surveyors Documents Working Party (2000) *Appointing a Building Surveyor*, RICS Guidance Note, 2nd edition. London: RICS Books.

Burgess, R.W. (2004) *Real Estate Home Inspection: Mastering the Profession*, 5th edition. Chicago: Dearborn Real Estate Education.

Burkinshaw, R. and Parrett, M. (2003) *Diagnosing Damp*. London: RICS Books.

Burridge, R. and Ormandy, D. (eds) (1995) *Unhealthy Housing: Research, Remedies and Reform*. London: E&FN Spon.

CAE (1999) *Designing for Accessibility: an essential guide for public buildings*. London: Centre for Accessible Environments.

Calder, A.J.J. (1986) Repair of cracked reinforced concrete: Assessment of injection methods, *Research Report 81*, Transport and Road Research Laboratory. London: Department of Transport.

Calder, A.J.J. and Thompson, D.M. (1988) Repair of cracked reinforced concrete: Assessment of corrosion protection, *Research Report 150*, Transport and Road Research Laboratory. London: Department of Transport.

Capper, P. (1987) *The Latent Damage Act 1986*. London: HMSO.

Carillion Services (2001) *Defects in Buildings: Symptoms, Investigation, Diagnosis and Care*, 3rd edition. London: The Stationery Office.

Carson, Dunlop & Associates Ltd (1993) *The Home Reference Book*. Toronto: Carson, Dunlop & Associates Ltd.

Cauldwell, R. (2000) *Inspecting a House* (For Pros by Pros Series). Newtown CT: The Taunton Press.

CCUS (2000) Scottish Slate: The potential for use in building repair and conservation area enhancement, *Research Report,* Technical Conservation, Research & Education Division. Edinburgh: Historic Scotland.

CEM (2003) The electrical installation survey, *Diploma in Surveying Revision Notes*. Reading: The College of Estate Management.

Chandler, I. (1989) *Building Technology 2 – Performance*. London: Mitchell/CIOB.

Chandler, I. (1991) *Repair and Refurbishment of Modern Buildings*. London: B.T. Batsford Ltd.

Chanter, B. and Swallow, P. (2000) *Building Maintenance Management*. Oxford: Blackwell Publishing.

Chappell, D. (1995) *Report Writing for Architects*. London: Blackwell Science.

Chew, M.L.Y. (1999) *Construction Technology for Tall Buildings*. Singapore: Singapore University Press/World Scientific.

Chynoweth, P. (2003) *The Party Wall Casebook*. Oxford: Blackwell Publishing.

CIB (1993) Building pathology: a state-of the-art report, *CIB Report Publication 155*, CIB Working Commission W86, June 1993. Holland: International Council for Building.

CIB (2003a) *Introduction to CIB*. Available online at:
http://www.cibworld.nl/pages/gi/Default.html.

CIB (2003b) 2nd International Symposium on Building Pathology, Durability and Rehabilitation: *Learning from Errors and Defects in Building*. Available online at:
http://www-ext.lnec.pt/cib_symposium_lisboa03/eng/main.htm.

CIB (2003c) *Working Commission W086: Building Pathology*. Available online at:
http://www.cibworld.nl/pages/gi/Default.html.

CIC (1997) *Definitions of Inspections and Surveys of Buildings* (an explanatory leaflet). London: Construction Industry Council.

CIRIA (2002) A Simple Guide to Controlling Risk, SP154. London: Construction Industry Research & Information Association.

Clancy, J. (1991) *Site Surveying and Levelling*, 2nd edition. London: Architectural Press.

Collins, J. (2001) *Old Houses Care and Repair*. Shaftsbury: Donhead Publishing.

Communities Scotland (2004) Housing and health in Scotland, *Scottish House Condition Survey 2002*, SHCS Working Paper No. 4. Edinburgh: Scottish Executive.

Communities Scotland (2004) Fuel poverty in Scotland, *Scottish House Condition Survey 2002*. Edinburgh: Scottish Executive.

Concrete Society (1984) Repair of concrete damaged by reinforcement corrosion, *Technical Report No. 26*. London: The Concrete Society.

Concrete Society (1989) Cathodic protection of reinforced concrete, *Technical Report No. 36*. London: The Concrete Society.

Concrete Society (1990) Assessment and Repair of Fire Damaged Concrete, *Technical Report No. 33*. London: The Concrete Society.

Concrete Society (1991) Patch repair of reinforced concrete – subject to reinforcement corrosion (model specification and method of measurement), *technical Report No. 38*. London: The Concrete Society.

Concrete Society (1992) Non-structural cracks in concrete, 3rd edition. *Technical Report No. 22*. London: The Concrete Society.

Concrete Society (1995) The relevance of cracking in concrete to corrosion of reinforcement, *Technical Report No. 44*. London: The Concrete Society.

Concrete Society (2000) Diagnosis of deterioration in concrete structures: Identification of defects, evaluation and development of remedial action, *Technical Report No. 54*. London: The Concrete Society.

Concrete Society and Corrosion Engineering Association (1989) Cathodic Protection of Reinforced Concrete, *Technical Report No. 36*. London: The Concrete Society.

Construction Audits Ltd and HAPM (1993) *Defects Avoidance Manual: New Build.* Garston: Building Research Establishment.

Cook, G.K. and Hinks, A.J. (1992) *Appraising Building Defects.* London: Longman.

Cook, R.U. and Gibbs, G.B. (1993) Crumbling heritage? (Studies of stone weathering in polluted atmospheres), *Report for National Power plc and PowerGen plc.* Wetherby: National Power plc.

Cornes, D.L. (1994) *Design Liability in the Construction Industry,* 4th edition. Oxford: Blackwell Science.

Council for the Care of Churches (1980) *A Guide to Church Inspection and Repair.* London: CIO Publishing.

Council for the Care of Churches (1991) *How to look after your church,* 3rd edition. London: CIO Publishing.

Council for the Care of Churches (1993) *The Conservation and Repair of Bells and Bellframes.* London: Church House Publishing.

Cox, J. and Letts, J. (2000) THATCH: Thatching in England 1940–1994, *English Heritage Research Transactions Research and Case Studies in Architectural Conservation,* Vol. 6. London: English Heritage.

Cozzi, G. (2000) *Home Inspection Business from A to Z,* 4th edition. New York: Nemmar Real Estate Training.

Cunnington, P. (1988) *How Old is Your House?* 2nd edition. Dorset: Marsdon House Publishers.

Cunnington, P. (1993) *How Old is that Church?* Dorset: Marsdon House Publishers.

Curtin, W. and Parkinson, E. (1989) *Structural Appraisal and Restoration of Victorian Buildings,* Conservation and Engineering Structures. London: Thomas Telford.

Curtin, W.G. (1994) Causes of defects. In: *Appraisal and Repair of Building Structures: Introductory Guide* (R. Holland, B.E. Montgomery-Smith and J.F.A. Moote, eds). London: Thomas Telford.

Curwell, S.R. and March, C.G. (eds) (1986) *Hazardous Building Materials: A Guide to Selection and Alternatives.* London: E & FN Spon.

Curwell, S.R., March, C.G. and Venables, R. (eds) (1990) *Buildings and Health: The Rosehaugh Guide to the Design, Construction, Use and Management of Buildings.* London: RIBA Publications.

Dallas, R. (ed.) (2003) Measured survey and building recording for historic buildings and structures, *Guide for Practitioners 4.* Edinburgh: Historic Scotland.

Davey, A., Heath, B., Hodges, D., Ketchin, M. and Milne, R. (1995) *The Care and Conservation of Georgian Houses,* 4th edition. London: Butterworth Architecture.

Desch, H.E. (1970) *Structural Surveying,* London: Griffin.

DETR (1998) *Housing fitness standard: consultation paper.* Department of the Environment, Transport and the Regions. London: HMSO.

DfEE (2000) *Asset Management Plans, Section 3: Condition Assessment.* London: The Department for Education and Employment.

Dickinson, P. and Thornton, N. (2004) *Cracking and Building Movement.* London: RICS Books.

DOE (1986) *Asbestos Materials in Buildings,* 2nd edition. London: HMSO.

Douglas, J. (2002) *Building Adaptation.* Oxford: Butterworth-Heinemann.

Driscoll, R.M.C. and Crilly, M.S. (2000) Subsidence damage to domestic buildings: lessons learned and questions remaining, *Foundation for the Built Environment Report 1.* London: Construction Communications Ltd.

Edwards, L. (1995) *Practical Risk Management in the Construction Industry.* London: Thomas Telford.

Egan, J. (1998) *Rethinking Construction*. London: Department of the Environment, Transport and the Regions.

Emmitt, S. (2002) *Architectural Technology*. Oxford: Blackwell Science.

Endean, K.F. (1995) *Investigating Rainwater Penetration of Modern Buildings*. Aldershot: Gower.

English Heritage (1995) *Easy Access to Historic Properties*. London: English Heritage.

English Heritage (1996) London terraced houses 1660–1860. In: *A Guide to Alterations and Extensions*. London: English Heritage.

Forsyth, M. (2004) *Bath*. Pevsner Buildings of England Series. London: Yale University Press.

Foster, L. (1997) *Access to the Historic Environment – Meeting the Needs of Disabled People*. Shaftsbury: Donhead.

Friedman, D. (2000) *The Investigation of Buildings: A Guide for Architects, Engineers, and Owners*. New York: W.W. Norton & Co.

Garratt, J. and Nowak, F. (1991) Tackling condensation: a guide to the causes of, and remedies for, surface condensation and mould in traditional housing, *BRE BR Report 174*. Garston: Building Research Establishment.

GB Geotechnics Ltd (2001) Non-destructive investigation of standing structures, *Technical Advice Note 23*. Edinburgh: Historic Scotland.

Gilbert, J. and Flint, A. (1992) *The Tenement Handbook: A Practical Guide to Living in a Tenement*. ASSIST Architects Ltd. Edinburgh: RIAS.

Glover, P. (2003) *Building Surveys*, 5th edition. Oxford: Butterworth-Heinemann.

Gowers, E. (1986) *The Complete Plain Words*, 3rd edition. (revised by S. Greenbaum and J. Whitcut). London: HMSO.

Graves, H.M. and Phillipson, M.C. (2000) *Potential Implications of Climate Change in the Built Environment*. London: BRE & Construction Communications Ltd.

Hall, G.T. (1984) *Revision Notes on Building Maintenance and Adaptation*. London: Butterworths.

Hall, N. (1988) *Thatching: A handbook*. London: Intermediate Technology Publications.

HAPM Ltd (1991) *Defects Avoidance Manual: New Build*. Garston: Building Research Establishment.

HAPM Ltd (2000) *Life Component Manual*, 2nd edition. London: E & FN Spon.

HAPM Ltd (2001) *Guide to Defects Avoidance*. London: Spon Press.

Harland, E. (1998) *Eco-Renovation: The ecological home improvement guide*, 2nd edition. Devon: Green Books.

Harper, D.R. (1978) *Building: The Process and the Product*. Lancaster: The Construction Press Ltd.

Harris, R. (1995) *Discovering Timber Framed Buildings*. Aylesbury: Shire Publications.

Harrison, H.W. (1996) *Roofs and Roofing: Performance, diagnosis, maintenance, repair and the avoidance of defects*, BRE Building Elements series. London: Construction Research Communications Ltd.

Harrison, H.W. and de Vekey, R.C. (1998) *Walls, windows and doors: Performance, diagnosis, maintenance, repair and the avoidance of defects*, BRE Building Elements series. London: Construction Research Communications Ltd.

Harrison, H.W. and Trotman, P.M. (2000) *Building services: Performance, diagnosis, maintenance, repair and the avoidance of defects*, BRE Building Elements series. London: Construction Research Communications Ltd.

Harrison, H.W. and Trotman, P.M. (2002) *Foundations, basements and external works: Performance, diagnosis, maintenance, repair and the avoidance of defects,*

BRE Building Elements series. London: Construction Research Communications Ltd.

Harrison, H.W., Mullin, S., Reeves, B. and Stevens, A. (2004) *Non-Traditional Housing Types – An Aid to Identification*. Garston: IHS Rapidoc and Building Research Establishment.

Hart, J.M. (1991) A practical guide to infra-red thermography for building surveys, *BRE Report BR 176*. Garston: Building Research Establishment.

Harvey, J.H. (1972) *Conservation of Building*. London: John Barker (Publishers) Ltd.

Haverstock, H. (1999) *Building Design Easibrief*, 3rd edition. Kent: Miller Freeman.

Heckroodt, R.O. (2002) *Guide to the Deterioration and Failure of Building Materials*. London: Thomas Telford Publishing.

Hilton, D.O. and Oliver, M.A. (1985) *Timber Framed Housing: Repairs, Maintenance and Extensions*. London: E & FN Spon.

Hinks, A.J. and Cook, G.K. (1997) *The Technology of Building Defects*. London: E & FN Spon.

Historic Scotland (2000a) Conservation laws: a guide to the preparation of conservation plans. *Heritage Policy*. Edinburgh: Historic Scotland.

Historic Scotland (2000b) Corrosion in masonry-clad early 20th century steel framed buildings, *Technical Advice Note 20*. Edinburgh: Historic Scotland.

Holden, T.G. (1998) The archaeology of Scottish thatch, *Technical Advice Note 13*. Edinburgh: Historic Scotland.

Holland, R. Montgomery-Smith, B.E. and Moore, J.F.A. (eds) (1994) *Appraisal and Repair of Building Structures*. London: Thomas Telford Services Ltd.

Hollis, M. (1988) *Surveying for Dilapidations*. London: Estates Gazette.

Hollis, M. (1990) *Cavity Wall Tie Failure*. London: Estates Gazette.

Hollis, M. (2005) *Surveying Buildings*, 5th edition. London: RICS Books.

Holmes, R. (1990) Property condition assessment – a cost effective approach, *Technical Information Service, No. 122*. Englemere, England: Chartered Institute of Building.

Holmes, S. and Wingate, M. (1997) *Building with Lime: A practical introduction*. London: Intermediate Technology Publications.

House Condition Surveys Team (2004) *Scottish House Condition Survey 2002* (Main Report). Edinburgh: Communities Scotland.

Howell, J. (1995) Moisture Measurement in Masonry: Guidance for Surveyors, In: *COBRA 95*, RICS Construction and Building Research Conference, at Heriot-Watt University, 8–9 September 1995. London: The Royal Institution of Chartered Surveyors.

Hoxley, M. (1997) Condition inspections of residential property: a procedural framework, *Structural Survey*, 20(1). Bradford: MCB University Press.

Hoxley, M. (2002) *Construction Companion to Building Surveys*. London: RIBA Publications.

HSE (1990) Evaluation and inspection of buildings and structures, *Health and Safety series booklet HS(G)58*. London: HMSO.

Hudson, R. (1984) *Corrosion of Structural Steelwork, 46 New Broad Street, London*. Swindon Laboratories: British Steel Corporation.

Hughes, P. (1986) The need for old buildings to 'breathe', *Information Sheet 4*. London: The Society for the Protection of Ancient Buildings (SPAB).

Hunt, R. Dyer, R.H. and Driscoll, R. (1991) Foundation movement and remedial underpinning in low-rise buildings, *BRE Report 184*. Garston: Building Research Establishment.

Hutchison, B.D. Barton, J. and Ellis, N. (1973) *Maintenance and Repair of Buildings and Their Internal Environment*. London: Newnes-Butterworths.

ICOMOS (1991) *Guide to Recording Historic Buildings*. London: Butterworth Architecture.

Illston, J.M. (ed.) (1993) *Construction Materials – Their Nature and Behaviour*. London: Chapman & Hall.

Imber, A. and Blake, L. (2004) *The Best of Mainly for Students*, Vol. 3. London: Estates Gazette.

Innocent, C.F. (1916) *The Development of English Building Construction*. Cambridge: Cambridge University Press (1999 edition reprinted by Donhead Publishing, Shaftesbury).

IRC (1991) *Protocols for building condition assessment*. Ottawa: Institute for Research in Construction.

Irvine, W. (1995) *Surveying for Construction*, 4th edition. New York: McGraw-Hill Publishing Co.

Irwin, R. (1995) *The Home Inspection Troubleshooter*. USA: Dearborn Trade Publishing.

Iselin, D.G. and Lemer, A.C. (eds) (1993) The fourth dimension in building: strategies for minimising obsolescence, *Committee on Facility Design to Minimize Premature Obsolescence*, Building Research Board. Washington DC: National Academy Press.

IStructE (1991) *Guide to surveys and inspections of buildings and similar structures*. London: The Institution of Structural Engineers.

IStructE (1992) *Appraisal of building structures*. London: Institution of Structural Engineers.

IStructE (2000) *Subsidence of low-rise buildings*, 2nd edition. London: Institution of Structural Engineers.

Kaminetzky, D. (1992) *Design and Construction Failures: Lessons from Forensic Investigations*. New York: McGraw-Hill Publishing Co.

Kelsall, F. (2001) Dating old buildings, In: *Building Conservation Directory*. Available online at: www.buildingconservation.com/articles.

Kibert, C.J. (ed.) (1999) *Reshaping the Built Environment*. Washington DC: Island Press.

Knight, J. (1995) *The Repair of Historic Buildings in Scotland*. Edinburgh: HMSO for Historic Scotland.

Kubal, M.T. (1993) *Waterproofing the Building Envelope*. New York: McGraw-Hill.

Latham, M. (1994) *Constructing the team*, Final Report. London: HMSO.

London Hazards Centre (1990) *Sick Building Syndrome (Causes, effects and control)*. London: London Hazards Centre Trust Ltd.

Loss Prevention Council (1993) *Asbestos use in buildings: the hazards and their limitations*, Report SHE 9:1993. Borehamwood: Loss Prevention Council.

Ltsiburek, J. and Carmody, J. (1993) *Moisture Control Handbook (Principles and Practices for Residential and Small Commercial Buildings)*. New York: Van Nostrand Reinhold.

Maltz, B. (2001) *Liability of the Negligent Surveyor*. London: Estates Gazette.

Marks, S. (ed.) (1996) *Concerning Buildings*. London: Architectural Press.

Marshall, D. (1996) Condition surveys for housing associations, *Construction Papers, No. 68* (P. Harlow, ed.). Englemere, England: Chartered Institute of Building.

Marshall, D., Worthing, D. and Heath, R. (1998) *Understanding Housing Defects*. London: Estates Gazette.

Marshall, D. and Worthing, D. (2000) *The Construction of Houses*, 3rd edition. London: Estates Gazette.

Massari, G. and Massari, I. (1993) *Damp Buildings, Old and New* (translated by C. Rockwell). Rome: ICCROM.

Matulionis, R.C. and Freitag, J.C. (eds) (1990) *Preventive Maintenance of Buildings*. New York: Van Nostrand Reinold.

Mayer, P. and Wornell, P. (eds) (1999) *HAPM Workmanship Checklists*, HAPM Publications Ltd. London: E&FN Spon.

McDonald, S. (ed.) (2002) *Building Pathology: Concrete*. Oxford: Blackwell Publishing.

McDonald, T.C. (1994) Understanding old buildings (the process of architectural investigation), *Preservation Brief 35*, Heritage Preservation Services, National Park Service. Available online from: http://www2.cr.nps/tps/briefs/brief35.htm.

McGregor, W. and Then, D.S.S. (1999) *Facilities Management and the Business of Space*. London: Arnold.

Melville, I.A. and Gordon, I.A. (1992) *Structural Surveys of Dwelling Houses*, 3rd edition. London: Estates Gazette.

Melville, I.A. and Gordon, I.A. (1998) *The Repair and Maintenance of Houses*, 2nd edition. London: Estates Gazette.

Melville, I.A. and Gordon, I.A. (2004) *Inspections and Reports on Dwellings: Assessing Age*. London: Estates Gazette.

Melville, I.A. and Gordon, I.A. (2005a) *Inspections and Reports on Dwellings: Inspecting*. London: Estates Gazette.

Melville, I.A. and Gordon, I.A. (2005b) *Inspections and Reports on Dwellings: Reporting for Buyers*. London: Estates Gazette.

Melville, I.A. and Gordon, I.A. (2005c) *Inspections and Reports on Dwelling: Reporting for Sellers*. London: Estates Gazette.

Mika, S.L.J. and Desch, S.C. (1988) *Structural Surveying*, 2nd edition. London: Macmillan.

Mills, E. (ed.) (1994) *Building Maintenance and Preservation*, 2nd edition. London: Architectural Press.

Moir, J. and Letts, J. (1996) THATCH: Thatching in England 1790–1940, *English Heritage Research Transactions Research & Case Studies in Architectural Conservation*, Vol. 5. London: English Heritage.

Moseley, L. and Curtis, M. (eds) (1973) *Goodin & Downing's Domestic Sanitation*. London: Estates Gazette.

Murdoch, J. and Murrells, P. (1996) *Law of Surveys and Valuations*. London: Estates Gazette.

Murdoch, J. (2002) *Negligence in Valuations and Surveys*, RICS Case in Point Series. London: RICS Books.

Myton-Davies, P. (1988) *A Practical Guide to Repair and Maintenance of Houses*. London: International Thomson Business Publishing.

NBA (1983) *Common Building Defects: Diagnosis and Remedy*. London: Construction Press.

NBA (1985) *Maintenance Cycles & Life Expectancies of Building Components & Materials: A guide to data and sources*. London: National Building Agency.

NCHA (1988) *Non-Traditional Housing in Northern England: Identification Booklet*. Newcastle: Northern Consortium of Housing Associations.

Newcomer, S. (1992) *Home Inspection Report*. USA: American Home Inspectors Training Institute Ltd.

NICEIC (2004) *Part P fact sheet*. Dunstable: National Inspection Council for Electrical Installation Contracting.

Nicholson, P. (1823) *New Practical Builder Vol. 1* (1993 Facsimile edition). London: Beacon Books.

Nicholson, P. (1823) *New Practical Builder Vol. 2* (1993 Facsimile edition). London: Beacon Books.

ODPM (2003a) *Factsheet 7: The home information pack (HIP)*, Office of the Deputy Prime Minister. London: HMSO.

ODPM (2003b) *Preparing for Floods: Interim guidance for improving the flood resistance of domestic and small business properties*, October 2003, Office of the Deputy Prime Minister. London: HMSO.

Oliver, A. (1997) *Dampness in Buildings*, 2nd edition revised by J. Douglas and J.S. Stirling). Oxford: Blackwell Science.

Orton, A. (1992) *The Way We Build Now: Form, scale, techniques*. New York: Van Nostrand Reinhold.

Oxley, R. (2003) *Survey and Repair of Traditional Buildings*. Shaftsbury: Donhead Publishing.

Oxley, T.A. and Gobert, E.G. (1994) *Dampness in Buildings (Diagnosis, Instruments, Treatment)*, 2nd edition. Oxford: Butterworth-Heinemann.

Palfreyman, J.W., Smith, D. and Low, G. (2002) Studies of the domestic dry rot fungus *Serpula lacrymans* with relevance to the management of decay in buildings, *Research Report*. Edinburgh: Historic Scotland.

Palfreyman, J.W. and Urquhart, D. (2002) The environmental control of dry rot, *Technical Advice Note 24*. Edinburgh: Historic Scotland.

Park, S. (1993) *Mothballing Historic Buildings*, Preservation Brief 31, National Parks Service (Preservation Assistance). Washington DC: US Department of the Interior.

Parkinson, K., Shaw, G. and Knowles, D. (1996) *Appraisal and Repair of Masonry*. London: Thomas Telford Services Ltd.

Parnham, P. (1996) *Prevention of Premature Staining of New Buildings*. London: Chapman & Hall.

Parnham, P. and Rispin, C. (2001) *Residential Property Appraisal*. London: E&FN Spon.

Party Wall Guidance Note Working Group (2003) *Party Wall Legislation and Procedure – RICS Guidance Note*. London: RICS Books.

Pickard, R.D. (1996) *Conservation in the Built Environment*. Harlow: Addison Wesley.

Pompeii, M.A. (2004) *Become a Home Inspector!* Fredericksburg, VA: Pompeii Engineers.

Powell, J.L. and Stewart, R. (eds) (2002) *Jackson and Powell's Professional Negligence*, 5th edition (Mr Justice Jackson, consulting ed.). London: Sweet & Maxwell.

Powys, A.R. (1995) *Repair of Ancient Buildings*, 3rd edition. London: SPAB.

Preiser, W.F. (1989) *Post Occupancy Evaluation*. New York: Van Nostrand Reinhold.

Preiser, W.F. and Vischer, J. (eds) (2004) *Assessing Building Performance*. New York: Butterworth-Heinemann.

Proverbs, D. and Soetanto, R. (2004) *Flood Damaged Property*. Oxford: Blackwell Publishing.

Pryke & Partners (1987) *Surveying Cracked Property: A Guide for Engineers and Surveyors*, 2nd edition. Essex: John Pryke & Partners.

Pye, P. and Harrison, H.W. (1997) *Floors and Flooring: Performance, diagnosis, maintenance, repair and the avoidance of defects*, BRE Building Elements series. London: Construction Research Communications Ltd.

Quiney, A. (1995) *Traditional Buildings of England*. London: Thames & Hudson.

Ransom, W. (1987) *Building Failures: Diagnosis, Remedy and Prevention*, 2nd edition. London: E&FN Spon.

Reeves, B.R. and Martin, G.R. (1989) The structural condition of Wimpey no-fines low-rise dwellings, *BRE Report 153*. Garston: Building Research Establishment.

Research, Analysis and Evaluation Division (2003) *English House Condition Survey 2001* (Main Report). London: Office of the Deputy Prime Minister.

Residential Property Faculty (2004) *Building Surveys of Residential Property*, 2nd edition. London: RICS Books.

Residential Property & Building Surveying Faculties (2004) *Residential Building Survey Examples Terms and Conditions of Engagement*. London: RICS Books.

Richardson, B.A. (2001) *Defects and Deterioration in Buildings*, 2nd edition. London: Spon Press.

Richardson, B.A. (2002) *Remedial Treatment of Buildings*, 2nd edition. Oxford: Butterworth-Heinemann.

Richardson, C. (1987) *AJ Guide to Structural Surveys*. London: Architectural Press.

Richardson, C. (1996) When the earth moves, *Architects Journal*, 22 February.

Richardson, C. (2000) Moving structures, *Architects Journal*, 14 September.

RICS (1995) *Flat Roof Covering Problems*, RICS Guidance Note. London: RICS Books.

RICS (1996) *Fire Damage Reinstatement*, RICS Information Paper. London: RICS Books.

RICS (1997a) *Stock Condition Surveys*, RICS Guidance Note. London: RICS Books.

RICS (1997b) *The Mundic Problem*, RICS Guidance Note, 2nd edition. London: RICS Books.

RICS (1998) *Building Surveys and Inspections of Commercial and Industrial Property: A Guidance Note for Surveyors*, 2nd edition. London: RICS Books.

RICS (2000) *Building Maintenance: Strategy, Planning and Procurement*, RICS Guidance Note. London: RICS Books.

RICS (2002) *The Mundic Problem Supplement – Stage 3 Expansion Testing*. London: RICS Books.

RICS (2003) *Dilapidations*, 2nd edition, RICS Guidance Note. London: RICS Books.

RICS (2004) *Surveying safely: Your guide to personal safety at work*. London: RICS Books. Available online at http://www.rics.org/Management/Healthandsafety/surv_safe_all.htm

RICS and EEO (1992) *Energy Efficiency in Buildings (Energy Appraisal of Existing Buildings – A Handbook for Surveyors)*. London: RICS Books.

Ridout, B. (1999) *Timber Decay in Buildings: The conservation approach to treatment*, English Heritage and Historic Scotland. London: E&FN Spon.

Robson, P. (1991) *Structural Appraisal of Traditional Buildings*. Shaftsbury: Donhead Publishing.

Robson, P. (1999) *Structural Repair of Traditional Buildings*. Shaftsbury: Donhead Publishing.

Ross, M. (1986) *Negligence in Surveying and Building*. London: Estates Gazette.

Ross, P. (2001) *Appraisal and Repair of Timber Structures*. London: Thomas Telford Services Ltd.

Rostron, J. (ed.) (1996) *Sick Building Syndrome*. London: E&FN Spon.

Rushton, T. (1992) Understanding Why Buildings Fail and Using 'Heir' Methodology in Practice, In: *Latest Thinking on Building Defects in Commercial Buildings: Their Diagnosis, Causes and Consequences*, one-day conference, Wednesday 15 July 1992, Henry Stewart Conference Studies, London.

Ryan P.A., Wolstenholme, R.P. and Howell, D.M. (1995) Durability of cladding, *A state of the art report*. London: Thomas Telford Services.

Scaddan, B. (2001) *IEE Wiring Regulations Illustrated and Explained*, 6th edition. Oxford: Newnes.

Scaduto, J.V. (1989) *What's It Worth: A Home Inspection and Appraisal Manual*, 2nd edition. Blue Ridge Summit, PA: TAB Books.

Scaduto, J.V. and Scaduto, M.J. (1988) *Keep Its Worth: Solving Most Common Building Problems*. Blue Ridge Summit, PA: TAB Books.

Scottish Executive Building Directorate (2001) *A Guide to Non-Traditional and Temporary Housing in Scotland 1923–1955*, 2nd edition. Edinburgh: The Stationery Office.

Scottish Executive Development Department (2004) Planning and Building Standards Advice on Flooding, *Planning Advice Note 69*. Edinburgh: Scottish Executive.

Scottish Executive (2003) 'Modernising Scotland's social housing', *Consultation Paper*, March 2003. Edinburgh: Scottish Executive.

Seeley, I.H. (1985) *Building Surveys, Reports and Dilapidations*. London: Palgrave Macmillan.

Seeley, I.H. (1987) *Building Maintenance*, 2nd edition. London: Palgrave Macmillan.

Singh, J. (ed.) (1994) *Building Mycology: Management of decay and health in buildings*. London: Chapman & Hall.

Singh, J. and Walker, B. (eds) (1996) *Allergy Problems in Buildings*. Cheltenham: Quay Books.

Smith, B.J. and Warke, P.A. (2004) *Stone Decay Its Causes and Controls*. Shaftsbury: Donhead Publishing.

Smith, D.J. (1994) *Underpinning and Repair of Subsidence Damage*. London: Camden Consultancy.

Smith, G. and May, P. (2001) *Blast Effects on Buildings*. London: Thomas Telford Services Ltd.

Smith, L. (1985) *Investigating Old Buildings*. London: Batsford Academic and Educational.

Smith, P. (ed.) (2004) *Rivington's Building Construction*, Vols. 1–3. Shaftesbury: Donhead Publishing.

Smith, P.F. (2001) *West & Smith's Law of Dilapidations*, 11th edition. London: Estates Gazette.

Son, T. and Yuen, G. (1996) *Building Maintenance Technology*. London: Macmillan.

Spada, M.D. (2001) *The Home Inspection Book: A Guide for Professionals*. USA: South-Western Educational Publishing.

Spain, B. (2000) *Spon's estimating costs guide to minor works, refurbishment and repairs*. London: Spon Press.

Staveley, H.S. (1998) Building Surveys, *Construction Papers, No. 89* (P. Harlow, ed.), Englemere, England: Chartered Institute of Building.

Stone, B. (2003) *The Consumer Advocate's Guide to Home Inspection.* Chicago: Dearborn Trade Publishing.

Swallow, P. (1998) Rehabilitation: Getting to know the building, *Building Engineer*, July/August. Northampton: Association of Building Engineers.

Swallow, P., Dallas, R., Jackson, S. and Watt, D. (2004) *Measurement and Recording of Buildings*, 2nd edition. Shaftsbury: Donhead Publishing.

Swindells, D.J. and Hutchings, M. (1991) *A Checklist for the Structural Survey of Period Timber Framed Buildings*, Building Conservation Group. London: RICS Business Books.

Switsur, R. (2001) Dating Technology. In: *Building Conservation Directory.* Available online at: www.buildingconservation.com/articles.

Taylor, G.D. (1998) *Materials of Construction*, 3rd edition. Essex: Longman Scientific & Technical.

Thomas, A.R. (1983) Treatment of damp in old buildings, *Technical Pamphlet No. 8.* London: SPAB.

Traister, J.E. (2000) *Home Inspection Handbook.* Carlsbad CA: Craftsman Book Co.

Trill, J. and Bowyer, J.T. (1972) *Problems in Building Construction: A Scientific Method Approach.* London: Architectural Press.

Trill, J. and Bowyer, J.T. (1973) *Problems in Building Construction: A Scientific Method Approach (Tutor's Guide).* London: Architectural Press.

Trotman P.M., Harrison, H.W. and Sanders, C. (2004) *Understanding Dampness: Effects, causes, diagnosis and remedies.* Garston: Building Research Establishment.

Vegoda, V. (1993) *A Practical Guide to Residential Surveys*, Need to Know Series No. 2. London: Lark Productions Ltd.

Vegoda, V. (1994) *An Introduction to Schedules of Dilapidations*, Need to Know Series No. 3. London: Lark Productions Ltd.

Vegoda, V. (2002) *The Dilapidations Handbook*, Need to Know Series No. 11. London: Lark Productions Ltd & Estates Gazette.

Ventolo, W.L. (1995) *Your Home Inspection Guide.* Chicago: Dearborn Financial Publishing.

Ventolo, W.L. (1996) *The Complete Home Inspection Kit.* Chicago: Longman Financial Services Publishing.

Vidal-Hall, J. (1977) *Report Writing*, Communication Skills Guides (revised by C. Wright). London: The Industrial Society.

Warszawski, A. (1999) *Industrialized and Automated Building Systems*, 2nd edition. London: E&FN Spon.

Watt, D. (1999) *Building Pathology: Principles and Practice.* Oxford: Blackwell Science.

Watt, D. and Swallow, P. (1996) *Surveying Historic Buildings.* Shaftsbury: Donhead Publishing.

Watts & Partners (2004) *Watts Pocket Handbook 2004.* London: Watts & Partners.

Weaver, R. (1999) *Conservation Techniques.* New York: Van Nostrand Reinhold.

Weeks, K.D. and Grimmer, A.E. (1995) *The Secretary of the Interior's Standards for the Treatment of Historic Properties with Guidelines for Preserving, Rehabilitating, Restoring and Reconstructing Historic Buildings.* Washington DC: US Department of the Interior.

Williams, A.R. (1993) *Domestic Building Surveys*. London: E & FN Spon.

Williams, A.W. and Ward, G.C. (1991) The renovation of no-fines housing, *BRE Report 191*. Garston: Building Research Establishment.

Williams, G. and Ashurst, N. (1986) An introduction to the treatment of rising damp in old buildings, *Information Sheet 6*. London: SPAB

Wood, B. (2003) *Building Care*. Oxford: Blackwell Publishing.

Wood, J. (ed.) (1994) *Buildings Archaeology: Applications in Practice*. Oxford: Oxbow Books.

Wordsworth, P. (2001) *Lee's Building Maintenance Management*, 4th edition. Oxford: Blackwell Science.

Yeomans, D. (1997) *Construction Since 1900: Materials*. London: BT Batsford Ltd.

Zunde, J.M. (1989) *Design Technology*. Sheffield: Pavic Publications.

Journals

The following technical journals are useful if not excellent sources of information on inspections and surveys of buildings and related matters.

Structural Survey

This is a key refereed journal dealing with the inspection and appraisal of buildings. It has been in existence since the early 1980s and is published by Emerald on a quarterly basis. Its editor is chartered building surveyor Dr Mike Hoxley, of Anglia Polytechnic University. The aims of this journal are to provide practical and up-to-date guidance and research for building surveyors and other construction professionals responsible for inspecting and refurbishing properties.

Journal of Building Appraisal

This is a recently published technical journal from Henry Stewart Publications. Its focus is very similar to that of the journal *Structural Survey*. Its publishers claim that it is designed to be the leading professional journal dealing with building condition, defects, repair and maintenance. Professor Malcolm Hollis, a well-respected chartered building surveyor and author of the book *Surveying Buildings*, is its editor.'

Building Research & Information

This is an international journal published six times a year by Spon Press for the CIB (the Conseil International du Bâtement – the French acronym for the International Council for Building). Richard Lorch edits this journal, which covers a wide range of topics relating to buildings and their services.

Construction and Building Materials

Elsevier publishes this refereed journal eight times a year. It contains several scientific papers on all kinds of materials used in the construction industry. Edinburgh University's Professor Mike Forde is its editor.

BRE publications

Preamble

The Building Research Establishment (BRE) produces an extensive range of very informative literature on all aspects of construction technology. The following *Digests, Good Building Guides, Good Repair Guides, Information Papers* and *Reports* are all relevant in some way to the inspection and surveying of building.

Digests

Digest 54 (1971) *Damp-proofing solid floors.*
Digest 110 (1972) *Condensation.*
Digest 125 (1988) *Colourless treatment of masonry.*
Digest 144 (1972) *Asphalt and built-up felt roofings: durability.*
Digest 145 (1972) *Heat losses through ground floors.*
Digest 152 (1973) *Repair and renovation of flood-damaged buildings.*
Digest 157 (1973) *Calcium silicate (sandlime, flintlime) brickwork.*
Digest 161 (1973) *Reinforced plastics cladding panels.*
Digest 163 (1974) *Drying out buildings.*
Digest 170 (1984) *Ventilation of internal bathrooms and WCs in dwellings.*
Digest 177 (1975) *Decay and conservation of stone masonry.*
Digest 180 (1986) *Condensation in roofs.*
Digest 190 (1976) *Heat losses from dwellings.*
Digest 192 (1976) *The acoustics of rooms for speech.*
Digest 196 (1976) *External rendered finishes.*
Digest 197 (1982) *Painting walls: Part 1 – choice of paint.*
Digest 198 (1977) *Painting walls: Part 2 – failures and remedies.*
Digest 200 (1977) *Repairing brickwork.*
Digest 217 (1978) *Wall cladding defects and their diagnosis.*
Digest 223 (1983) *Wall cladding: designing to minimise defects due to inaccuracies and movements.*
Digest 235 (1989) *Fixings for non-loadbearing precast concrete cladding panels.*
Digest 236 (1980) *Cavity insulation.*
Digest 238 (1992) *Reducing the risk of pest infestation: design recommendations and literature review.*
Digest 245 (1981) *Rising damp in walls – diagnosis and treatment.*
Digest 251 (1983) *Assessment of low-rise structures.*
Digest 268 (1988) *Common defects in low-rise traditional housing.*
Digest 270 (1983) *Condensation in insulated domestic roofs.*
Digest 280 (1983) *Cleaning external surfaces of buildings.*

Digest 282 (1983) *Structural appraisal of buildings with long spans.*

Digest 293 (1985) *Improving the sound insulation of separating walls and floors.*

Digest 296 (1985) *Timbers – their natural durability and resistance to preservative treatments.*

Digest 297 (1985) *Surface condensation and mould growth in traditionally-built dwellings.*

Digest 298 (1999) *Low-rise building foundations: the influence of trees in clay soils.*

Digest 299 (1985) *Dry rot: its recognition and control.*

Digest 300 (1985) *Toxic effects of fires.*

Digest 301 (1985) *Corrosion of metals by wood.*

Digest 304 (1985) *Preventing decay in external joinery.*

Digest 306 (1986) *Domestic draughtproofing: ventilation considerations.*

Digest 307 (1986) *Identifying damage by wood-boring insects.*

Digest 308 (1983) *Unvented domestic hot water systems.*

Digest 311 (1986) *Wind scour of gravel ballast on roofs.*

Digest 312 (1986) *Flat roof design: the technical options.*

Digest 313 (1986) *Mini piling for low-rise housing.*

Digest 320 (1987) *Fire doors.*

Digest 321 (1987) *Timber for joinery.*

Digest 324 (1987) *Flat roof design: insulation.*

Digest 327 (1992) *Insecticidal treatments against wood-boring insects.*

Digest 329 (2000) *Installing wall ties in existing construction.*

Digest 330 (1988) *Alkali aggregate reactions in concrete.*

Digest 331 (1983) *GRC.*

Digest 334 (1988) *Sound insulation of separating walls and floors, Part 1: walls.*

Digest 334 (1988) *Sound insulation of separating walls and floors, Part 2: floors.*

Digest 336 (1988) *Swimming pool roofs: minimizing the risks of condensation.*

Digest 337 (1983) *Sound insulation: basic principles.*

Digest 345 (1989) *Wet rots: recognition and control.*

Digest 351 (1990) *Recovering old timber roofs.*

Digest 352 (1990) *Underpinning.*

Digest 354 (1990) *Painting of exterior wood.*

Digest 355 (1990) *Energy efficiency in dwellings.*

Digest 358 (1991) *CFCs in buildings.*

Digest 359 (1991) *Repairing brick and block masonry.*

Digest 361 (1983) *Why do buildings crack?*

Digest 362 (1991) *Building mortar.*

Digest 363 (1991) *Sulphate and acid resistance of concrete in the ground.*

Digest 364 (1991) *Design of timber floors to prevent decay.*

Digest 365 (1991) *Soakaway design.*

Digest 366 (1991) *Structural appraisal of existing buildings for change of use.*

Digest 369 (1992) *Interstitial condensation and fabric deterioration.*

Digest 370 (1992) *Control of lichens, moulds and similar growths.*

Digest 371 (1992) *Remedial wood preservatives: use them safely.*

Digest 372 (1992) *Flat roof design: waterproof membranes.*

Digest 373 (1992) *Wood chipboard.*

Digest 375 (1992) *Wood-based products: their contribution to the conservation of forest resources.*

Digest 378 (1993) *Wood preservatives: application methods.*

Digest 379 (1993) *Double glazing for heat and sound insulation.*

Digest 380 (1993) *Damp-proof courses.*
Digest 387 (1993) *Natural finishes for exterior wood.*
Digest 389 (1999) *Concrete – cracking and corrosion of reinforcement.*
Digest 392 (1994) *Assessment of existing high alumina cement construction in the UK.*
Digest 396 (1993) *Smoke control in buildings.*
Digest 398 (1993) *Continuous mechanical ventilation in dwellings: design, installation and operation.*
Digest 401 (1995) *Replacing wall ties.*
Digest 403 (1996) *Damage to structure from ground-borne vibration.*
Digest 404 (1996) *PVC-U windows.*
Digest 405 (1993) *Carbonation of concrete and its effects on durability.*
Digest 406 (1998) *Wind actions on buildings.*
Digest 410 (1994) *Cementitious renders for external walls.*
Digest 412 (1996) *Desiccation in clay soils.*
Digest 415 (1996) *Reducing the risk of pest infestations in buildings.*
Digest 418 (1996) *Bird, bee and plant damage to buildings.*
Digest 422 (1983) *Painting exterior wood.*
Digest 428 (1983) *Protecting buildings against lightning.*
Digest 429 (1999) *Timbers: their natural durability and resistance to preservative treatment.*
Digest 430 (2000) *Plastics external glazing.*
Digest 432 (1999) *Corrosion of reinforcement in concrete: electrochemical monitoring.*
Digest 437 (1999) *Industrialised and platform floors: mezzanine and raised storey.*
Digest 440 (1999) *Weathering and white PVC-U.*
Digest 443 Part 1 (1999) *Termites and UK buildings: biology, detection and diagnosis.*
Digest 443 Part 2 (1999) *Termites and UK buildings: control and management of subterranean termites.*
Digest 444 Part 1(2000) *Corrosion of steel in concrete: Investigation and assessment.*
Digest 444 Part 2 (2000) *Corrosion of steel in concrete: Durability of reinforced concrete structures.*
Digest 444 Part 3 (2000) *Corrosion of steel in concrete: Protection and remediation.*
Digest 448 (2000) *Cleaning buildings: legislation and good practice.*
Digest 449 Part 1 (2000) *Cleaning external masonry: developing and implementing a strategy.*
Digest 449 Part 2 (2000) *Cleaning external masonry: methods and materials.*
Digest 452 (2000) *Whole life costing and life cycle assessment for sustainable building design.*
Digest 455 (2001) *Corrosion of steel in concrete: service life design and prediction.*
Digest 478 (2003) *Building performance feedback: getting started.*

Special digests

SD3 (2002) *HAC in the UK: Assessment, durability management, maintenance, repair and refurbishment.*

Good building guides (GBG)

GBG 1 (1990) *Repairing or replacing lintels (Revision 1).*
GBG 2 (1990) *Surveying masonry chimneys for repair or rebuilding.*

GBG 3 (1990) *Damp proofing basements.*

GBG 4 (1990) *Repairing or rebuilding masonry chimneys.*

GBG 5 (1990) *Choosing between cavity, internal and external wall insulation.*

GBG 6 (1990) *Outline guide to assessment of traditional housing for rehabilitation.*

GBG 7 (1991) *Replacing failed plasterwork.*

GBG 8 (1991) *Bracing trussed rafter roofs.*

GBG 9 (1991) *Habitability guidelines for existing housing.*

GBG 10 (1992) *Temporary support: assessing loads above openings in external walls,* 2nd edition.

GBG 11 (1993) *Supplementary guidance for assessment of timber framed houses, Part 1: Examination.*

GBG 12 (1993) *Supplementary guidance for assessment of timber framed houses, Part 2: Interpretation.*

GBG 13 (1992) *Surveying brick or block freestanding walls.*

GBG 14 (1992) *Building brick and block freestanding walls (Revision 1).*

GBG 15 (1992) *Providing temporary support during work on openings in external walls.*

GBG 17 (1993) *Freestanding brick walls – repairs to copings and cappings.*

GBG 18 (1994) *Choosing external rendering.*

GBG 20 (1999) *Removing internal loadbearing walls in older dwellings.*

GBG 21 (1996) *Joist hangers.*

GBG 22 (1995) *Maintaining exterior wood finishes.*

GBG 23 (1995) *Assessing external rendering for replacement or repair.*

GBG 24 (1995) *Repairing external rendering.*

GBG 25 (1995) *Buildings and radon.*

GBG 26 (1995) *Minimising noise from domestic fan systems.*

GBG 27 (1996) *Building brickwork and blockwork retaining walls.*

GBG 28 Part 1 (1997) *Domestic floors: construction, insulation and damp-proofing.*

GBG 28 Part 2 (1997) *Domestic floors: assessing them for replacement or repair – concrete floors, screeds and finishes.*

GBG 28 Part 3 (1997) *Domestic floors: assessing them for replacement or repair – timber floors.*

GBG 28 Part 4 (1997) *Domestic floors: repairing or replacing floors and flooring – magnesite, tiles, slabs and screeds.*

GBG 28 Part 5 (1997) *Domestic floors: repairing or replacing floors and flooring – wood blocks and suspended timber.*

GBG 29 Part 1 (1999) *Connecting walls and floors – a practical guide.*

GBG 29 Part 2 (1999) *Connecting walls and floors – design and performance.*

GBG 30 (1999) *Carbon monoxide detectors.*

GBG 31 (1999) *Insulated external cladding systems.*

GBG 32 (1999) *Ventilating thatched roofs.*

GBG 33 (1999) *Building damp-free cavity walls.*

GBG 34 (1999) *Building in winter.*

GBG 35 (1999) *Building without cold spots.*

GBG 36 (1999) *Building a new felted flat roof.*

GBG 37 (2000) *Insulating roofs at rafter level.*

GBG 38 (2000) *Disposing of rainwater.*

GBG 39 Part 1 (2000) *Simple foundation design for low rise housing: site investigation.*

GBG 39 Part 2 (2000) *Simple foundation design for low rise housing: rule of thumb design.*

GBG 39 Part 3 (2000) *Simple foundation design for low rise housing: groundworks – getting it right.*

GBG 40 (2000) *Protecting pipes from freezing.*
GBG 41 (2000) *Installing wall ties.*
GBG 42 (2000) *Reed beds (two part set).*
GBG 43 (2000) *Insulated profiled metal roofs.*
GBG 44 Part 1 (2001) *Insulating masonry cavity walls: techniques and materials.*
GBG 44 Part 2 (2001) *Insulating masonry cavity walls: principal risks and guidance.*
GBG 45 (2001) *Insulating ground floors.*
GBG 46 (2001) *Domestic chimneys for solid fuel: flue design and installation.*
GBG 47 (2001) *Level external thresholds: reducing moisture penetration and thermal bridging.*
GBG 48 (2001) *Installing domestic automatic door controls.*
GBG 49 (2001) *Installing domestic automatic window controls.*
GBG 50 (2001) *Insulating solid masonry walls.*
GBG 51 (2001) *Ventilated and unventilated cold pitched roofs.*
GBG 52 Part 1 (2002) *Site cut pitched timber roofs: design.*
GBG 52 Part 2 (2002) *Site cut pitched timber roofs: construction.*
GBG 53 (2001) *Foundations for low-rise building extensions.*
GBG 54 Part 1 (2001) *Construction site communications: general.*
GBG 54 Part 2 (2001) *Construction site communications: masonry.*
GBG 55 (2002) *Quality mark scheme.*
GBG 56 (2002) *Off-site construction: an introduction.*
GBG 57 (2002) *Construction and demolition waste.*

Good repair guides (GRG)

GRG 1 (1996) *Cracks caused by foundation movement.*
GRG 2 (1996) *Damage to buildings caused by trees.*
GRG 3 (1996) *Repairing damage to brick and block walls.*
GRG 4 (1997) *Repairing masonry wall ties.*
GRG 5 (1997) *Diagnosing the causes of dampness.*
GRG 6 (1997) *Treating rising damp in houses.*
GRG 7 (1997) *Treating condensation in houses.*
GRG 8 (1997) *Treating rain penetration in houses.*
GRG 9 (1997) *Repairing and replacing rainwater goods.*
GRG 10 Part 1 (1997) *Repairing timber windows: investigating defects and dealing with water leakage.*
GRG 10 Part 2 (1997) *Repairing timber windows: dealing with draughty windows, condensation in sealed units, operating problems, and deterioration of frames.*
GRG11 Part 1 (1997) *Repairing flood damage: immediate action.*
GRG11 Part 2 (1997) *Repairing flood damage: ground floors and basements.*
GRG11 Part 3 (1997) *Repairing flood damage: foundations and walls.*
GRG11 Part 4 (1997) *Repairing flood damage: services, secondary elements, finishings and fittings.*
GRG12 (1997) *Wood rot: assessing and treating decay.*
GRG 13 Part 1 (1998) *Woodboring insect attack: identifying and assessing damage.*
GRG 13 Part 2 (1998) *Woodboring insect attack: treating damage.*
GRG 14 (1998) *Re-covering pitched roofs.*
GRG 15 (1998) *Repairing chimneys and parapets.*

GRG 16 Part 1 (1998) *Flat roofs: assessing bitumen felt and mastic asphalt roofs.*
GRG 16 Part 2 (1998) *Flat roofs: making repairs to bitumen felt and mastic asphalt roofs.*
GRG 17 (1998) *Repairing and replacing ground floors.*
GRG 18 (1998) *Replacing plasterwork.*
GRG 19 (1998) *Internal painting: hints and tips.*
GRG 20 Part 1 (1998) *Repairing frost damage: roofing.*
GRG 20 Part 2 (1998) *Repairing frost damage: walls.*
GRG 21 (1998) *Improving ventilation in housing.*
GRG 22 Part 1 (1999) *Improving sound insulation.*
GRG 22 Part 2 (1999) *Improving sound insulation.*
GRG 23 (1999) *Treating dampness in basements.*
GRG 24 (1999) *Repointing external brick and block walls.*
GRG 25 (1999) *Supporting temporary openings.*
GRG 26 Part 1 (2000) *Improving energy efficiency: thermal insulation.*
GRG 26 Part 2 (2000) *Improving energy efficiency: boilers and heating systems, draughtproofing.*
GRG 27 Part 1 (2000) *Cleaning external walls of buildings: cleaning methods.*
GRG 27 Part 2 (2000) *Cleaning external walls of building: removing dirt and stains.*
GRG 28 (2000) *Repairing brick and block freestanding walls.*
GRG 29 (2000) *Refixing ceramic tiles to internal walls.*
GRG 30 (2000) *Remedying condensation in domestic pitched tiled roofs.*
GRG 31 (2001) *Hot air repair of PVC-u window and door frames.*
GRG 32 (2001) *Dealing with noisy plumbing.*
GRG 33 Part 1 (2002) *Assessing moisture in building materials – sources of moisture.*
GRG 33 Part 2 (2002) *Assessing moisture in building materials – measuring moisture content.*
GRG 33 Part 3 (2002) *Assessing moisture in building materials – interpreting moisture data.*

Information papers (IP)

IP 4/81 (1981) *Performance of cavity wall ties.*
IP 11/81 (1981) *Maintenance of flat roofs.*
IP 13/81 (1981) *Conversions of older property to house single young people.*
IP 26/81 (1981) *Solar reflective paints.*
IP 10/84 (1984) *The structural condition of prefabricated reinforced concrete houses designed before 1960.*
IP 6/86 (1986) *Spacing of wall ties in cavity walls.*
IP 16/87 (1987) *Maintaining paintwork on exterior timber.*
IP 6/88 (1988) *Methods of improving sound insulation between converted flats.*
IP 9/88 (1988) *Methods of reducing impact sounds in buildings.*
IP 13/88 (1988) *Energy assessment for dwellings using BREDEM worksheets.*
IP 16/88 (1988) *Ties for cavity walls: new developments.*
IP 17/88 (1988) *Ties for masonry cladding.*
IP 2/89 (1989) *Thermal performance of lightweight inverted warm deck flat roofs.*
IP 20/89 (1989) *An energy-efficient refurbishment of electrically heated high-rise flats.*
IP 21/89 (1989) *Passive stack ventilation in dwellings.*
IP 7/90 (1990) *An introduction to infra-red thermography for building surveys.*

IP 8/90 (1990) *Retail warehouses: potential for increasing energy efficiency.*
IP 12/90 (1990) *Corrosion of steel wall ties: history, occurrence, background and treatment.*
IP 13/90 (1990) *Corrosion of steel wall ties: recognition and inspection.*
IP 15/90 (1990) *Defects in local authority housing: results of a building problems survey.*
IP 1/91 (1991) *Durability of non-asbestos fibre-reinforced cement.*
IP 4/91 (1991) *Improving the energy-efficient performance of high-rise housing.*
IP 13/91 (1991) *Energy-efficient refurbishment of pre-1919 housing.*
IP 20/92 (1992) *Energy use in office buildings.*
IP 10/93 (1993) *Avoiding latent mortar defects in masonry.*
IP 11/93 (1993) *Ecolabelling of building materials.*
IP 12/93 (1993) *Industrial building refurbishments: opportunities for energy efficiency.*
IP 12/94 (1994) *Assessing condensation risk and heat loss at thermal bridges round openings.*
IP 1/96 (1996) *Management of construction and demolition wastes.*
IP 12/97 (1997) *Plastics recycling in the construction industry.*
IP 9/99 (1999) *Cleaning exterior masonry: pretreatment assessment of a stone building.*
IP 7/00 (2000) *Reclamation and recycling of building materials: industry position report.*
IP 12/00 (2000) *Positive input ventilation in dwellings.*
IP 16/00 (2000) *Low-solvent primers: performance in construction steelwork.*
IP 17/00 (2000) *Advanced technologies for 21st century building services.*
IP 18/00 (2000) *Ammonia refrigerant in buildings: minimising the hazards.*
IP 4/01 (2001) *Reducing impact and structure-borne sound in buildings.*
IP 3/03 (2003) *Dynamic insulation for energy saving and comfort.*

BRE leaflets

BR 156 (1989) *Cast rendered no-fines houses.*
BR 160 (1989) *No-fines houses.*

Other reports

BR 113 (1987) *Steel-framed and steel-clad houses: inspection and assessment.*
BR 143 (1989) *Thermal insulation: avoiding risks.*
BR 148 (1989) *Atholl steel-framed, steel clad houses.*
BR 166 (1990) *Rehabilitation – a review of quality in traditional housing.*
BR 167 (1990) *Assessing traditional housing for rehabilitation.*
BR 168 (1990) *Surveyor's checklist for rehabilitation of traditional housing.*
BR 170 (1990) *Energy use in buildings and carbon dioxide emissions.*
BR 184 (1991) *Foundation movement and remedial underpinning in low-rise buildings.*
BR 185 (1991) *Overroofing: especially for large panel system dwellings.*
BR 238 (1993) *Upgrading floors – sound control for homes.*
BR 250 (1993) *Surveying dwellings with high indoor radon levels: a BRE guide to radon remedial measures in existing dwellings.*
BR 262 (2002) *Thermal insulation: avoiding risks.*
BR 267 (1994) *Radon: major alterations and conversions.*
BR 299 (1996) *Indoor air quality in homes, Part 1.*

BR 349 (1999) *Impact of climate change on building.*

BR 364 (1999) *Solar shading of buildings.*

BR 366 (1999) *Sustainable retail premises: an environmental guide to design, refurbishment and management of retail premises.*

BR 389 (2000) *EcoHomes: the environmental rating for homes.*

BR 390 (2000) *The Green Guide to Housing Specification: an environmental profiling system for building materials and components.*

BR 402 (2000) *A risk assessment procedure for health and safety in buildings.*

BR 456 (2001) *Deconstruction and reuse of construction materials.*

BR 456 (2003) *Control of dust from construction and demolition activities.*

BWPDA (British Wood Preserving and Damp-Proofing Association) publications

Timber leaflets

T1 (1995) *Fungal decay in buildings – dry rot and wet rot.*

T2 (1993) *Preservative treatment of timber.*

T3 (1993) *Methods of applying preservatives.*

T4 (1993) *The preservation of window joinery.*

T5 (1993) *Preservative treatment against wood borers.*

T6 (1990) *The use of creosote oil for wood preservation.*

T7 (1993) *The treatment of timber with water borne preservatives.*

T8 (1993) *Water repellent wood preservatives.*

T9 (1990) *Preserving wood in home and garden.*

FR1 (1995) *The treatment of solid timber and panel products with flame retarders.*

Dampness literature

DP1 (1990) *The use of moisture meters to establish the presence of rising damp.*

DP 2 (1990) *Plastering in association with damp-proof coursing.*

DP 3 (1990) *Condensation.*

DP 4 (1990) *Methods of analysis for damp-proof course fluids.*

DP 5 (1990) *Chemical damp-proof course insertion – the attendant problems.*

DP 6 (1990) *Safety in damp-proofing.*

DP 7 (1993) *Chemical damp-proof course in walls – detection techniques and their limitations.*

DP 8 (1993) *Damp-proof barriers (tanking) in association with chemical damp-proof coursing.*

DP 9 (1993) *Guidelines to survey report writing.*

DP 10 (1990) *Hygroscopic salts and rising damp.*

Codes of Practice

Code of Practice for Remedial Timber Treatment (1990)

Code of Practice for the Installation of Chemical Damp-proof Courses (1986)

Code of Practice for Safe Design and Operation of Timber (1993)

WPA (Weatherproofing Advisers Ltd) technical publications (see website address in next section)

Technical Guide (c.2004) *Defects associated with organically coated steel sheeting.*
Technical Guide (c.2004) *Treating asbestos cement roof products.*

Some useful internet sites

The number of Internet sites related to construction and surveying is growing rapidly. The following list, therefore, is a sample of some of the most prominent sites available in this area.

www.ashi.com (ASHI is the oldest, largest, and most recognised organisation for home inspectors in North America).
www.bcis.co.uk (for information on building costs).
www.bre.co.uk (website of the Building Research Establishment).
www.buildingpreservation.com/ (Graham Coleman's informative website on dampness and timber treatment).
www.bwpda.co.uk (website of the British Wood Preserving and Damp-proofing Association).
www.cbppp.org.uk/cbpp (for information on the Construction Best Practice Programme).
www.cibworld.nl (website of the International Council for Research and Innovation in Building and Construction (CIB) based in Holland).
www.cic.org.uk (Construction Industry Council website).
www.cisti.nrc.ca/irc (website of the Canadian Institute for Research in Construction, which contains an extensive list of useful articles on a whole range of building problems).
www.clasp.gov.uk (for full details of the various types of CLASP system buildings – including case studies of refurbished properties built using this method).
www.concreterepair.org.uk (website of the Concrete Repair Association).
www.direct.gov.uk (for information on a wide range of sources relating to general and local government).
www.dryrotdogs.com/ (Hutton + Rostron's informative website that contains articles on moisture problems as well as fungal attack in buildings).
www.dti.gov.uk (for information relating to the construction industry).
www.fogit.co.uk (Rentokil's property care website).
www.hmso.gov.uk (for details of government publications).
www.housedustmite.org.uk (for useful information on the risks and remedies associated with house dustmite infestation in dwellings).
www.leanconstruction.org (website of the Lean Construction Institute, USA).
www.planningportal.gov.uk (UK government's website on planning matters).
www.odpm.gov.uk (website of the Office of the Deputy Prime Minister, which contains information on the Building Regulations and other technical matters related to construction).
www.publications.parliament.uk (for details of parliamentary committee reports).
www.safeguardchem.com/Downloads (for downloads on damp proofing and timber treatments).

www.scotland.gov.uk (for full access to the Scottish Building Regulations).

www.ukonline.gov.uk (for information on a wide range of sources relating to general and local government).

www.weatherproofing.co.uk (website of Weatherproofing Advisers Ltd).

www2.cr.nps.gov/tps/briefs/presbhom.htm (website of the US National Park Service heritage preservation services containing an excellent range of articles and briefs on the inspection, appraisal and conservation of historic buildings).

Index